International Business and the Eclectic Paradigm

The eclectic paradigm has become the dominant theoretical basis in the study of international business, multinational corporations and internationalisation over the last two decades. However, developments such as economic globalisation and the subsequent growth of global and alliance capitalism have fundamentally affected the way in which MNC activities are undertaken and organised. The contributions to this volume evaluate the eclectic paradigm in the global economy and its validity as a theoretical basis to understand these developments. Contributions by leading scholars – including John H. Dunning, Stephen Guisinger, John Cantwell, Terutomo Ozawa, Gabriel R.G. Benito, Lars Oxelheim, Rajneesh Narula, Timothy M. Devinney and Anoop Madhok – examine the legacy of the eclectic paradigm.

This book systematically evaluates how the eclectic paradigm has survived the test of time, from a wide variety of perspectives. Contributors discuss whether and how the Ownership–Location–Internalisation (OLI) framework continues to be applicable to the perspective of their own particular analytical concerns. Has the explanatory usefulness of the paradigm diminished? To what extent is it a useful tool from the perspective of, for instance, trade theory, finance, evolutionary economics, resource-based theory or strategic management? Can it be utilised to explain new developments in international business and economics? Do these require new ideas and concepts to be integrated within the eclectic paradigm? What are the new challenges to which international business theorising must respond? This volume includes contributions by leading scholars from a variety of perspectives, disciplines and subject areas in international business, and represents the state of the art in international business theory.

John Cantwell is Professor of International Business at Rutgers University, USA (from 2002), while on leave of absence as Professor of International Economics at the University of Reading, UK. His main research areas are the analysis of corporate technological change and international business. He is the author of *Technological Innovation and Multinational Corporations* (1989). **Rajneesh Narula** is Professor in the Department of International Economics and Management (INT) at the Copenhagen Business School, and at the Centre for Technology, Innovation and Culture (TIK), University of Oslo. His research interests include foreign investment and economic development and globalisation and technology. He is the author of *Multinational Investment and Economic Structure* (1996).

Routledge Studies in International Business and the World Economy

1 **States and Firms**
Multinational enterprises in institutional competition
Razeen Sally

2 **Multinational Restructuring, Internationalization and Small Economies**
The Swedish case
Thomas Andersson, Torbjörn Fredriksson and Roger Svensson

3 **Foreign Direct Investment and Governments**
Catalysts for economic restructuring
Edited by John H. Dunning and Rajneesh Narula

4 **Multinational Investments and Economic Structure**
Globalization and competitiveness
Rajneesh Narula

5 **Entrepreneurship in a Global Context**
Edited by Sue Birley and Ian Macmillan

6 **The Global Structure of Financial Markets**
An overview
Edited by Dilip K. Ghosh and Edgar Ortiz

7 **Alliance Capitalism and Global Business**
John H. Dunning

8 **Multinational Enterprises from the Netherlands**
Edited by Roger van Hoesel and Rajneesh Narula

9 **Competition, Growth Strategies and the Globalization of Services**
Real estate advisory services in Japan, Europe and the United States
Terrence LaPier

10 **European Integration and Foreign Direct Investment in the EU**
The case of the Korean consumer electronics industry
Sang Hyup Shin

11 **New Multinational Enterprises from Korea and Taiwan**
Beyond export-led growth
Roger van Hoesel

12 **Competitive Industrial Development in the Age of Information**
The role of co-operation in the technology sector
Edited by Richard J. Braudo and Jeffrey G. MacIntosh

13 **The Global Restructuring of the Steel Industry**
Innovations, institutions and industrial change
Anthony P. D'Costa

14 **Privatisation and Liberalisation in European Telecommunications**
Comparing Britain, the Netherlands and France
Willem Hulsink

15 **Multinational Corporations**
Emergence and evolution
Paz Estrella Tolentino

16 **Foreign Direct Investment in Emerging Economies**
Corporate strategy and investment behaviour in the Caribbean
Lou Anne A. Barclay

17 **European Integrations and Global Corporate Strategies**
Edited by François Chesnais, Grazia Ietto-Gillies and Roberto Simonetti

18 **The Globalisation of Corporate R&D**
Implications for innovation systems in host countries
Prasada Reddy

19 **Globalization of Services**
Some implications for theory and practice
Edited by Yair Aharoni and Lilach Nachum

20 **A Century of Foreign Investment in the Third World**
Michael J. Twomey

21 **Global Capitalism at Bay**
John H. Dunning

22 **Foreign Direct Investment**
Research issues
Edited by Bijit Bora

23 **Ford and the Global Strategies of Multinationals**
The North American auto industry
Isabel Studer Noguez

24 **The World Trade Organization Millennium Round**
Freer trade in the next century
Klaus Deutsch/Bernhard Speyer

25 **Consultancy and Innovation**
The business service revolution in Europe
Edited by Peter Wood

26 **Knowledge Economies**
Clusters, learning and co-operative advantage
Philip Cooke

27 **The Role of Resources in Global Competition**
John Fahy

28 **Globalization, Employment and the Workplace**
Diverse impacts
Edited by Yaw A. Debrah and Ian G. Smith

29 **Transnational Corporations**
Fragmentation amidst integration
Grazia Ietto-Gillies

30 **Growth Theory and Growth Policy**
Edited by Harald Hagemann and Stephan Seiter

31 **International Business and the Eclectic Paradigm**
Developing the OLI framework
Edited by John Cantwell and Rajneesh Narula

32 **Regulating Global Trade and the Environment**
Paul Street

International Business and the Eclectic Paradigm

Developing the OLI framework

Edited by John Cantwell and Rajneesh Narula

Routledge
Taylor & Francis Group
LONDON AND NEW YORK

First published 2003 by Routledge
2 Park Square, Milton Park, Abingdon, Oxon OX14 4RN

Simultaneously published in the USA and Canada
by Routledge
711 Third Avenue, New York, NY 10017

Routledge is an imprint of the Taylor & Francis Group

Typeset in Baskerville by Exe Valley Dataset Ltd, Exeter

British Library Cataloguing in Publication Data
A catalogue record for this book is available from the British Library

Library of Congress Cataloging in Publication Data
International business and the eclectic paradigm: developing the OLI framework / edited by John Cantwell and Rajneesh Narula.
p. cm.
Includes bibliographical references and index.
1. International business enterprises. 2. Strategic alliances (Business). 3. Investments, Foreign. 4. Industrial location. 5. Technology transfer–Economic aspects. 6. Technological innovations–Economic aspects. 7. Globalization–Economic aspects. I. Cantwell, John. II. Narula, Rajneesh, 1963–

HD2755.5.I5354 2003
338.8´8–dc21 2003043134

ISBN 0–415–31678–2

Contents

Figures

Tables

Contributors

Gabriel R.G. Benito is Professor at the Department of Strategy, Norwegian School of Management BI. His research interests are in international business, strategic management and corporate governance.

John Cantwell is Professor of International Business at Rutgers University in the US (from 2002), while on leave of absence as Professor of International Economics at the University of Reading in the UK.

Sergio Castello is an Associate Professor of Global Business and Economics, the University of Mobile, Alabama, USA. His publications include *Globalization of Small Economies as a Strategic Behavior in International Business* (co-authored with T. Ozawa), New York: Garland, 1999.

Timothy M. Devinney is Professor of Strategy at the Australian Graduate School of Management (AGSM) and Director of the Centre for Corporate Change. Before joining the AGSM he held positions on the faculties of Vanderbilt University and the University of California, Los Angeles (UCLA).

John H. Dunning is Emeritus Professor of International Business at the University of Reading, UK, and State of New Jersey Professor of International Business at Rutgers University.

Lorraine Eden is Professor of Management and a Texas A&M Faculty Fellow at Texas A&M University, College Station, Texas. Her research on multinational enterprises focuses on transfer pricing, international taxation, MNE–state relations and regional integration.

Stephen Guisinger was Professor of Organizations, Strategy, and International Management at the University of Texas at Dallas. He passed away on 3 July 2001, just before the original special issue was published.

Anoop Madhok is Associate Professor of Management at the David Eccles School of Business, University of Utah. His research interests include multinational strategy, interfirm collaborations and the theory of the firm.

Elizabeth Maitland is Senior Lecturer in the School of International Business at the University of New South Wales. She researches and teaches in the areas of international business and strategy, with particular research focuses on corruption, Australian firms in Asia and relations between sovereign states and multinational enterprises.

David F. Midgley is Professor of Marketing at INSEAD, Fontainebleau, France. Previously he held positions at the Anderson School, University of California, Los Angeles and the Australian Graduate School of Management.

Rajneesh Narula is Professor at the Department of International Economics and Management (INT) at the Copenhagen Business School, and at the Centre for Technology, Innovation and Culture (TIK), University of Oslo.

Stephen Nicholas is Sesqui Professor of International Business at the University of Sydney, Australia. Stephen Nicholas researches in international business and consults for international companies in Australia, Thailand and Malaysia, as well as the Australian government.

Lars Oxelheim is Professor of International Business and Finance, Lund Institute of Economic Research, Lund University. He is also affiliated with the Research Institute of Industrial Economics (IUI), Stockholm.

Terutomo Ozawa is a Professor of Economics, Colorado State University, Fort Collins, CO, USA. He has published extensively, including the latest book, *Business Restructuring in Asia: Cross-Border M&As in the Crisis Period* (co-authored with J. Zhan), Copenhagen: Copenhagen Business School Press, 2001.

Anupama Phene is currently Assistant Professor of Strategy at the David Eccles School of Business, University of Utah. Her research primarily focuses on firm strategy in a multinational context, evolution of firms and subsidiaries and development of firm capabilities in knowledge creation and transfer.

Lucia Piscitello is Associate Professor in Industrial Economics and Management in the Department of Management, Economics and Industrial Engineering at the Politecnico di Milano. Her main research areas are the economics of technological change and the theory of multinational enterprise.

Trond Randøy is Associate Professor of International Business and Management, Agder University College, Norway, and Agder Research, Norway. School of Management, Kristiansand.

Arthur Stonehill is a Visiting Professor of Finance and International Business, University of Hawaii at Manoa and Professor Emeritus, Oregon State University.

Paz Estrella Tolentino is Lecturer in International Business at the School of Management and Organizational Psychology, Birkbeck College, University of London.

Sverre Tomassen is lecturer in International Business at the Norwegian School of Management BI. His main line of research deals with the relationship between headquarters and foreign subsidiaries, especially with regard to performance assessment and evaluation.

Sunil Venaik is Assistant Professor of Marketing at the Indian Institute of Management, Ahmedabad. Previously he held a position at the School of International Business, University of New South Wales, Sydney.

Cliff Wymbs is an Assistant Professor of Marketing at the Zicklin School of Business, Baruch College/CUNY. His research interests include the study of international business organizational modalities, the evolution of e-commerce and the digital enterprise, and the formation of service-related knowledge clusters.

Acknowledgements

Editing an academic volume may not be on the same scale as producing *Ben Hur*, but it also requires a cast of unsung heroes who toil behind the scenes, giving cheerfully of their time and resources for little reward and very little acknowledgement. Unlike Hollywood, they do not number in the thousands, so it is feasible to actually list these gentle folk.

First, we'd like to thank Eleanor Morgan, editor of the *International Journal of the Economics of Business* (IJEB). Eleanor helped to organise the special issue on which this volume is based, and it is safe to say that without her encouragement this project would never have happened. We would also like to thank Taylor & Francis, publishers of the *International Journal of the Economics of Business* (www.tandf.co.uk), for allowing us to reproduce material here.

Second, we are grateful to our various colleagues who cheerfully agreed to review the various contributions to the special issue and this volume, providing copious comments that have helped shape the final product in one way or the other. In alphabetical order, these include Christian Bellak, Gabriel Benito, Julian Birkinshaw, Pontus Braunerhjelm, Tine Bruland, Peter Buckley, Laurance Capron, Mark Casson, Richard Caves, Peter Dicken, Yves Doz, John Dunning, Pervez Ghauri, Peter Gray, Lars Håkanson, Neil Kay, Anoop Madhok, Klaus Meyer, John McGee, Rolf Mirus, Ram Mudambi, Terutomo Ozawa, Nigel Pain, Bent Petersen, Tom Pugel, David Ricks, Alan Rugman, Frédérique Sachwald, Francesca Sanna-Randaccio, Art Stonehill, Katharine Wakelin and Udo Zander.

Third, we would like to thank Marlene Larsen and Jeanne Albrecht at the Copenhagen Business School for secretarial support, and Sudha Menon for her assistance in indexing.

Last, but by no means least, we would like to acknowledge our intellectual and personal debt to John Dunning. Both of us were his PhD students, and he has been nothing if not supportive of our individual and joint endeavours over the years, as anyone who has had the good fortune to work with him will attest to.

Rajneesh Narula
John Cantwell
15 December 2002

1 Revisiting the eclectic paradigm

New developments and current issues

John Cantwell and Rajneesh Narula

Introduction

The eclectic paradigm has been the leading explanation for the growth of multinational activity over the past two decades. Its broad appeal has made it a mainstay in explaining various sectors and types of activities. Its simplicity and general nature makes it compatible with a number of schools of economic and managerial thought. It has been applied by management scholars, economic geographers, evolutionary economists, resource-based theorists and development economists, to mention a few. It has also been the object of considerable criticism.

Yet it faces fresh challenges to its continued applicability in the light of new developments associated with the process of globalisation. Economic globalisation and the subsequent growth of global and alliance capitalism have fundamentally affected the way in which international and domestic business activities are undertaken and organised.

Some have pointed to its limited appeal in predicting trends and developments. New forms of activity such as asset-augmenting investment (by multinational corporations (MNCs) of all nationalities and varying levels of multinationality), in a variety and combination of modes of operation such as alliances and outsourcing, have required new ideas and approaches to be integrated within the eclectic paradigm. This chapter systematically discusses these current issues and related relevant theoretical developments, and revisits the eclectic paradigm within this context. The various contributions to the book are then placed within the resulting framework.

We highlight the fact that globalisation has increased the interactive dynamics between, and amongst 'O', ' L' and 'I' characteristics at firm, industry and country level, in at least two ways. First, a knowledge-based society has meant that the efficient exploitation of MNCs' ownership advantages and the continual need to augment and sustain their competitive advantages is ever more crucial, leading to a complex interdependence between ownership and location advantages. Second, globalisation has affected how MNCs seek to coherently organise their cross-border activities in response to changing boundaries of the firm.

The objective of this volume is to systematically evaluate how the eclectic paradigm has survived the test of time, from a wide variety of perspectives. Contributors were specifically given a remit to discuss whether and how the Ownership-Location-Internalisation (OLI) framework continues to be applicable to the perspective of their own particular analytical concerns. Has the explanatory usefulness of the paradigm diminished? To what extent is it a useful tool from the perspective of (for instance) trade theory, economic geography, evolutionary economics, resource based theory or strategic management? Can it be utilised to explain new developments in international business and economics? Do these require new ideas and concepts to be integrated within the eclectic paradigm? What are the new challenges to which international business theorising must respond?

The continuing usefulness of the OLI framework with the evolution of the MNC

The eclectic paradigm grew out of a desire to synthesise elements from the transaction cost (internalisation) and market power theories of the individual firm in its relationship to markets with macroeconomic approaches to international production (such as the original product cycle model) at the country level. However, it was not intended to be a complete synthesis as it is not possible to fully encompass a set of theories which may address rather different questions or rely on different views of the world. It was therefore soon acknowledged that it was not itself another theory. It was instead intended to provide an overall analytical framework for empirical investigations which would draw the attention of the analyst to the most important theories for the problem at hand. It also provided a framework for a comparison between theories, by establishing the common ground or the points of contact between them, and clarifying the relationship between different levels of analysis and the different questions which theorists have been concerned to address. Thus, for example, internalisation theory may be the most relevant under certain circumstances or when answering certain kinds of questions (such as those related to backward vertical integration into resource extraction, as in the early work of Hennart (1982) or Read (1986)), while the determinants of the competitive strategy of firms in their final product market may be more pertinent in other cases (such as issues of technological competition or cooperation).

These points have sometimes been missed by commentators on and especially by critics of the eclectic paradigm, who have tended to subsume the generality of its framework into one particular theoretical line of argument (some have treated it as a mere extension of transaction cost or internalisation theory) or to one particular level of analysis (such as the firm level); or who at the other extreme suppose that it somehow amounts to an analytical fudge since it doesn't firmly 'nail its theoretical colours to the mast'. Rather, the eclectic paradigm provides a means of assessing evidence on foreign-owned production in a variety of empirical contexts, as a means of determining which theories and which level of analysis are most appropriate to any given set of circumstances. The paradigm is not therefore a substitute for making judgements over how best to synthesise relevant complementary theories, or how to choose between potentially competing theories. Instead, it provides a framework which facilitates just such judgements, and helps to operationalise them.

In the eclectic paradigm it is contended that MNCs have competitive or 'ownership' advantages *vis-à-vis* their major rivals, which they utilise in establishing production in sites that are attractive due to their 'location' advantages (Dunning 1977, 1979, 1988, 1995). According to Dunning, two types of competitive advantage can be distinguished: the first is attributable to the ownership of particular unique intangible assets (such as firm-specific technology), and the second is due to the ownership of complementary assets (such as the ability to create new technologies, or the ability to effectively coordinate cross-border activities). MNCs retain control over their networks of assets (productive, commercial, financial and so forth) because of the 'internalisation' advantages of doing so. Internalisation advantages arise both from the greater ease with which an integrated firm is able to appropriate a full return on its ownership of distinctive assets such as its own technology, as well as directly from the coordination of the use of complementary assets, subject to the costs of managing a more complex network.

Frameworks and theories in international business, as in other social sciences, attempt to explain events and developments that have already occurred, and it is axiomatic that things do not stay the same: The behaviour of firms in the future cannot be accurately predicted by their actions in the past. Firms are *path-dependent*, but not entirely so. To put it another way, there is a vintage effect. The rest of the actors in the economy (institutions, competitors, customers and suppliers) evolve largely independently of a given firm, but changes in their behaviour impinge on its behaviour.

The last few decades have seen a fundamental change in the activities of MNCs. MNC activity has increased not just in its extent, but also in its intensity, and variety. These developments are often associated with the process of globalisation. Economic globalisation refers to the increasing cross-border interdependence and integration of production and markets for goods, services and capital. This process has led to a widening of the extent and form of international transactions, and to a deepening of the interdependence between the actions of economic actors located in one country and those located in others (Dunning 1997). *Inter alia*, one of the primary consequences of globalisation has been the growing convergence of income levels, consumption patterns and institutional structures, both among the industrialised countries and between them and the more advanced developing countries; and also the increasing significance of intra-firm trade in goods and services. The influence of globalisation on international business activity has ushered in fundamental changes in which MNCs undertake cross-border activities, in what Dunning (1995) has described as 'alliance capitalism'.

It is hard, if not impossible, to establish clear causalities between developments relating to globalisation, but their interdependence is unmistakeable (Narula 2003). For instance, the kinds of technologies across countries have shown to have converged because of, *inter alia*, increasing cross-border competition and the increasing interdependence of economic actors in different locations. These developments have changed the way firms organise their innovative activity both spatially and organisationally. There is also an increasing international aspect of R&D activity, and a growth in the use of collaborative R&D, both within and across borders. Competition is global in nature, and this affects the way in which firms sustain their competitiveness. Cantwell and Sanna-Randaccio (1990) have shown, for instance, that firms seek to emulate the technological advantages of leading competitors in the same industry, regardless of their national location. Similarly, Narula and Hagedoorn (1999) have shown that firms seek to engage in R&D alliances with technological leaders in the same industry, irrespective of their national origins.

Firms increasingly seek to invest abroad for a growing variety of reasons, and while asset-exploiting activities still predominate as a motivation, the tendency for firms to invest abroad in order to augment their existing assets is now also substantial, and forces scholars of international business to rephrase their enquiries.

Thus, while the early question in the international business literature was 'why do technologically advantaged firms go abroad to exploit their advantage (and transfer their technology internally to be able to do so)?', with the rapid expansion of activity in now-established MNCs from a wider range of home countries a second question has arisen, namely, 'why do existing MNCs source technology creation internationally through an internal network of geographically dispersed affiliates?' The eclectic paradigm is equally applicable to both these kinds of question. Thus, in the 1980s those of us that continued to insist on the concept of ownership advantages as a condition for international corporate expansion reformulated the concept more precisely. Critics of the concept had focused their attack on an interpretation of Hymer's earliest discussion of ownership advantages, as a net cost advantage of foreign-owned over indigenous firms in the relevant local market (and this is still often the version that critics prefer to disparage today, seemingly unaware that the discussion, like the real world, has moved on!).

The first and obvious revision is that ownership advantages must be thought of in relation to the international competition mainly from other MNCs rather than relative to domestic companies in a particular host country. Today MNCs are generally competing with one another in international markets, they are usually not in the earliest stages of internationalisation and their investments are not all of a local market-oriented kind (unlike in Hymer's case, in which he had addressed the more specific question of why firms initially go abroad and begin to engage in outward direct investment). Indeed, beyond the notion that firms evolve in their intensity of overseas activity, the internationalisation sequence argument whereby firms first export, and then engage in foreign direct investment (FDI) has now largely been abandoned. Firms in certain sectors skip exporting, and proceed directly to FDI, still others seek to augment assets in several locations, sometimes without any other prior asset-exploiting activities overseas. It is not unusual for MNCs to use several organisational modes simultaneously in the same location, and with the same partners (Benito and Welch 1994; Petersen and Welch 2001).

Second, innovation and hence innovative advantages are differentiated and relative concepts, not indicative of some notional technology frontier. All surviving MNCs have some distinctive competitive edges, and it is these differentiated firm-specific strengths that constitute each firm's ownership advantages rather than some overall absolute cost advantage. Hence, MNCs that are not world leaders or do not hold an overall absolute cost advantage over most indigenous firms in the countries in which they invest may still have ownership advantages especially in operating in certain differentiated kinds of environment, and some of them have been able to upgrade these advantages more rapidly than in the past, encouraging and facilitating a faster internationalisation. Third, therefore, there is a complementarity between the initial ownership advantages of the firm and its ability to consolidate and extend these advantages through an international network of competence-creating subsidiaries. MNCs with greater initial ownership advantages have a greater absorptive capacity to be able to extract and utilise the potential for new innovation to be found in foreign centres of excellence. So it is rather facile to counterpose the initial holding of ownership advantages and the ability to build advantages through asset-seeking foreign direct investment, as has sometimes been done by would-be critics of the eclectic paradigm, since actually the two tend to go together. Likewise, in a successful process of international network creation in the MNC, ownership and location advantages are cumulatively developed together, but this does not imply that the conceptual distinction between them is not useful. For a further discussion of the role of ownership advantages and a review of the debate over their necessity see Cantwell (2000a).

Concerning the relationship between ownership advantages and internalisation advantages in the light of the rise of internationally integrated networks in MNCs, the eclectic paradigm provides a suitable framework for an interchange between the evolutionary theory of technological change applied to MNCs (Cantwell 1989, 1994; Hagedoorn and Narula 2001) and the transaction cost approach applied to the theory of the existence of the MNC (Buckley and Casson 1976). Whereas the theory of cumulative technological change, like the work of the classical economists, is a theory of production (and the changing technology of production), as it stands the Coasian theory of the firm, like the neoclassical economic thinking of which it is a criticism, is a pure theory of exchange. Exchange takes place under a variety of institutional arrangements, in markets or within the firm. However, in order to make the Coasian theory of the firm itself evolutionary it would be necessary to specify how transaction costs are themselves influenced by the growth and technological innovation of firms.

As Casson (1986) pointed out, while transaction cost theory specifies conditions under which non-market institutional arrangements will obtain (for example, within the firm), it does so at present to the exclusion of any active role for managerial strategy. In other words, while a theory of the MNC, couched in an exchange framework, can explain the existence of the MNC or the firm, it still left the firm itself as a passive reactor to transactional circumstances. It relates to the external influences on growth, and not to the internal sources of growth which make the firm inherently dynamic when under competition in its final product market. Changes in the organisation or control of production are merely a response to changes in the costs of various exchange relationships in markets or otherwise. This procedure may be justified if one is concerned only with the internalisation of intermediate product markets actively replacing trade between independent parties, but in the evolutionary or resource-based view this is not the main source of growth of the firm.

It seems reasonable to suppose that the accumulation of technology and the growth of production within the firm will affect the transaction costs of exchange. The transaction cost theorist has instead tended to start from exchange in a market, which gives way to more consciously organised control where it is relatively inefficient. Coase (1937) stressed the market conditions which lead to a reorganisation or an extension in the organisation of the firm. However, the nature and extent of a firm's transactions and cooperative arrangements with other firms, as well as its market share, also depend upon its innovative capacity *vis-à-vis* other firms. As paths of technological development become established within each firm this is likely to affect the conditions of technology transfer between them, which has been the focus of much of the internalisation literature.

The collection of papers we have brought together here for this volume can be grouped under at least one of three headings. The first two represent a focus on particular elements of the eclectic paradigm. The first group are concerned with the increasing interaction between ownership advantages and location advantages, as referred to earlier. This has been a central concern for macroeconomic theories of trade and development (the eclectic paradigm applied to the level of countries or national systems, and their inward and outward investment), and for microeconomic theories of the changing internal division of labour within the firm (the increasing significance of competence-creating subsidiaries in favoured locations and asset-seeking types of investment, as discussed above), as well as analyses of localised inter-company spillovers. The second group are concerned with the shifting characteristics of internalisation advantages, and with new ways of viewing the organisational coherence of economic activities. This is the centre of attention for work on the changing boundaries of the firm in the form of a proliferation of inter-firm alliances and mergers and acquisitions, and for discussions that contrast the transaction cost and resource-based perspectives on the firm, or which develop the organisational strategy perspective on the firm. The third group of contributions seek to apply the eclectic paradigm. Some of these have an empirical or industry-specific focus. Others have sought to use the eclectic paradigm as a tool to apply the eclectic paradigm to particular questions, or to propose new avenues for future research. We now turn to a brief review of contributions under these three headings.

Increasing interaction between ownership and location advantages

Recent research applying the framework of the competence-based approach to multinational firms (Cantwell 1989) has attempted to trace out the technological evolution of large multinational corporations over time as a path-dependent learning process following distinct corporate technological trajectories (Dosi 1982). In the course of this process, MNCs move into new technological fields and they establish innovative activities in multiple geographical sites as a reflection of the development of the underlying capability of firms. In the internationalisation field, new theoretical and empirical models have been devised of the process by which multinational companies access locationally dispersed technological assets, through their own international operations and through alliances with other firms (Cantwell 1989; Kogut and Chang 1991; Dunning 1995; Dunning and Narula 1995; Pugel *et al.* 1996; Almeida 1996; Frost 1996; Cantwell and Barrera 1998; Kuemmerle 1999; Cantwell and Janne 1999; Pearce 1999; Zander 1999; Narula 2002a). We have suggested (e.g. Cantwell and Piscitello 2000) that in the more recent internationally integrated or 'globalised' MNC, the geographical dispersion of innovation may come to facilitate the technological development of the firm, since the MNC can tap into alternative streams of innovation in different centres, and establish favourable cross-border interactions between them (Cantwell 1995; Zander 1997; Dunning 1996).

The firms of each country tend to embark on a path of technological accumulation that has certain unique characteristics and sustains a distinct profile of national technological specialisation (Rosenberg 1976; Pavitt 1987; Cantwell 2000b). The kinds of linkages that grow up between competitors, suppliers and customers in any regional district or country are also, to some extent, peculiar to that location, and imbue the technology creation of its firms with distinctive features (Mariotti and Piscitello 2001). For these reasons, other MNCs often need to be on-site with their own production and their innovatory capacity if they are to properly benefit from the latest advances in geographically localised technological development, to feed their innovation (Cantwell 1989; Kogut and Chang 1991). Moreover, due to the complexity of technological learning, and the significance of maintaining face-to-face contacts, the localisation of technological contacts tends to occur at a regional level within host countries (Jaffe *et al.* 1993; Almeida 1996; Cantwell and Iammarino 1998, 2000; Verspagen and Schoenmakers 2000). By contrast, where the technological capacity of a host country is weak in the sector concerned, the investments of MNCs may drive out local competition and reduce local technological capability still further (Cantwell 1987). It is therefore typically when there is already a strong existing domestic technological presence that the R&D of foreign-owned affiliates is most likely to become substantial, and to gain a creative role with respect to the global technological development strategy of the MNC as a whole. In other words, it is where the ownership advantages of investing firms and the location advantages of the host region or country are strongest that we find the greatest potential scope for a process of the mutual reinforcement of these advantages through the two-way spillover effects of internationalisation.

It is self-evident that the interaction between ownership and location advantages has important implications for policies relating to economic growth and development. Countries seek to attract MNC activity as a means of improving their location advantages (and consequently the ownership advantages of local firms), through spillovers and linkages due to MNC activity. However, the quality and extent of the externalities due to MNC activities depends on the motivation of their investment, which is itself dependent on the kinds of location advantages available to them (Narula and Dunning 2000). Even where the 'right kinds' of FDI (i.e. those which provide optimal potential spillovers and linkages) are located in the host country, the ownership advantages of domestic firms need to have the necessary absorptive capability to benefit from them. In other words, there is a powerful dynamic interaction and interdependence between the ownership advantages and location advantages that underlies FDI-supported economic development (Narula 1996).

This ownership-location advantage dynamic and its policy implications has been tackled from several perspectives. There has been a more aggregate and macro-view (see, e.g., Dunning 1981, 1988; Lall 1993; Tolentino 1993; Narula 1996; Ozawa 1992, 1996), while the work of others has focused on particular countries or groups of countries (e.g. Lall (ed.) 1983; Kokko 1992; Blomstrom 1989; van Hoesel 1999).

The chapter by Terutomo Ozawa and Sergio Castello (Chapter 4) falls firmly within this tradition. This chapter examines the role of the MNC in endogenous growth. In particular, they explore the proactive role of governments in promoting FDI-assisted growth, building on previous work on FDI and governments (see the various contributions to Dunning and Narula 1996). Their chapter fleshes out some of the processes that underlie the virtuous circle of technology accumulation (Cantwell 1987), but particularly emphasising the importance of macro-organizational policy, and relating this conceptualisation to received theory on endogenous growth.

This subject is also touched on by Dunning in his contribution here (Chapter 2). He reminds us, *inter alia*, that while the tenets of the eclectic paradigm have remained the same, it is its application to particular issues that has led to its evolution. The economic and socio-political realities within which international business is conducted is continually evolving, and so too has the way in which the eclectic paradigm applied, if it is to serve as a yardstick by which to understand the complexity of the phenomena in question. Dunning – ever-cognisant of new developments – provides a glimpse towards the end of his chapter of the issues that he feels still challenge the IB scholar, and those that might be expected to arise.

Elizabeth Maitland and Stephen Nicholas (Chapter 3) flesh out an important under-theorised 'boundary problem' between ownership and location advantages: the concept of location-specific ownership advantages. They trace the evolution of the literature on ownership advantages, and argue that the growing emphasis in the IB literature on resource-based approaches has tended to focus attention on firm-specifc ownership assets, and away from country-specific (yet also firm-specific) assets. Maitland and Nicholas suggest that concepts from new institutional economics may help better explain why location-specific advantages exist, as it takes a systems view of economic activity. MNCs may derive ownership advantages that are non-replicable in another locations because of the unique nature in which they are linked through formal and informal institutions to other economic and non-economic actors. Their line of reasoning shares much in common with recent work by scholars interested in 'systems' view of economies, where the importance of institutions is also being acknowledged (Edquist and Johnson 1997; Cantwell and Kosmopoulou 2002; Narula 2002a; Coriat and Weinstein 2002).

New perspectives on internalisation advantages

Alliance capitalism implies that the favoured mode of cross-border value-adding activity has begun to shift away from an emphasis on hierarchies towards a richer variety of organisational modes. This has occurred along with a systematic shift towards the dis-integration of the vertically integrated firm. This trend has occurred throughout the value chain, and across industries, but nowhere is this development more starkly observed than in the area of innovation and technology development.

In particular, there has been a tremendous growth in the use of external networks by firms of all sizes (Hagedoorn 1996). Indeed, Duysters *et al.*

(1999) note that alliances have shifted from being regarded as a peripheral aspect to being a cornerstone of the firm's technological strategy. In addition to the declining costs of monitoring and exploiting networks, there has also been a growing need for firms to possess multiple technological competences (Granstrand *et al.* 1997). This trend has largely been a result of the increased knowledge content of products in general, and the cross-fertilisation of previously distinct technological areas. Firms have sought to utilise 'non-internal' means to undertake innovation, and we refer specifically to the use of strategic alliances and outsourcing. This trend has been noted by Veugelers and Cassiman (1999), Archibugi and Iammarino (1999) and Narula (2001) among others. The facilitating role of globalisation has expanded firms' use of external resources to reduce, *inter alia*, innovation time spans, costs and risks, and acquire greater flexibility in their operations (Hagedoorn 1993). The improved enforceability of contracts and declining transaction and monitoring costs resulting from developments associated with globalisation have made it easier for firms of all sizes to monitor, identify and establish collaborative ventures than previously had been the case (Narula 1999, 2001). In other words, hierarchical control and full internalisation is no longer always a first-best option to MNCs, especially where innovatory activities are concerned. Even where this is so, full internalisation may simply not be a choice available to the MNC. Take the case where firm A seeks technological competences that are firm-specific to B, and these are a small aspect of the technological assets of firm B. Further assume that while there are many variations of this technology available from other firms, firm B's design is the dominant one, or is most compatible with firm A's other technological assets. To buy the technology from another firm would mean higher costs. If the technology is unique, firm B is unlikely to want to license. Besides, the technology may be largely uncodifiable. It is in neither firm's interest to engage in arm's-length transaction, and short of acquiring the entire firm B – another expensive option, especially if the technology sought forms only a small part of firm B's assets – it must seek some sort of an alliance. In other words, firms that undertake R&D alliances often have fewer organisational options than alliances in production or sales.

Even smaller technology-based MNCs are nowadays involved in a web of such agreements, and their growing significance raises numerous theoretical conundrums (Narula 2002b). There are sets of issues that raise new perspectives for internalisation advantages, because cooperation is inherently complex as an activity. First, because such increasingly complex linkages, both of networks internal to the firm, and those between external networks and internal networks, require complex coordination if they are to provide optimal benefits (see Zanfei 2000 for a discussion). Such networks are not only difficult to manage, but also require considerable resources (both managerial and financial). It is no surprise, therefore, that external technology development is primarily the domain of larger MNCs with greater resources, and more experience in trans-national activity (Hagedoorn and Schakenraad 1994). But technological assets represent only one aspect of an MNC's ownership advantages. The firm-specific ownership advantages that derive from its ability to manage and coordinate intra-firm and inter-firm transactions (Ot advantages) are central to understanding the competitiveness of firms. Indeed, the lack of technological advantages can be offset by superior Ot advantages, and may be used to develop and acquire technological assets. Managing a web of different types of agreements across borders is not without its price, and highlights the role of Ot advantages in the success of the MNC. On the one hand, allowing for differences in the motivation to locate overseas (which may themselves derive from simple firm- and industry-specific differences), geographical proximity to host locations is important. On the other, the knowledge-based firm needs to seek *internal* proximity between strategically important functions located abroad (such as R&D) and the rest of the MNE (Blanc and Sierra 1999). This acts as a centripetal force on R&D, and accounts for a tendency of firms to locate the most strategically significant aspects of their activities closer to headquarters. A dispersion of activities across the globe also requires extensive coordination between them – and particularly with headquarters – if they are to function in an efficient manner with regards to the collection and dissemination of information.

The contribution of Anoop Madhok and Anupama Phene (Chapter 5) avers that the management of intangible assets is potentially a core, inimitable advantage of the firm. The MNC, in their view, is a sub-economy in its own right. Different forms of organisation act as complements to one another, rather than as substitutes for traditional hierarchical modes. Madhok and Phene suggest that there is an increasing need for the economics and strategic management literatures to converge. While one branch of economic theory (the transaction cost school) has asked why the firm exists, strategic management theory (like the economic theory of the growth of the firm, which derives from Penrose's resource-based approach) is more interested in explaining the different performance among firms, but these, they argue, are co-dependent, and co-evolutionary.

A second major issue is the 'fuzzy border problem' because it becomes increasingly difficult to clearly identify where the boundaries of the firm lie. Internalisation advantages by themselves provide only a partial explanation for the growth in alliances, and only suggest why one group may derive greater benefits from collaboration than other groups. It does not answer why firms increasingly prefer quasi-hierarchical arrangements to fully internalised ones. Cost issues are not always central to explaining the growth of cooperation. If this were so, the decline in monitoring and transaction costs should lead to at least the similar extent of benefits for traditional hierarchical arrangements. In answering this, it is important to reflect on the presence of the word 'strategic' in strategic alliances. What differentiates a strategic alliance from a customer–supplier network is the underlying motive of the cooperation (Narula and Hagedoorn 1999). The primary motivation for a customer–supplier network (essentially an outsourcing agreement, where joint work is minimal) is that it is primarily cost-economising in nature, while strategic alliances embody a second motivation, which is strategic in nature. 'Strategic' suggests that such agreements are aimed at *long-term profit-optimising* objectives by attempting to enhance the value of the firm's assets.

Several reasons exist for the growth in popularity of cooperative agreements which embody a strategic element. One explanation is based on the increased competition due to liberalisation of markets and the globalised nature of the operations of firms. Such increased competition has led to a low-growth scenario over the past two decades or so, and firms need to seek cheaper sources of inputs or divert sales from slow or negative growth markets (Buckley and Casson 1998). Such changes often need to be undertaken with rapidity. Declining transaction costs associated with contractual or quasi-internalised relationships in addition to falling profits margins has led to a disintegration of certain firms in particular industries, as they seek flexibility and lower risk.

Other more strategic reasons exist for preferring alliances to hierarchies. First, firms do not always have recourse to patenting as a means to protect their intellectual property, and must rely on secrecy or co-invention instead (Levin *et al.* 1987). Second, by co-invention, alliances allow firms to monitor competitors, and allow firms that are engaged in conducting similar research. As Narula and Dunning (1998) note, firms may also engage in alliances in order to co-opt the competition. Take the situation where two firms in the same industry are pursuing an important new breakthrough. Neither can be certain that they will win the race to innovate. As such, it may be in their best interest to collaborate, thus ensuring both that they are jointly 'first': half a pie may be considered better in conditions of uncertainty while there is a probability that there may be none at all.

Third, MNCs do not only seek to partner with 'technology leaders'. If this were the case, asset-augmenting activities would remain the exclusive domain of only a handful of firms. Why would a potential partner wish to collaborate with another which has limited or as-yet-undemonstrated resources to offer? First, because of the nature of innovation, the only way to determine the nature of a potential partner's research efforts is to examine them. One way it can do so is by engaging in some form of mutual hostage exchange, which an alliance provides. Second, even where the partner's resources prove to be of a limited or inappropriate nature, and the alliance is terminated prematurely, information about its former partner's competencies are then available to either firm in future periods, should it require competencies similar to those on offer by its ex-partner. Third, as Hagedoorn and Duysters (1997) have argued, while selecting partners that are well-established players in existing technologies may represent a profit maximising situation, it is optimal only in a static environment. In a dynamic environment, where there is a possibility of technological change (or even a change in technological trajectories), having ties to a wide group of companies, including companies that have yet to demonstrate their value, represents a higher learning potential. At the technology frontier where dominant technological designs have not yet been determined and several potential options exist, it pays to have a number of overlapping, redundant agreements. It may be optimal to partner with all sorts of companies, even those without a demonstrated track record. Finally, as Kay (1997) explains, 'it is necessary to engage in networks with certain firms not because they trust their partners, but *in order* to trust their partners'.

Limitations in the applicability of economic theory-based explanations to strategic management issues forms the basis of Stephen Guisinger's contribution (Chapter 6). He suggests that the theoretical perspectives based on an economic framework such as the eclectic paradigm tend to focus on explaining how firms accommodate themselves to their environment. However, its focus has been more aggregated, suggesting that it is insufficient – if it is to continue to be useful to organisation theory – for the OLI framework to merely explain in general terms why a firm exists (using the transaction cost approach) or why it grows successfully relative to other firms (as in the resource-based or technological accumulation approach), but it must also relate to the details of its internal organisational characteristics. The OLI, according to Guisinger, needs to take into account 'more finely grained firm structures such as business processes' from the organisational theory literature, which help determine profitability. MNCs need to both adapt to their environment, and to accommodate themselves in an environmentally complex environment. He proposes a modification of the OLI framework to OLMA (Ownership, Location, Mode of entry and Adjustment).

Inter-firm cooperation is by no means a simple cure to a firm's ills. Considerable risks and costs are associated with such a strategy. As a general rule, firms find it extremely costly and difficult to access competencies from other firms or locations in fields which are unrelated to their own capabilities, and with which they have little initial familiarity; while the internalisation advantages of in-house combinations of activities derive (*inter alia*) from the technological coherence of these activities (Teece *et al.* 1994). This is essentially due to the need for 'absorptive capacity' when the firm acquires knowledge from its external environment or one knowledge-creating part of the firm interacts with another, and which requires the recipient to have some innovative potential of its own to be able to learn and effectively adapt the technologies to which it may wish to have access (Cohen and Levinthal 1989). Thus, inter-firm technological alliances tend to develop in areas in which partner companies share some complementary capabilities, and these alliances create a greater degree of interaction between the partners' respective paths of learning and innovation

(Cantwell and Colombo 2000). However, the extent and form of interaction between learning activities depend upon the organisational form of co-operation. Learning may either continue to be organised relatively autonomously in each partner company in the case of licensing (when the agreement is restricted to an essentially quasi-market exchange of knowledge), or cooperation may be extended to the coordination of learning activities themselves (which is sometimes a feature of joint ventures) (Cantwell and Barrera 1998). This varies considerably by industry. Non-equity forms of agreements tend to be more efficient for undertaking activity in more research-intensive industries, and where technological change is rapid since they promote negotiation and can lead to more intensive cooperation than equity forms. However, where firms seek to learn *and transfer* tacit knowledge back to the parent firm, such as market-specific knowledge when entering a new market, or are engaged in production as well as research, equity forms of agreement may be more appropriate. Equity agreements are preferred in relatively mature sectors, while non-equity agreements are utilised in high-tech sectors (Hagedoorn and Narula 1996).

Inter-company cooperation sometimes involves a coordination of learning processes, but sometimes involves their greater separation or differentiation. Thus, from the competence-based perspective, the issue may not be whether economic activity is organised by hierarchy within the firm, or by markets between firms, or between these two extremes by inter-company cooperative agreements (the transaction cost view); but the issue is rather the extent of cooperative linkages between the learning processes that lie behind economic activities (whether such cooperation occurs under common ownership within the firm, or through the joint development activities of independent companies). From this viewpoint, the central concern is the motivation and objectives of different forms of corporate organisation in the context of the potential (or lack of potential) that exists for learning, and not the differences in the various types of business organisation in themselves.

To reiterate what was emphasised earlier, the eclectic paradigm does not need to be wedded to one particular theory of the firm (in this case the transaction cost approach) to the exclusion of any other. Internalisation advantages or disadvantages in situations of combined or interactive learning are better explained by a competence-based approach to the firm, but this is not necessarily at odds with the use of the transaction cost theory in relation to more stable established activities. Hence, the eclectic paradigm facilitates judgements over how to choose between these theories of the firm where they offer alternative views, and then how they might be synthesised or related to one another once the proper position of each has been recognised.

Part of the answer to the question of why it is that technology is developed in cross-border networks within the MNC is provided by internalisation theory, which focuses on why MNCs as opposed to purely national firms have come into existence. That is, if the initiating firm is to appropriate a full return on its technological advantage, and if it is to coordinate the successful introduction of its new technology elsewhere, then it must exercise direct control over the network as a whole. However, this may be not so much a feature of the market for technological knowledge which is the focus of internalisation theory, as a feature of the very nature of technological development itself. In the alternative evolutionary or resource-based view, technological knowledge is not an immediately usable intermediate product in its own right, but is rather an input into the collective corporate learning process by which tacit capability and hence technology as a whole is generated. As such, it is an input that normally has its greatest relevance to the learning process of the firm that created it and set the problem-solving agenda to which it represents a response, and thus it is likely to be of the greatest value to the originating company (Cantwell 1991, 1994).

Suppose for a moment that the act of exchanging technological knowledge between firms does not present a problem, in that a reasonable price for such an exchange can always be readily agreed, such as in a framework of cross-licensing agreements. Now consider an international industry in which constituent firms produce more or less identical products for the same international markets. However, each firm has its own quite specific process technology, derived from a distinct technological tradition (say, different chemical processes with a similar end result). In this situation, if technological accumulation is continuous in each firm, raising its productivity or lowering its costs along a given line of technological development, then no existing firm would abandon its existing pattern of innovation and buy in all its technological knowledge from a competitor. It would be far more costly, and perhaps even infeasible, for an existing firm to switch into a completely new line of technological development, by comparison with the costs of the potential seller of technology simply extending its own network. It is because technology is differentiated across countries even within the same firm, but especially between different firms, that technology transfer is a costly process (as demonstrated by Teece 1977). Some exchanges of technological knowledge between existing firms will take place, since alternative lines of technological accumulation in the same industry are often complementary to one another, and so spillovers occur and may be facilitated through inter-company alliances in which knowledge is exchanged and occasionally jointly developed. However, where technological knowledge is brought in it must be adapted to the specific context of the firm's own tacit capability (the other necessary component of any operational technology) and then incorporated into an existing stream of innovation, and this adaptation becomes part and parcel of the on-going process within an established firm of generating its own technology.

In the case outlined, the retention of technology within each firm has little to do with any failure or malfunctioning of the market for technological knowledge, but everything to do with the close association between the generation and the utilisation of a distinctive type of technology within each firm. By extending its own network, each firm extends the use of its own unique line of technological development, and by extending it into new environments it increases the complexity of this development. The expansion of international production thereby brings gains to the firms as a whole, as the experience gained from adapting its technology under new conditions feeds back new ideas for development to the rest of its system. For this reason, once they have achieved a sufficient level of technological strength in their own right, firms are particularly keen to produce in the locations from which their major international rivals have emanated, which offer them access to alternative sources of complementary innovation. Thus, on this evolutionary interpretation of the eclectic paradigm internalisation advantages are attributable to the conditions for improved organisational learning and technology creation, rather than to the conditions for a more efficient cost-minimising organisation of an established set of transactions.

In Chapter 7, Paz Estrella Tolentino attempts to bridge the schism between the interpretations of the concept of internalisation between those that claim internalisation (the transaction cost approach) is itself a general theory, and the concept of internalisation which allows for other potential theoretical explanations of firm behaviour within the eclectic paradigm. She does so by proposing that firms may exercise common control over assets used in geographically dispersed locations either because the markets for these assets are internalised to reduce the transaction costs of exchange (in which case internalisation is a sufficient theory of the multinational dimension of the firm), or instead because the distant use of the assets is connected to the internal ability of the firm to continue to generate endogenously such assets (in which case a resource-based or competence-based approach is relevant to the specific combination of ownership and internalisation).

Applications of the paradigm: future directions and challenges

As the various chapters in this book demonstrate, the eclectic paradigm is a fundamental tool for analysis at several levels, and from numerous perspectives. Its versatility is consistent with a number of prevailing theoretical approaches, including transaction costs theory, resource based theory of the firm, evolutionary theory, the product life cycle theory, new institutional economics, to mention a few.

That it has become the underpinning of a vast literature attests to its staying power. It is not a theory in itself, but through its application to particular managerial and economic issues, it has inspired various theories and other frameworks which fill this void. We feel that much of the criticism that the eclectic paradigm faces is a result of misunderstanding the fundamental difference between the eclectic paradigm and eclectic paradigm-*inspired* theories.

A colleague of ours once described the eclectic paradigm as an intellectual coat-hanger. Just as a coat-hanger is not restricted in its use to coats but to all manner of clothing, so too, she remarked, was it with the eclectic paradigm. Its generality allows it to be applied to a wider context of issues, some that only peripherally relate to the MNC.

In applying the OLI framework to a particular line of inquiry, theorists have merged other frameworks and disciplines to 'flesh out' and contextualise the subject under investigation. Occasionally these theories fail the test of time, or prove otherwise wanting, but their inapplicability does not in itself negate the usefulness of the eclectic paradigm. Other 'marriages' between the eclectic paradigm and external frameworks and theories have survived and continue to evolve in their new, 'fused' identity.

Chapter 8 by Timothy Devinney, David Midgley and Sunil Venaik – possibly the most ambitious contribution theoretically – follows in this tradition. Devinney *et al.* seek to clarify the eclectic paradigm so that it might be more structured, and empirically verifiable. They argue the OLI framework – when applied to understanding the evolution of MNCs' strategy and structure – lacks dynamic character, because it is ill-equipped to deal cause-and-effect on at least two levels. First, the O, L and I pillars are not independent constructs and affect each other. Second, the relationship between firm, industry and country level variables has not been systematically developed.

Devinney *et al.*'s approach to understanding an MNC's entry strategy rests on a separation of endogenous effects from exogenous effects. They introduce concepts from contingency theory as well as strategic management, arguing that 'what matters is the fit between structure, environment and managerial beliefs about structure and environment'.

Chapter 9 by Gabriel Benito and Sverre Tomassen also applies the OLI framework to systematically analyse the under-researched area of MNC performance, in the sense in particular of inter-MNC variety and hence differentiated types and degrees of success, rather than performance in the more usual but less challenging sense of a uniform criterion that can be applied across MNCs in general. While they too – like Devinney *et al.* – seek to apply the eclectic paradigm to expand our theoretical understanding of individual MNCs, they do not seek so much to integrate the O, L and I variables, but rather to examine the relationship of each to performance independently. Their analysis follows a rich tradition of bringing out and comparing or integrating the sub-theories that underlie the eclectic paradigm.

One of the primary motivations of this book (and the special issue on which this volume is partly based) has been to assess the eclectic paradigm in light of changing ways in which business – and particularly international business – is conducted. These are primarily associated, but not restricted to, the effects of economic globalisation, and the consequent growth of global capitalism, alliance activity and asset-augmenting FDI.

These developments are significant, as they have potentially changed fundamental definitions. For instance, transactions are not always associated with material goods, and are no longer always *physically* traded in the traditional sense. The concept of location has become more tenuous, and the seemingly immutable idea of national borders brought into question, as nation states are affected by the commercial activities of firms outside their sovereignty. Firms, likewise, are influenced by policies of political entities outside their production base. The intervention of the European Commission in the merger of Boeing and McDonnell Douglas (both with production almost wholly within the borders of the US) illustrates this well. Internalisation issues have received the most attention, as the conventional belief that full internalisation is the preferred mode of MNCs has been challenged by the growing use of strategic alliances.

These developments inevitably raise the 'fuzzy border problem'. Borders have become increasingly unclear, be they of nation states, or of firms. The growing reliance upon inter-firm cooperation – both through outsourcing and through strategic alliances – underlie firms' 'fuzzy' boundaries (Narula 2003). The fuzziness arises from lack of clarity in what constitutes an economic entity. At what point can relations between economic entities, which are physically independent and geographically disparate, be seen to be inter-firm (or at the country level, international), and when is it regarded as intra-firm (domestic)? But at what point do transactions become inter-firm? What constitutes an MNC?

Most of the chapters in this volume address the rising complexity of these processes, and the responsiveness of the eclectic paradigm to these challenges. By and large, the eclectic paradigm proves able to handle these developments.

Although most of the Triad (Europe, North America and Japan) countries are increasingly post-industrial in the sense that a dominant and growing share of value-adding activities in these economies are in the tertiary sector, the role of the MNC tends to be under-researched. Two chapters in this book take a specifically non-manufacturing view of the eclectic paradigm. In Chapter 12, Lucia Piscitello undertakes a study of the banking sector, and demonstrates the applicability of the eclectic paradigm. She also provides an excellent general literature review of the eclectic paradigm to the services sector. Piscitello's objective is not theory-building, but to empirically test the applicability of the eclectic paradigm using a unique data set on the internationalisation of Italian banks. Chapter 10, by Lars Oxelheim, Trond Randøy and Arthur Stonehill also deals with the services sector. They delve into the specific and peculiar nature of finance-specific O, L and I factors.

The chapter by Cliff Wymbs and John Dunning (Chapter 11) deals squarely with a topic at the heart of the 'new economy'. On the one hand, e-commerce is a facilitator of international business, creating new virtual markets that both substitute and complement existing markets and hierarchies. This raises new perspectives on internalisation advantages. On the other hand, e-commerce is an industry by itself, and this has created alternative markets and hierarchies devoted to the generation of economic rent. In other words, it is a value-generating activity, just as any other. Activities within the e-commerce industry – allowing for differences in context – can broadly be explained by the eclectic paradigm. Firms involved in e-commerce, according to Wymbs and Dunning, seek to leverage the Internet to exploit or augment existing ownership advantages by choosing locations that best suit their purposes in much the same way as conventional MNCs. At the same time, they highlight the unique features of electronic markets as markets *per se*. Electronic markets provide an alternative means of conducting transactions when traditional markets fail, without resorting to hierarchical activity.

Relative to the other contributions, Chapters 9, 10, 11 and 12 stand out in that they do not focus much on the interaction of the OLI variables. Most of the other chapters evaluate the dynamics between, and amongst 'O', ' L' and 'I', and to understanding their applicability to particular issues. They seek to relate the eclectic paradigm with other frameworks and supplementary theories to address specific questions.

In the concluding Chapter 13, Lorraine Eden revisits the essential elements or building blocks of the OLI paradigm, and traces out how each of these 'O', ' L' and 'I' constructs came to be refined and depicted increasingly purposefully over time. She focuses in particular on the gradual development of John Dunning's own thinking in this regard, and on how he has reformulated the eclectic paradigm in response both to critics, and to changes in the real world environment and the issues it raises for discussion. The chapter goes back to questions that are brought up by various of our other authors, most notably as to how well the OLI paradigm can accommodate or has accommodated firm-specific strategy variables, and the shift from early internationalisation to the modern cross-border management of knowledge networks. Eden's overall assessment is that it may have become the case that the range of international business issues which concern scholars has extended so much that we cannot always rely on a single overarching framework to act as the starting point for the examination of all the questions or issues which may be raised. However, she concludes that we do still very much need such an overall analytical framework to provide our discipline with coherence in its wider and most ambitious theoretical approaches, and that the eclectic paradigm is still easily the best such construction that we have available to us, not least because of the way in which it has been continuously updated and reconsidered in line with contemporary events.

Although there is some disagreement about what might be the most appropriate level of analysis, the validity of particular findings, and how the eclectic paradigm may be best applied, it serves its fundamental purpose, as a framework *per se*. It seems to us that with debates such as those within the covers of this volume, only the most foolhardy would disagree with our prognosis that the eclectic paradigm is alive and well.

References

Almeida, P. (1996) 'Knowledge sourcing by foreign multinationals: patent citation analysis in the US semiconductor industry', *Strategic Management Journal*, 17: 155–65.

Archibugi, D. and Iammarino, S. (1999) 'The policy implications of the globalisation of innovation', *Research Policy*, 28: 317–36.

Benito, G. and Welch, L. (1994) 'Foreign market servicing: beyond choice of entry mode', *Journal of International Marketing*, 2(2): 7–27.

Blanc, H. and Sierra, C. (1999) 'The internationalisation of R&D by multinationals: a trade-off between external and internal proximity', *Cambridge Journal of Economics*, 23: 187–206.

Blomstrom, M. (1989) *Foreign Investment and Spillovers: A Study of Technology Transfer to Mexico*, London: Routledge.

Buckley, P.J. and Casson, M.C. (1976) *The Future of the Multinational Enterprise*, London: Macmillan.

Buckley, P. and Casson, M. (1998) 'Models of the multinational enterprise', *Journal of International Business Studies*, 29(1): 21–44.

Cantwell, J.A. (1987) 'The reorganisation of European industries after integration: selected evidence on the role of multinational enterprise activities', *Journal of Common Market Studies* (reprinted in J.H. Dunning and P. Robson, P. (eds), 1988, *Multinationals and the European Community*, Oxford: Basil Blackwell), 26: 127–51.

Cantwell, J.A. (1989) *Technological Innovation and Multinational Corporations*, Oxford: Basil Blackwell.

Cantwell, J.A. (1991) 'The theory of technological competence and its application to international production', in D. McFetridge (ed.), *Foreign Investment, Technology and Economic Growth*, Calgary: University of Calgary Press.

Cantwell, J.A. (1994) 'Introduction', in J.A. Cantwell (ed.), *Transnational Corporations and Innovatory Activities*, London: Routledge.

Cantwell, J.A. (1995) 'The globalisation of technology: What remains of the product cycle model?', *Cambridge Journal of Economics*, 19: 155–74.

Cantwell, J.A. (2000a) 'A survey of theories of international production', in C.N. Pitelis and R. Sugden (eds), *The Nature of the Transnational Firm*, London: Routledge, Second Edition.

Cantwell, J.A. (2000b) 'Technological lock-in of large firms since the interwar period', *European Review of Economic History*, 4: 147–74.

Cantwell, J.A. and Barrera, M.P. (1998) 'The localisation of corporate technological trajectories in the interwar cartels: cooperative learning versus an exchange of knowledge', *Economics of Innovation and New Technology*, 6: 257–90.

Cantwell, J.A. and Colombo, M.G. (2000) 'Technological and output complementarities, and inter-firm cooperation in information technology ventures', *Journal of Management and Governance*, 4: 117–47.

Cantwell, J.A. and Iammarino, S. (1998) 'MNCs, technological innovation and regional systems in the EU: some evidence in the Italian case', *International Journal of the Economics of Business*, 5: 383–408.

Cantwell, J.A. and Iammarino, S. (2000) 'Multinational corporations and the location of innovation in the regions', *Regional Studies*, 34: 317–32.

Cantwell, J.A. and Janne, O.E.M. (1999) 'Technological globalisation and innovative centres: the role of corporate technological leadership and locational hierarchy', *Research Policy*, 28: 119–44.

Cantwell, J.A. and Kosmopoulou, E. (2002) 'What determines the internationalisation of corporate technology?' in V. Havila, M. Forsgren, and H. Håkanson, (eds), *Critical Perspectives on Internationalisation*, Oxford: Pergamon.

Cantwell, J.A. and Piscitello, L. (2000) 'Accumulating technological competence – its changing impact on corporate diversification and internationalisation', *Industrial and Corporate Change*, 9: 21–51.

Cantwell, J. and Sanna-Randaccio, F. (1990) 'The growth of multinationals and the catching up effect', *Economic Notes*, 19 (July): 1–23.

Casson, M.C. (1986) 'General theories of the multinational enterprise: a critical examination', in G. Jones and P. Hertner (eds), *Multinationals: Theory and History*, Farnborough: Gower.

Coase, R.H. (1937) 'The nature of the firm', *Economica*, 4: 386–405.

Cohen, W.M. and Levinthal, D.A. (1989) 'Innovation and learning: the two faces of R&D', *Economic Journal*, 99: 569–96.

Coriat, B. and Weinstein, O. (2002) 'Organizations, firms and institutions in the generation of innovation', *Research Policy*, 31: 273–90.

Dosi, G. (1982) 'Technological paradigms and technological trajectories: a suggested interpretation of the determinants and directions of technical change', *Research Policy*, 2: 147–62.

Dunning, J.H. (1977) 'Trade, location of economic activity and the MNE: a search for an eclectic approach', in B. Ohlin, P.O. Hesselborn and P.M. Wijkman (eds), *The International Allocation of Economic Activity*, London: Macmillan.

Dunning, J.H. (1979) 'Explaining changing patterns of international production: in defence of the eclectic theory', *Oxford Bulletin of Economics and Statistics*, 41: 269–95.

Dunning J.H. (1981) 'Explaining the international direct investment position of countries: towards a dynamic or developmental approach', *Weltwirtschaftliches Archiv*, 119: 30–64.

Dunning, J.H. (1988) *Explaining International Production*, London: Unwin Hyman.

Dunning, J.H. (1995) 'Reappraising the eclectic paradigm in an age of alliance capitalism', *Journal of International Business Studies*, 26: 461–91.

Dunning, J.H. (1996) 'The geographical sources of the competitiveness of firms: some results of a new survey', *Transnational Corporations*, 5: 1–29.

Dunning, J.H. (1997) *Alliance Capitalism and Global Business*, London: Routledge.

Dunning, J.H. and Narula, R. (1995) 'The R&D activities of foreign firms in the United States', *International Studies of Management & Organization*, 25(1–2): 39–73.

Dunning, J.H. and Narula, R. (1996) *Foreign Direct Investment and Governments: Catalysts for Economic Restructuring*, London: Routledge.

Duysters, G., Kok, G. and Vaandrager, M. (1999) 'Crafting successful strategic partnerships', *R&D Management*, 29: 343–51.

Edquist, C. and Johnson, B. (1997) 'Institutions and organisations in systems of innovation', in C. Edquist (ed.), *Systems of Innovation: Technologies, Institutions and Organisations*, London and Washington: Frances Pinter.

Frost, T.S. (1996) 'From exploitation to exploration: the geographic sources of subsidiary innovations and the evolutionary theory of the multinational enterprise', *mimeo,* University of Western Ontario.

Granstrand, O., Patel, P. and Pavitt, K. (1997) 'Multi-technology corporations: why they have "distributed" rather than "distinctive core" competencies', *California Management Review*, Summer, 39(4): 8–25.

Hagedoorn, J. (1993) 'Understanding the rationale of strategic technology partnering: interorganizational modes of cooperation and sectoral differences', *Strategic Management Journal,* 14: 371–85.

Hagedoorn, J. (1996) 'Trends and patterns in strategic technology partnering since the early seventies', *Review of Industrial Organization,* 11: 601–16.

Hagedoorn, J. and Duysters, G. (1997) 'Satisficing strategies in dynamic inter-firm networks – the efficacy of quasi-redundant contacts', *MERIT working paper series* 97–106.

Hagedoorn, J. and Narula, R. (1996) 'Choosing modes of governance for strategic technology partnering: international and sectoral differences', *Journal of International Business Studies*, 27: 265–84.

Hagedoorn, J. and Narula, R. (2001) 'Evolutionary understanding of corporate foreign investment behaviour: US Foreign Direct Investment in Europe', in R. Narula, (ed.), *International Trade, Investment in a Globalising World,* New York: Pergamon.

Hagedoorn, J. and Schakenraad, J. (1994) 'The effect of strategic technology alliances on company performance', *Strategic Management Journal,* 15: 291–311.

Hennart, J.-F. (1982) *A Theory of Multinational Enterprise*, Ann Arbor, MI: University of Michigan Press.

Hoesel, R. van (1999) *New Multinational Enterprises from Korea and Taiwan*, London: Routledge.

Jaffe A.B., Trajtenberg, M. and Henderson, R. (1993) 'Geographic localization of knowledge spillovers as evidenced by patent citations', *Quarterly Journal of Economics*, 63: 577–98.

Kay, N. (1997) *Pattern in Corporate Evolution*, Oxford: Oxford University Press.

Kogut, B. and Chang, S.J. (1991) 'Technological capabilities and Japanese foreign direct investment in the United States', *Review of Economics and Statistics*, 73: 401–13.

Kokko, A. (1992) *Foreign Direct Investment, Host Country Characteristics, and Spillovers*, Stockholm: Stockholm School of Economics.

Kuemmerle, W. (1999) 'The drivers of foreign direct investment into research and development – an empirical investigation', *Journal of International Business Studies*, 30: 1–24.

Lall, S. (1983) *Third World Multinationals*, Chichester: John Wiley.

Lall, S. (1993) 'Transnational corporations and economic development', in S. Lall (ed.), *Transnational Corporations and Economic Development*, United Nations Library on Transnational Corporations, London: Routledge.

Levin, R., Klevorick, A., Nelson, R. and Winter, S. (1987) 'Appropriating the returns from industrial research and development', *Brookings Papers on Economics Activity*, 3: 783–820.

Mariotti, S. and Piscitello, L. (2001) 'Localised capabilities and internationalisation

of manufacturing activities by SMEs', *Entrepreneurship & Regional Development*, 13: 65–80.
Narula, R. (1996) *Multinational Investment and Economic Structure*, London: Routledge.
Narula, R. (1999) 'Explaining strategic R&D alliances by European firms', *Journal of Common Market Studies*, 37(4): 711–23.
Narula, R. (2001) 'Choosing between modes of non-internal technological activities by firms: some technological and economic factors', *Technology Analysis and Strategic Management*, 13: 152–70.
Narula, R. (2002a) 'Innovation systems and "inertia" in R&D location: Norwegian firms and the role of systemic lock-in', *Research Policy*, 31: 795–816.
Narula, R. (2002b) 'R&D collaboration by SMEs: Some analytical issues and evidence', in F. Contractor and P. Lorange (eds), *Cooperative Strategies and Alliances*, Oxford: Pergamon Press.
Narula, R. (2003) *Globalization and Technology*, Cambridge: Polity Press.
Narula, R. and Dunning, J.H. (1998) 'Explaining international R&D alliances and the role of governments', *International Business Review*, 7: 377–97.
Narula, R. and Dunning, J.H. (2000) 'Industrial development, globalisation and multinational enterprises: new realities for developing countries', *Oxford Development Studies*, 28(2).
Narula, R. and Hagedoorn, J. (1999) 'Innovating through strategic alliances: Moving towards international partnerships and contractual agreements', *Technovation*, 19: 283–94.
Ozawa, T. (1992) 'Theory of FDI as a dynamic paradigm of economic development', *Transnational Corporations*, 1(1), February.
Ozawa, T. (1996) 'Japan: The macro-IDP, meso-IDPs and the Technology Development Path (TDP)', in J.H. Dunning and R. Narula (eds), *Foreign Direct Investment and Governments: Catalysts for Economic Restructuring*, London: Routledge.
Pavitt, K.L.R. (1987) 'International patterns of technological accumulation', in N. Hood and J.E. Vahne (eds), *Strategies in Global Competition*, London: Croom Helm.
Pearce, R.D. (1999) 'Decentralised R&D and strategic competitiveness: globalised approaches to generation and use of technology in MNEs', *Research Policy*, 28: 157–78.
Petersen, B. and Welch, L. (2001) 'Foreign operation mode combinations and internationalisation', *Journal of Business Research*, 55: 157–62.
Pugel, T.A., Kragas, E.S. and Kimura, Y. (1996) 'Further evidence on Japanese direct investment in US manufacturing', *Review of Economics and Statistics*, 78: 208–13.
Read, R. (1986) 'The banana industry: oligopoly and barriers to entry', in M.C. Casson *et al.* (eds), *Multinationals and World Trade*, London: Allen and Unwin.
Rosenberg, N. (1976) *Perspectives on Technology*, Cambridge: Cambridge University Press.
Teece, D.J. (1977) 'Technology transfer by multinational firms: the resource costs of transferring technological know-how', *Economic Journal*, 87: 242–61.
Teece D.J., Rumelt, R., Dosi, G. and Winter, S.G. (1994) 'Understanding corporate coherence: theory and evidence', *Journal of Economic Behavior and Organization*, 23: 1–30.
Tolentino, P.E. (1993) *Technological Innovation and Third World Multinationals*, London: Routledge.

Verspagen, B. and Schoenmakers, W. (2000) 'The spatial dimension of knowledge spillovers in Europe: evidence from firm patenting data', *MERIT Working Paper* no. 2/20–16.

Veugelers, R. and Cassiman, B. (1999) 'Make and buy in innovation strategies: evidence from Belgian manufacturing firms', *Research Policy*, 28: 63–80.

Zander I., (1997) 'Technological diversification in the multinational corporation: historical evolution and future prospects', *Research Policy*, 26: 209–28.

Zander, I. (1999) 'How do you mean "global"? An empirical investigation of innovation networks in the multinational corporation', *Research Policy*, 28: 195–214.

Zanfei, A. (2000) 'Transnational firms and the changing organisation of innovative activities', *Cambridge Journal of Economics*, 24: 515–42.

2 The eclectic (OLI) paradigm of international production

Past, present and future

*John H. Dunning**

Its origins

Although the eclectic paradigm (or the eclectic theory as it was initially called) of international production was first put forward by the present author at a Nobel Symposium in Stockholm in 1976, its origins can be traced back to the mid-1950s. At that time, I was writing my PhD thesis, later to be published as a book (Dunning 1958), on US direct investment in British manufacturing industry. Earlier research by Rostas (1948), Frankel (1955) and some Anglo-American study teams[1] had shown that the labour productivity in US manufacturing industry was, on average, 2-5 times higher than that in UK industry. The question this fact posed in my mind was: 'was this difference in productivity a reflection of the superior indigenous (and immobile) resources of the US (cf. the UK) economy; or was it due to the more proficient way in which the managers of US firms (cf. UK firms) harnessed and organised these resources?' – a capability which, I argued, might be transferable, at least to some extent across national boundaries.

The hypothesis of my thesis then was, if the superior productivity was entirely managerially related, US manufacturing affiliates in the UK should perform at least as well as their parent companies, and fare considerably better than their indigenous competitors. This I identified as the *ownership* specific effect, as the productivity differences were presumed to rest on the spatially transferable intangible assets of the parent companies. If, however, the US affiliates in the UK recorded no better performances than their UK competitors, and hence, much poorer than that of their parent companies, I hypothesised that this would be due to the non-transferable (i.e. immobile) characteristics of the US economy. This I called the *location-specific* component of any productivity differential.

As might be expected, I discovered that US affiliates were not as productive as their parent companies, but were more productive than their local

* This chapter draws on various past contributions of the author, but most particularly on those of Dunning (2000a and b).

competitors. This then suggested that, in the 1950s, at least, Anglo-American productivity differences were partly explainable by location (L) and partly by ownership (O) specific characteristics. However, my study omitted to ask a follow-up question, namely, to what extent was the origin of the O advantages of US firms itself home country-specific? Neither did it attempt to distinguish between those O advantages that arose *as a consequence* of US direct investment in the UK, and those that the US firms possessed *prior* to engaging in foreign production.

I took up the theme of ownership and location advantages again in two papers written in the early 1970s. The first (Dunning 1972), concerned the likely impact of Britain's membership of the European Common Market (ECM). In it, I suggested that while the removal of tariff barriers would cause some realignment of the location of economic activity within the ECM, it would also be likely to affect the competitive position of firms of different national origins, and, in consequence, the ownership of production in the ECM.[2]

The second paper (Dunning 1973) was an attempt to review the various attempts which had been made to explain the activities of firms outside their national boundaries over the past decade. In that contribution I tried to integrate the industrial organisational and locational determinants of international production. I argued (like Hymer 1960, 1976) that while the first was necessary to explain why the affiliates of foreign firms could compete successfully with domestic firms in supplying the latter's markets, the second was relevant to explaining why the former firms chose to supply their markets from a foreign, rather than from a domestic, base.

In 1975, I was asked to present a paper at a Nobel Symposium on the International Location of Economic Activity which was held in Stockholm in June 1976. This symposium was organised by Bertil Ohlin and attended by leading international economists, economic geographers and regional scientists. For the most part, the seminar was oriented towards an evaluation of country-specific factors influencing the changing distribution of international economic activity, but the starting-point of my paper was that a country's economic space could be considered in two ways. The first was the value of output produced within its national boundaries independently of the ownership of that production. The second was the output produced by its own firms, including that part produced outside its national boundaries. This further distinguished between the competitive advantages of countries and that of firms which I have been at pains to stress in several of my writings.[3]

In explaining the activity of firms outside their national boundaries, I extended the O and L advantages identified in my earlier research to include another set of choices available to firms, which related to the way the firms organised the generation and use of the resources and capabilities within their jurisdiction and those they could access in different locations. In other words, I acknowledged that, to fully explain the extent and pattern of the foreign value-added activities of firms, one also had to explain why such firms opted to internally generate and/or exploit their O-specific advantages, rather than to acquire and/or sell these, or their rights, through the open market. Such advantages I referred to as *internalisation* (I) advantages; and these became the third leg of the ownership, location and internalisation (OLI) tripod in explaining the scope and geography of value-added activities by multinational companies (MNCs).

I would be the first to admit that, in my work on I advantages, I was considerably influenced by my colleagues Peter Buckley and Mark Casson of the University of Reading, who were in the process of writing *The Future of the Multinational Enterprise* (1976), although my first exposure to the concept of internalisation, as applied to the MNC, came in 1972 when I read an article by J.C. McManus (1972) in Guy Paquet's edited volume on *The Multinational Firm and the Nation State.* A year later, on a visit to Uppsala in Sweden, I had conversations with Nils Lundgren, a Swedish economist who was thinking along the same lines in his attempt to explain the growth of Swedish foreign direct investment (FDI).[4] Nevertheless, I regarded this new insight as a useful addition to my own approach in explaining the determinants of the foreign production, and not a replacement of it, a view I still hold today.

Over the past two and one-half decades, I have benefited enormously from the comments of friends and colleagues on the eclectic paradigm of international production.[5] I accept that in my earlier work I did tend to look at these advantages more as those that arose from the way O advantages were exploited rather than as a market replacement activity which conferred its own hierarchical advantages. While, as set out in my book *Multinational Enterprises and the Global Economy* (Dunning 1993a), I still prefer to think of O advantages as any kind of income-generating asset that allow firms to engage in foreign production, I readily acknowledge that these may arise as a direct consequence of cross-border market-replacing activities. But, even where this is the case, I believe that a firm's ability to benefit from such activities must be related to the assets which it possesses *prior* to the act of internalisation.

The economies of common governance arise because a firm integrates its existing activities with new activities. For example, a firm that is currently producing in country A, and believes it will benefit from the economies of scope and by the diversification of risks if it produces in country B, will gain from such diversification only if it produces in both country A and country B! A firm that benefits from the cross-border economies of scope or scale will only do so if the new investment is *in addition* to its existing investment. A firm that makes a foreign acquisition to obtain new and up to date technology or managerial capabilities presumably does so because it believes it can use such assets along with its existing core competences in a way which will protect or augment its competitive position.[6] This may seem an obvious point, but, to me at any rate, the distinction between the benefits that accrue from the gains to be had from internalising the market of an existing asset and those that arise from coordinating existing assets with new assets, *vis-à-vis* some alternative use which might be made of those assets, is an important one.

The key propositions of the eclectic paradigm

Let me now reiterate the propositions of the eclectic paradigm.[7] The subject to be explained is the extent and pattern of international production, i.e. production financed by FDI and undertaken by MNCs. The paradigm avers that, at any given moment of time, this will be determined by the configuration of three sets of forces:

1 The (net) competitive advantages which firms of one nationality possess over those of another nationality in supplying any particular market or set of markets. These advantages may arise either from the firm's privileged ownership of, or access to, a set of income-generating assets,[8] or from their ability to coordinate these assets with other assets across national boundaries in a way that benefits them relative to their competitors, or potential competitors.
2 The extent to which firms perceive it to be in their best interests to internalise the markets for the generation and/or the use of these assets, and by so doing add value to them.
3 The extent to which firms choose to locate these value-adding activities outside their national boundaries.

The eclectic paradigm further avers that the significance of each of these advantages and the configuration between them is likely to be context specific, and in particular, is likely to vary across industries (or types of value-added activities), regions or countries (the geographical dimension) and among firms. Thus there are likely to be country-specific differences in the ownership advantages of (say) Korean firms compared with (say) Canadian firms. The extent of market failure influencing whether or not the market for technology is internalised is likely to be different in (say) the wood and pulp industry than in (say) the semi-conductor industry; while the relationship to the comparative locational advantages of Thailand and Taiwan as a manufacturing base for motor vehicles may be differently regarded by (say) the Toyota than (say) the Honda Corporation.

In my more recent writings, I have argued that the eclectic paradigm is best regarded as a framework for analysing the determinants of international production rather than as a predictive theory of the MNC *qua* MNC. I have frequently asserted that no single theory can be expected to satisfactorily encompass all kinds of foreign-owned value-added activity simply because the motivations for, and expectations from, such production vary a great deal. The variables necessary to explain import-substituting

based upon the O advantages of the investing firms in time t may well affect the L advantages of the host country in time $t + 1$; while the response of firms, by use of either a 'voice' or an 'exit' strategy, to market failure (Hirschman 1970) and/or their choice of location for their innovating activities, might critically affect the shape of their future O advantages. Indeed, I would go further and suggest it is the successful co-ordination of the O advantages of foreign and domestic firms with their own L advantages, and how each affects and is affected by the modality of resource deployment, that determines the extent to which a particular country is able to sustain, or upgrade its wealth-creating capacities over a period of time.[11]

No role for strategy: a static approach?

It has also been argued that the eclectic paradigm insufficiently allows for differences in the strategic response of firms to any given configuration of OLI variables. This criticism may be coupled with another that suggests the paradigm is couched in static (or comparatively static) terms and offers little guidance as to the dynamics of the internationalisation process of firms (or of countries). In my book *The Globalization of Business* (Dunning 1993b), I took on board this criticism (which, incidentally, may also be levelled at the internalisation theory of the MNCs). My reasoning is as follows. At a given moment of time, the extent and pattern of MNC activity represents a point on a set of trajectories towards (or, for that matter, away from) their internationalisation path. That trajectory itself is set by the continuous and iterative interaction between the OLI configuration over successive time periods and the strategy of firms in response to these configurations, that, in turn, will influence the OLI configuration in a subsequent moment of time. Let OLI_{t0} be the OLI configuration in time $_{t0,}$ OLI_{t1} the OLI configuration in time $_{t1}$, S_{t-n} the past (i.e. pre t_0) strategies of firms still being worked out, and be any change in the strategic response of firms to that configuration between time t_0 and t_1. Then, *ceteris paribus,*

$$OLI_{t1}=f(OLI_{t0}S_{t-n}\Delta S_{t1\rightarrow t2}) \tag{1}$$

If we extend the analysis to a second time period t_2, then

$$OLI_{t2}=f(OLI_{t1}S_{t-n}\Delta S_{t1\rightarrow t2}) \tag{2}$$

This analysis further suggests that and determine the path of the movement from OLI_{t0} to OLI_{t2}.

Strategic response is, of course, just one of the many *endogenous* variables which might affect the OLI configuration of firms (mainly by its impact on O and I advantages). Others include technological and/or organisational innovations, changes in the composition of senior management, increases in labour productivity, new marketing techniques, mergers and acquisitions, and so on. No less significant are *exogenous* changes, such as changes in: population; raw material prices; exchange rates; national government policies; actions taken by international agencies; and so on. If we take all endogenous variables other than strategy, to be *EN* and all exogenous variables to be *EX*, and we assume that changes in *EN* and *EX* do not affect the firms' strategies, then we can rewrite equation (1) as

$$OLI_{t1}=f(OLI_{t0}S_{t-n}\Delta S_{t0\rightarrow t1}\Delta EN_{t0\rightarrow t1}\Delta EX_{t0\rightarrow t1}) \quad (3)$$

Equation 2 can be similarly reconstructed and it is easy to incorporate any change in strategy which embraces the response to ΔEN and ΔEX if it occurs before $_{t1}$ is reached by adding * to $\Delta S_{t0\rightarrow t1}$ in the equation.

Of course, it may be argued that this drives a coach and horses through the generality of eclectic paradigm; as the nature of the interaction between the value of most exogenous and endogenous variables likely to affect international production and the strategy of firms are difficult to predict. Yet, from the time of Vernon (1966) onwards, economists and business analysts have been trying to do just that. In one of his later writings, Vernon (1974), for example, suggested that both the strategy of firms and the locational advantages of at least some value-added activities associated with the production of a product change as that product moves through its cycle. Thus a firm's O and L position affecting investment in time $t + 1$ (e.g. the mature stage of the product cycle) is affected both by its OL configuration in the early (i.e. the innovating) stage of the cycle, and by the changes in the exogenous variables, e.g. demand by the foreign customers, and endogenous variables, e.g. the presence (or absence) of economies of plant size, and any changes in the strategy of firms consequential upon these eventualities.

Later scholars have more explicitly introduced a time- and strategy-related dimension into their analysis. Again, reinterpreting Knickerbocker's analysis in terms of the OLI paradigm, we may say that firms are prompted to go overseas, in part at least, because they consider their O advantages are (or could become) threatened, if they do not follow their competitors' lead or because their advantages would be less without their presence.[12] In other words, the strategy followed by firms in response to a given OLI configuration in time t_0 is governed by their desire to protect or influence that configuration in t_1. (This incidentally does not necessarily mean that all firms will engage in more FDI.)

The Kojima criticism of the eclectic paradigm

Let me next turn to Kiyoshi Kojima's criticism of the eclectic paradigm (Kojima 1982). To Kojima, my approach, and that of the internalisation scholars, is purely a micro-economic phenomenon. Indeed, he seems to assume that the internalisation and eclectic paradigms are trying to explain the same phenomenon. They are not. As far as I am aware, no one from the internalisation school has sought to explain the changing propensity of *countries* to invest, or be invested in, over time.

Nevertheless, Kojima is right in supposing my macro-economic perspective is different from his. Let me give an analogy: suppose the subject for explanation is the trade in goods. Kojima would be interested in answering the question 'Why does one country export certain types of goods and import other kinds of goods?', whereas I would be concerned with explaining whether a particular country was a net importer or exporter of particular types of goods or of all goods. And, I admit that, at a macro level, the latter is a somewhat meaningless question, as in the last resort, and over a sufficiently long period of time, the balance of payments must balance. But, this is *not* the case with the stocks and flows of international investment.

Moreover, most investment owned and controlled by MNCs is a different phenomenon from foreign portfolio investment.[13] So, indeed, is trade conducted *within* MNCs different from trade between independent parties. In other words, as I have elaborated on elsewhere (Dunning 1993a), organisational issues do inject the need for a set of analytical tools different from those offered by traditional trade theory.

This is where I think Kojima's criticism of the eclectic paradigm falls down. He insists upon applying a strictly neo-classical framework of thought to explain a phenomenon that is outside that framework of thought. Moreover, like neo-classical theory, his approach to explaining FDI is more normative than mine. However, in various of my writings (see, for example, Chapter 10 of Dunning 1988, and Chapter 13 of Dunning 1993a), I have attempted to give some normative content to the eclectic paradigm by suggesting the conditions for optimising the benefits which host countries might secure from inbound FDI.

Extending and reconfiguring the eclectic paradigm

The investment development path (IDP)

One of my first applications of the eclectic paradigm was to examine its relevance in explaining the changing international position of countries as they passed through different stages of development. The concept of the investment development cycle (or path) was first put forward in 1975[14] and has since gone through various iterations (e.g. Dunning 1981, 1988, 1993a; Dunning and Narula 1996; Narula 1996; Dunning *et al.* 2001). The basic hypothesis of the IDP is that as a country develops, the configuration of the OLI advantages facing foreign-owned firms that might invest in that country and of its own firms that might invest overseas, undergoes change, and that it is possible to identify both the conditions making for the change and their effect on the trajectory of the country's development. The concept also suggests the ways in which the interaction between foreign and domestic firms might itself influence the country's investment path; but only recently has this aspect been incorporated in the literature.[15] The IDP identifies several stages of development a country might pass through. The first stage is one of pre-industrialisation, in which a country is presumed to have no inbound or outbound investment, in the first case because it has insufficient locational attractions, and in the second because its own firms possess few or no ownership advantages. Depending on its resources, government policy, the organisation of activity, and the strategy of firms, the OLI configuration changes so as first to attract inward investment in resource-based sectors, in the traditional and labour-intensive manufacturing sectors, in trade and distribution, in transport and communications, construction and perhaps in tourism.

Depending on the extent to which the country is able to create a satisfactory legal system, commercial infrastructure and business culture, and to provide the business sector with the transport and communications facilities and human resources they need; and depending on its government's policy toward inward direct investment (cf. Japan, which largely disallowed such investment in the 1960s, with Germany, which adopted an open-door policy toward it), its locational attractions will increase, and because foreign firms are likely to have more experience in manufacturing the goods and services now likely to be demanded (and have probably penetrated the local market by imports in any case) inward investment will continue to grow. Gradually it, and any investment by indigenous firms, will affect both supply and demand conditions for the products supplied by foreign firms and their desire to internalise their markets for the competitive advantages.

The improvement in the L advantages of countries may also help indigenous firms to upgrade their own competitive advantages. The growth of Japanese outward investment and, more recently, that of several developing countries is entirely consistent with a reconfiguration of the OLI advantages of indigenous firms brought about by the development process.[16] Once again, changes in the value of both exogenous and endogenous variables affect each of these components. In this early stage, the role of the home government is especially important. In various of his writings (see, for example, Ozawa 1989, 1992, 1996), Terutomo Ozawa has demonstrated the critical role of the Japanese government in influencing the ability of Japanese firms both to generate competitive advantages relative to their competitors, and to locate their value-added activities outside of Japan. It has also affected the strategy of the Japanese companies themselves.

As countries move along their development path, the OLI configuration facing outward and inward investors continues to change. Some foreign (and domestic) firms, which earlier found a country attractive to invest in because of its low labour costs or plentiful natural resources, no longer do so. In other cases, its L advantages have become more attractive as an indigenous technological infrastructure and pool of skilled labour is built up. This, in turn, makes it possible for domestic firms to develop their own O advantages and begin exporting capital.

Next, as countries reach some degree of economic maturity, the OLI configuration facing their own firms may be such that their propensity to engage in outward direct investment exceeds that of foreign-based firms to engage in inward investment. Again, whether or not this happens rests on the strategy of firms and the policies of national governments to generate the competitive (and especially innovatory) advantages of their own firms and to make their own locations attractive to both domestic and foreign investors.

The literature is replete with examples of the kinds of variables likely to influence the OLI configuration over time and the determinants of the value of these variables. Predictions for individual countries are difficult because they require forecasting the behaviour of governments. Different countries at the same stage of their development paths seem to display different propensities to engage in outward and inward FDI. Others may display similar propensities for different reasons. Thus, in the late 1980s, both Sweden and Japan were significant net outward investors, but whereas the Japanese push outwards represented a positive restructuring to make way for the upgrading of its domestic industry, in the Swedish case, it was more symptomatic of the falling competitiveness of the domestic economy. By the mid-1990s, however, Swedish inbound investment was rising quite rapidly, in part, because of Swedish accession to the European Union in 1995 (Zander and Zander 1996).

The final stage of the IDP occurs when there is a fluctuating balance between outward and inward direct investment. This arises when there is some degree of convergence between the level of development and the economic structure of countries, and also where firms engage in FDI, not only to exploit their existing O advantages in a foreign location, but also to augment these advantages by acquiring complementary assets or new markets. In the mid-1990s, this stage has been reached by the more advanced industrial economies, whose wealth creation and productivity growth are increasingly based on their ability to harness and effectively utilise all forms of knowledge or intellectual capital. At this stage too, the role of government is often of critical importance in influencing the quality of L specific advantages; and in setting the competitive environment for their own firms to effectively exploit the opportunities offered by the global economy (Dunning and Narula 1996; Narula 1996).

I have illustrated at some length from the IDP, because it does introduce (albeit at a macro level) a dynamic element into the eclectic paradigm. Moreover, it confirms that the equations 1–3 set out in a previous section do seem to make sense. The configuration of OLI variables affecting the (say) Japanese firms in the world economy in 1997 is a function of the OLI configuration facing them in (say) the mid-1980s and the changes in the endogenous and exogenous variables which have affected their behaviour in the intermediate period. Of these, there is strong evidence that the way in which these two sets of variables interact is, itself, an important factor determining the movement towards a new OLI configuration. We also believe that the concept outlined is very relevant in explaining the recent growth of outward investment from Third World countries, especially from South Korea, Singapore, Taiwan, and Mexico (Van Hoesel 1999; Dunning *et al.* 2001).

Acquiring a competitive advantage through foreign direct investment

Over the last decade or more foreign direct investment intended to augment the existing O or competitive advantages of firms has become an increasingly important form of cross-border economic activity, as, indeed, has the growth of inter-firm strategic alliances. Both forms of trans-border economic involvement reflect the perceived need by firms domiciled in one country not only to capture the technological and marketing synergies offered by firms in other countries, but also, more generally, to harness or tap into the created assets of foreign competitors, suppliers, customers and those offered by national educational and innovatory systems.

Recent technological advances, more intensive inter-firm competition, the opening up of new markets, and the increasing mobility of some kinds of firm-specific assets have, then, led to new motives for foreign production not only as a means of exploiting the existing O specific advantages of the investing firms, but also as a vehicle for augmenting these advantages. In the second half of the 1990s, cross-border acquisitions and mergers accounted for by far the greater part of new MNC activity (UNCTAD 2000); while non-equity alliances, particularly those geared towards innovatory activities, have become an increasingly important component of corporate strategy. Nowhere is this more clearly demonstrated than in the sourcing of technological assets. Several recently published studies[17] have shown that MNCs from all countries are increasingly reaching beyond their national borders to create or gain access to resources and capabilities which complement their existing core competencies. These same studies have also suggested that the locational requirements of strategic asset-seeking FDI are different from those of natural resource-seeking, market-seeking or efficiency-seeking FDI, inasmuch as the former is attracted less by the need to reduce production costs, overcome trade barriers, and exploit economies of scale, but more by the presence of high-quality physical and human infrastructure and a favourable political and commercial ethos towards M&As and cooperative alliances.

Other research seeking to explain the outward FDI by MNCs from developing into developed countries has questioned the ability of traditional theories to explain this phenomenon. This is case where the O advantages of the investing firms transferable outside their home countries.[18] While I would goal of asset-seeking investment is to protect or augm firm's core competences I would still contend that i between these and those which directly arise from the access to new resources and capabilities, and the ability of the firm to manage these which make up the O advantage. Without som contribution from the acquiring MNC, the FDI would, in effect, be a foreign portfolio investment.[19]

The theory of asset-augmenting (cf. asset-exploiting) MNC activity is still in its infancy, but it seems likely to challenge researchers in the next decade. Some useful insights have already been made by scholars such as Tom Wesson (1993, 1997, 2002) and Shige Makino (1998); while other parallels between asset-seeking FDI and foreign portfolio (FPI) investment are explored later in this chapter. Certainly it would be difficult to deny that the variables determining the current wave of intra-Triad M&As are the same as determining inbound greenfield FDI into China or Vietnam; or indeed the spate of acquisition of Korean by US and European firms since 1997. But does this mean that the tenets of the eclectic paradigm are no longer relevant?

For reasons I've set out elsewhere (Dunning 2000b), I believe that although the search for newly created assets adds a new dimension to our thinking about the rationale for FDI, and can only be explained by a reconfiguration of traditional OLI variables, the essential propositions of the eclectic paradigm still remain intact and valid. Irrespective of the motive for MNC activity, its extent, pattern and form still rest on the interaction between the O-specific advantages of investing firms – including the willingness and ability of such firms to access new assets and coordinate these with their existing assets – and the advantages of countries and also on the relative costs and benefits of engaging in this interaction by alternative modes of governance and noticeably that of administrative fiat (i.e. I advantages).

Extending the eclectic paradigm to embrace non-equity alliances

This brings us neatly to an extension to the eclectic paradigm which I first set out in an article published in the *Journal of International Business Studies* in 1995 (Dunning 1995b). The title of the article was 'Reappraising the eclectic paradigm in an age of alliance capitalism', and its main theme was that as cooperation and competition are increasingly becoming complementary modalities for resource creation and allocation in market-based economies, so the concept of the individual firm as the sole or independent source of intellectual capital is no longer sustainable. Rather, it is better viewed as an organiser of a collection of created assets, some of which it

.erates internally and others which it accesses from other firms, yet over ɪe deployment of which it exercises some kind of influence or control.

Accepting this view, then, the O-specific advantages of MNCs will depend not only upon those internally generated, but also upon their competence to seek out, harness and influence the innovation, price and quality of assets of other institutions with which they have an on-going cooperative relationship. Such a relationship may take various forms, such as a strategic technological or marketing alliance between two or more competitors; a subcontracting agreement between a firm and one (or more) of its suppliers; or a licensing or franchising agreement between a firm and one (or more) of its customers.

Similarly, in its choice of a foreign site for its value-added activities, an MNC will be influenced not only by how location-bound resources and/or markets affect its direct costs, but also by how they affect its ability to acquire and exploit the O-specific assets of related firms with which it has some kind of coalition. The costs and benefits of accessing these latter assets by some form of cooperative agreement, rather than by a direct acquisition or merger, will determine the modality by which the O advantages of firms of one country are coordinated with those of another and also with the L advantages of that country; hence the 'I' component of the eclectic paradigm, as initially applied to the markets versus hierarchies choice, needs to be widened to embrace more 'voice'-oriented strategies of firms, which are directed to capturing the benefits of quasi-integration offered by trans-border coalitions and cooperative relationships.

In short, the content and significance of the OLI configuration affecting the determinants of international production need to be reconsidered in the light of the emergence of alliance capitalism and contemporary technological developments, all of which are pointing to the need of firms to embrace a plurality of intra- and inter-firm cooperative relationships if they are to be successful competitors in the global marketplace.

The relevance of the paradigm to explaining patterns of trade and portfolio investment

In my 1976 Nobel Symposium presentation, I argued that the contents of the eclectic theory could be equally used to explain the level and pattern of trade. This was taken up further in 1995 in a paper entitled 'What's wrong and right with trade theory' (Dunning 1995a). There I suggested that firms would export goods and services from a production base in their home country whenever the L advantages of creating or utilising their O-specific advantages were greater than servicing the foreign markets from a foreign location. The extent to which exports were internalised within the firm (*intra*-firm exports) or sold to third parties (*inter*-firm exports) would reflect the relative transaction costs of the two modes of servicing the foreign markets.

Similarly, the extent to which firms imported goods and services, as compared with producing them in a domestic location, would depend on relative location-bound assets offered by the exporting and importing countries; and the relative O advantages of the importing firm (and/or its affiliates).

I believe that by acknowledging the role of the O advantages of firms – in addition to the L advantages of countries, together with the relative costs and benefits of accessing or exploiting these two sets of advantages by way of intra-firm rather than inter-firm cross-border transactions – trade theory can be, and is to some extent, considerably enriched.[20]

More recently I have attempted to use the eclectic paradigm to help our understanding about the determinants of foreign portfolio investment (FPI). In a paper written jointly with John Dilyard, and published in *Transnational Corporations* (1999) I argued that while the major explanatory differences between the two kinds of capital exports rested on the kind of O advantages possessed by the two groups of investors and the extent to which such advantages were coordinated with those of the potential host countries, via internal fiat (in the case of FDI) or the external market (in the case of FPI), there were others – notably those to do with locational choice – which were very similar. Thus, for example, the geography of intra-Triad cross-border M&As over the past decade has been closely paralleled by that of FDI; while the same risk factors that are important in determining FDI in the developing countries are also those which explain FPI. Moreover there is increasing evidence (set out in Dunning and Dilyard 1999) that the two kinds of foreign investment are complementary to, rather than substitutable for, each other, with FDI tending to lead (private) FPI, at least in the early stages of a country's IDP.

E-commerce and relational assets

Finally I wish to make brief mention of two further extensions of the eclectic paradigm which take particular cognisance of contemporary trends in the globalising economy. The first by Cliff Wymbs and myself has been to consider how one of the most dramatic and far-reaching technological innovations of the 1990s, namely, the advent of e-commerce and the Internet, is affecting our explanations for MNC activity. In this chapter we identify both the specific attributes of this new means of information harnessing and communication, and how they are affecting our understanding about the OLI variables affecting firms, and their strategic response to them. Second, and moving along a rather different trajectory of thought, but as a natural extension of our earlier writings on asset-seeking FDI and alliance capitalism, we have begun to explore the likely impact of the growing importance of relational assets (R-assets)[21] both at the corporate and social level, on the OLI configuration affecting MNC activity (Dunning 2002). In particular, we have argued that as the *access* to exogenous resources and capabilities and the organisation of these with the internally *owned* resources and capabilities becomes a more important determinant of commercial success, so the willingness and ability of firms to conduct harmonious value-adding and/or exchange relationships is becoming a more critical advantage. Such advantages are often cumulative, and arise from previous or current dyadic or network relationships.[22] It is also apparent that at a macro-level, social relational capital measured, *inter alia*, by the lack of crime, bribery, corruption and terrorism is becoming a more important factor influencing the location of economic activity by MNCs; while the balance of costs and benefits in owning or accessing resources and capabilities is affecting the way in which they are organised.

In examining the implications of these two developments – the one relating to the innovation and deployment of new technological assets, and the other, that of human assets – I believe the eclectic paradigm provides a powerful analytical framework. Within that framework it also offers up a number of new contextually related hypotheses, as a result of which received theories of the firm and the location of economic activity may need reappraisal.

Concluding remarks: a look towards the future

Let me conclude by re-emphasising a number of points. The first is that, although I have sometimes illustrated the eclectic paradigm by reference to the individual firm, my main focus of interest is in explaining the international production of all firms from a particular country or group of countries. Because of this, I contend that it is inappropriate to compare the merits and demerits of the eclectic paradigm with that of internalisation and other theories of the firm.

Second, I accept that some O-specific advantages are directly the result of firms internalising the market for their intermediate products across national borders. However, since this very act of internalisation puts the internalising firms at an advantage relative to non-internalising firms, I think it appropriate to refer to this benefit as an advantage and to internalisation as the modality by which this advantage is realised.

Third, I acknowledge that the eclectic paradigm as originally conceived is uncomfortable in dealing with the dynamics of international production. However, I would argue that it can help to explain why an industry's or country's international investment profile may be different in two points of time. To link these two points one needs to introduce changes in the exogenous or endogenous variables, including strategy, and how these, in turn, affect the OLI configuration. I have illustrated from the IDP how this may be done at a macro level. At an industry or micro level, only a detailed examination of the profile of individual firms can resolve this problem. The reclassification of firms into strategic groups (McGee and Thomas 1986) is helping to show us that the type of strategic behaviour is not an idiosyncratic variable but can be related to certain characteristics of firms (or groups of firms).

Fourth, I have endeavoured to explain differences between my approach and that of Kiyoshi Kojima, and I hope I have made it clear that this is primarily a difference in emphasis and perspective rather than of reasoning between us.

Fifth, I have emphasised that as strategic asset-acquiring FDI and non-equity alliances have become more important forms of international economic involvement, so the OLI configuration of the eclectic paradigm requires some reappraisal. In particular, I acknowledge that, without knowing whether a firm is contemplating a FDI to exploit a competitive strength or to overcome or counteract a competitive weakness, it is difficult to offer any prediction. Only by treating the cumulative process of sustaining and advancing the core-competences of firms (rather than a discrete and once-and-for-all transaction) can this conundrum be resolved. This then suggests that in future, the eclectic paradigm might better address itself to explaining the changing characteristics of international production than to its level and composition at a particular moment of time.

Finally, while accepting there are other paradigms which seek to offer general explanations of the internationalisation process of firms and/or their international management strategies, we do not consider these to be competing paradigms to our own. Management-related paradigms, for example, are interested in explaining the behaviour of managers in harnessing and utilising scarce and non-imitable resources, not the overall level and pattern of FDI or MNCs activity (or changes to same). Organisational paradigms are directed to evaluating the costs and benefits of alternative institutional mechanisms for organising a given set of resources and capabilities, independently of the location of these assets. Paradigms offered by marketing scholars usually focus on the process and/or form of international market entry and/or growth.

Technological- and network-related paradigms of international production come nearest to our own approach, but cannot comfortably explain some kinds of FDI in developing countries and in some service sectors. With a few exceptions modern paradigms of international trade ignore or downplay the significance of firm-specific advantages. Finance-related paradigms can offer only limited insights into the growth of corporate networks and cross-border strategic alliances.[23]

We conclude then, that an add-on dynamic component to the eclectic paradigm, an extension of its constituent parts to embrace asset-augmenting FDI and cross-border non-equity ventures, and a more explicit acknowledgement of the increasing role of the access of ownership of resources and capabilities can do much to uphold its position as the dominant analytical framework for examining the determinants of MNC activity. We believe that recent technological and economic events, and the emergence of new explanations of MNC activity have added to, rather than subtracted from, the robustness of the paradigm. While accepting that, in spite of its eclecticism (*sic*), there may be some kinds of foreign-owned value-added activities which do not fit comfortably into its construction, we do believe that it continues to meet most of the criteria of a good paradigm and that it is not yet approaching its own 'creative destruction' (Foss 1996).[24]

Notes

1 As reviewed by Graham Hutton (1953).
2 Which, indeed, is exactly what has transpired. For an examination of the impact of European economic integration on transatlantic FDI flows see Dunning (1993b)
3 See especially Chapter 4 of Dunning (1993a).
4 The only English source of Lundgren's thoughts on this matter is his comment on my paper for the 1976 Nobel Symposium in Ohlin *et al*. Birgitta Swedenborg took up and extended the theme in her excellent study on Swedish MNCs (Swedenborg 1979).
5 For an explanation of the difference between a paradigm and a theory see e.g. Dunning (1988: Chapters 1 and 2, 1993a: Chapter 4, and 2000b).
6 If there is no synergy between a firm's existing assets and those it acquires, it is difficult to see how this can be thought of as a direct investment; although I readily admit there are such investments which are classified in this way.
7 For a full exposition see Dunning (1988, 1993a, 1995a, 2000b).
8 It is worth noting that such advantages may stem from the forces of monopoly or of (dynamic) competition. Most references to the competitive advantages of firms embrace both types of advantages, and it is in this sense we use ownership advantages.
9 As set out in more detail in Dunning (1988 and 2000a).
10 Profitable, that is, from the viewpoint of the investing companies. We also use the word 'profitable' in a generic sense to embrace the long-term commercial goals of these companies.
11 Witness, for example, the case of Singapore and the way in which the Singaporean government has used the O advantages of foreign investors in conjunction with those of its own firms and with the L attractions of its own immobile resources and capabilities, to advance its postwar economic prosperity (Haley *et al.* 1996).
12 For example, by their competitors capturing markets which might otherwise be theirs.
13 This is discussed later in this chapter.
14 By Peter Buckley and me at a conference of the UK chapter of the Academy of International Business at Manchester.
15 See Tolentino (1993), Narula (1996), Buckley and Castro (1998), Bellak (2000) and Dunning *et al*. (2001) and various contributions in Dunning and Narula (1996).
16 But see our remarks about asset-seeking FDI in the following sub-section.
17 See, for example, Dalton and Serapio (1995), Dunning and Narula (1995), Almeida (1996), Dunning (1996), Kogut (1996), Kuemmerle (1997, 1999).
18 That is to say where they are based on the L advantages of the home country.
19 Which, *de facto* is sometimes the case, with MNCs behaving as holding companies.

20 See especially the writings of Helpman (1985), Markusen (1995, 1998 and 2001) and Markusen and Venables (1988), Gray (1999) and Helpman (1984).
21 At a corporate level R-assets are defined as the willingness and capacity of a firm – or more correctly persons within a firm – to conduct (on behalf of that firm) beneficial relations, both with other persons within the firm and between themselves and persons in other institutions.
22 The contribution of spatial and other networks to the O advantages of firms is now being closely studied by scholars. See, for example, Chen (2000) and Enright (2000).
23 References relating to the above paradigms are given in Dunning (2000b). For a recent attempt to more explicitly incorporate finance specific variables into the eclectic paradigm see Stonehill and Oxelheim (2001) and Chapter 10 of this volume.
24 For a somewhat different, and highly refreshing, approach to some of the concepts dealt with in this chapter, see a recently published article by Boddewyn and Iyer (1999).

References

Aliber, R.Z. (1970) 'A theory of foreign direct investment', in C.P. Kindleberger (ed.), *The International Corporation*, Cambridge, MA: MIT Press.

Aliber, R.Z. (1971) 'The multinational enterprise in a multiple currency world', in J.H. Dunning (ed.), *The Multinational Enterprise*, London: Allen & Unwin, 49–56.

Almeida, P. (1996) 'Knowledge sourcing by foreign multinationals: patent citation analysis in the US semi-conductor industry', *Strategic Management Journal*, Winter, special issue, 17: 155–65.

Bellak, C. (2000) *The Investment Development Path of Austria*, Vienna, Vienna University of Economics, mimeo.

Boddewyn, J. and Lyer, G. (1999) 'International business research', *Management International Review*, 39(2): 161–84.

Buckley, P.J. and Casson, M.C. (1976) *The Future of the Multinational Enterprise*, London: Macmillan.

Buckley, P.J. and Casson, M.C. (1988) 'Models of the multinational enterprise', *Journal of International Business Studies*, 29(1): 21–44.

Buckley, P.J. and Castro, F.B. (1998) 'The investment development path: the case of Portugal', *Transnational Corporations*, 7(1): 1–15.

Chen, T.J. (2000) *Network Resources for Internationalisation, The Case of Taiwan's Electronic Firms*, Taipei National Taiwan University, mimeo.

Dalton, D.H. and Serapio, M.G. (1995) *Globalizing Industrial Research and Development*, Washington, DC: US Department of Commerce, Office of Technology Policy, Asia Pacific Technology and Programs, October.

Dunning, J.H. (1958) *American Investment in British Manufacturing Industry*, London: Allen & Unwin.

Dunning, J.H. (1972) *The Location of International Firms in an Enlarged EEC. An Exploratory Paper*, Manchester: Manchester Statistical Society.

Dunning, J.H. (1973) 'The determinants of international production', *Oxford Economic Papers*, 25: 289–336.

Dunning, J.H. (1981) 'Explaining the international direct investment position of countries: towards a dynamic or developmental approach', *Weltwirtschaftliches Archiv*, 119: 30–64.

Dunning, J.H. (1988) *Explaining International Production*, London: Unwin Hyman.

Dunning, J.H. (1993a) *Multinational Enterprises and the Global Economy*, Wokingham, UK and Reading, MA: Addison Wesley.

Dunning, J.H. (1993b) *The Globalization of Business*, London and New York: Routledge.

Dunning, J.H. (1995a) 'What's wrong and right with trade theory', *International Trade Journal*, 9(2): 153–202.

Dunning, J.H. (1995b) 'Reappraising the eclectic paradigm in the age of alliance capitalism', *Journal of International Business Studies*, 26(3): 461–93.

Dunning, J.H. (1996) 'The geographical sources of competitiveness of firms: the results of a new survey', *Transnational Corporations*, 5(3), December.

Dunning, J.H. (1997) *Alliance Capitalism and Global Business*, London and New York: Routledge.

Dunning, J.H. (2000a) 'The eclectic paradigm of international production; a personal perspective', in C.N. Pitelis and R. Sugden (eds), *The Nature of the Transnational Firm*, London and New York: Routledge: 119–39.

Dunning, J.H. (2000b) 'The eclectic paradigm as an envelope for economic and business theories of MNE activity', *International Business Review*, 9(1): 163–90.

Dunning, J.H. (2002) 'Relational assets, networks and international business activity', in F. Contractor and P. Lorange (eds) *Cooperative Strategies and Alliances*, Oxford: Elsevier: 569–95.

Dunning, J.H. and Dilyard, J. (1999) 'Towards a general paradigm of foreign direct and foreign portfolio investment', *Transnational Corporations*, 8(1): 1–52.

Dunning, J.H. and Narula, R. (1995) 'The R&D activities of foreign firms in the US', *International Studies in Management and Organisation*, 25: 39–73.

Dunning, J.H. and Narula, R. (eds) (1996) *Foreign Direct Investment and Governments*, London and New York: Routledge.

Dunning, J.H., Kim, C. and Lin, J.D. (2001) 'Incorporating trade into the investment development path: a case study of Korea and Taiwan', *Oxford Development Studies*, 29: 145–54.

Enright, M. (2000) 'Globalization, regionalization and the knowledge-based economy in Hong Kong', in J.H. Dunning (ed.), *Regions, Globalization and the Knowledge-Based Economy*, Oxford: Oxford University Press: 391–406.

Foss, N.J. (1996) 'Research in strategy, economics and Michael Porter', *Journal of Management Studies*, 33(1): 1–24.

Frankel, M. (1955) 'Anglo-American productivity differences – their magnitude and some causes', *American Economic Review*, 45 (May): 94–112.

Gray, H.P. (1999) *Global Economic Involvement*, Copenhagen: Copenhagen Business Press.

Haley, C.V., Lan, L. and Toh, M.H. (1996) 'Singapore incorporated: reinterpreting Singapore's business environment through a corporate metaphor', *Management Decision*, 34(9): 17–28.

Helpman, E. (1984) 'A simple theory of trade with multinational corporations', *Journal of Political Economy*, 92: 451–71.

Helpman, E. (1985) 'Multinational corporations and trade structure', *Review of Economic Studies*, 52: 443–58.

Hirschman, A. (1970) *Exit, Voice and Loyalty: International Differences in Work-Related Values*, Cambridge, MA: Harvard University Press.

Hoesel, R. van (1999) *New Multinational Enterprises from Korea and Taiwan*, London and New York: Routledge.

Hutton, G. (1953) *We Too Can Prosper*, London: Allen & Unwin.

Hymer, S. (1960) *The International Operations of National Firms: A Study of Direct Investment*, Cambridge, MA: MIT Press.

Hymer, S. (1976) *The International Operations of National Firms: A Study of Direct Investment*, Cambridge, MA: MIT Press.

Knickerbocker, F.T. (1973) *Oligopolistic Reaction and the Multinational Enterprise*, Cambridge, MA: Harvard University Press.

Kogut, B. (1996) 'Platform investments and volatile exchange rates: direct investment in the US by Japanese electronic companies', *Review of Economics and Statistics*, May, 78(2): 221–31.

Kojima, K. (1978) *Direct Foreign Investment: A Japanese Model of Multinational Business Operations*, London: Croom Helm.

Kojima, K. (1982) 'Macro economic versus international business approaches to foreign direct investment', *Hotosubashi Journal of Economics*, 23: 1–19.

Kuemmerle, W. (1997) 'Building effective capabilities abroad', *Harvard Business Review*, March/April, 61–70.

Kuemmerle, W. (1999) 'The drivers of foreign direct investment into research and development: an empirical investigation', *Journal of International Business Studies*, 30(1): 1–24.

Makino, S. (1998) *Towards a Theory of Asset Seeking Foreign Direct Investment*, Hong Kong: The Chinese University, mimeo.

Markusen, J.R. (1995) 'The boundaries of multinational enterprises and the theory of international trade', *Journal of Economic Perspectives*, 9(2): 169–89.

Markusen, J.R. (1998) 'Multinational firms, location and trade', *The World Economy*, 21: 733–56.

Markusen, J.R. (2001) 'International trade theory and international business', in A.R. Rugman and T. Brewer (eds), *The Oxford Handbook of International Business*, Oxford: Oxford University Press: 69–87.

Markusen, J.R. and Venables, A.J. (1988) 'Multinational firms and the new trade theory', *Journal of International Economics*, 46: 183–303.

McGee, J. and Thomas, H. (1986) 'Strategic groups: theory, research and taxonomy', *Strategic Management Journal*, (7): 141–60.

McManus, J.C. (1972) 'The theory of the multinational firm', in G. Paquet (ed.), *The Multinational Firm and the Nation State*, Toronto: Collier-Macmillan: 66–93.

Narula, R. (1996) *Multinational Investment and Economic Structure*, London and New York: Routledge.

Ohlin, B., Hesselborn, P.-O. and Wijkman, P.M. (eds) *The International Allocation of Economic Activity*, London and Basingstoke: Macmillan.

Ozawa, T. (1989) *Japan's Strategic Investment Policy Towards Developing Countries: From the Ad Hoc to a New Comprehensive Approach*, Fort Collins University of Colorado, mimeo.

Ozawa, T. (1992) 'Foreign direct investment and economic development', *Transnational Corporations*, 1(1): 27–54.

Ozawa, T. (1996) 'Japan: the macro-IDP, meso-IDPs and the technology development path (TDP)', in J.H. Dunning and R. Narula (eds), *Foreign Direct Investment and Governments*, London and New York: Routledge: 142–73.

Rostas, L. (1948) *Comparative Productivity in British and American Industry*, Cambridge: Cambridge University Press.

Rugman, A.M. (1980) 'Internalization theory and corporate international finance', *California Management Review*, 13: 73–9.

Rugman, A.M. (1997) *The Scientific Papers of Alan Rugman*, Vol. 1, London: Edward Elgar.

Stonehill, A. and Oxelheim, L. (2001) *On the Treatment of Finance Specific Factors Within the OLI Paradigm*, Hawaii, University of Hawaii, mimeo.

Swedenborg, B. (1979) *The Multinational Operations of Swedish Firms: An Analysis of Determinants and Effects*, Stockholm: Almquist & Wiksell.

Tolentino, P.E. (1993) *Technological Innovation and Third World Multinationals*, London: Routledge.

UNCTAD (2000) World Investment Report, *Cross Border Mergers and Acquisitions and Development*, New York and Geneva: UN.

Vernon, R. (1966) 'International investment and international trade in the product cycle', *Quarterly Journal of Economics*, 80: 90–207.

Vernon, R. (1974) 'The location of economic activity', in J.H. Dunning (ed.), *Economic Analysis and the Multinational Enterprise*, London: Allen & Unwin.

Wesson, T.J. (1993) *An alternative motivation for foreign direct investment*, PhD dissertation, Harvard University, Cambridge, MA.

Wesson, T.J. (1997) 'A model of asset seeking foreign direct investment', *Proceedings International Business Division, The Administrative Sciences Association of Canada*, 18(8): 110–20.

Wesson, T.J. (2002) *Foreign Direct Investment and Competitive Advantage*, Cheltenham: Edward Elgar.

Zander, I. and Zander, U. (1996) 'Sweden: a latecomer to industrialization', in J.H. Dunning and R. Narula (eds), *Foreign Direct Investment and Governments*, London and New York: Routledge: 101–41.

3 New institutional economics

An organising framework for OLI

Elizabeth Maitland and Stephen Nicholas

New institutional economics and international business

New institutional economics (NIE) and international business (IB) research share a similar inception: the inability of prevailing theories to explain readily observable phenomena. For early IB theorists, the elusiveness of the multinational enterprise in orthodox economics spurred the development of new models and conceptualisations of cross-border economic activity to explain both the emergence of multinational firms and the industrial and geographic distribution of foreign direct investment (FDI). For new institutional economists, the aridness of neoclassical economics lay in the abstraction of its black box treatment of the firm, and its inability to explain the differential growth and development of societies across time and across geographic space.

NIE and IB research form part of a much broader thrust in the social sciences to prise open the complexities of organisations – the firm, the household and government – and inject greater realism into the modelling of economic behaviour and activity. In evolutionary economics, organisational theory, business history and law, the reformulation of the cognitive and behavioural assumptions of the neoclassical framework has formed the basis for new theories of organisation and economic behaviour. Increasingly, these theories of organisation are converging into a single synthesis that seeks the source and renewal of competitive advantage in the distinctive resources and capabilities of the firm and explains the boundaries of the firm. The strength of this emerging synthesis is its emphasis on modelling the complexity of the environment within which the firm makes these strategic decisions. As Buchanan (1995: 171) observed: '[o]nce attention is drawn to a structure, to a process, and away from resources, goods and services, many of the trappings of orthodox economic theory fall away.'

For IB researchers, these are critical developments. Building a rigorous theoretical framework for IB research requires consistency in the core assumptions – the building blocks – of the theories assembled. While constituent elements of Dunning's (1973, 1977, 1979, 1995, 2000a, 2000b,

Chapter 2) Ownership–Location–Internalisation (OLI) paradigm rely on powerful theoretical models, the theories assembled have not always been consistent or complementary. OLI's application by IB researchers has often reflected this underlying lack of theoretical consistency. NIE provides a consistent analytical framework for understanding the evolution and operation of societies, their institutional environment and their economic organisations, including the international firm.

Dunning has shown that the power of OLI lies in its ability to focus attention on a range of IB research problems – as Williamson (1993: 115) has observed, '[l]isting researchable problems requires a good nose'. This chapter shows how NIE provides an analytical framework for organising many of the elements in OLI in a theoretically consistent manner. In particular, the chapter identifies the troublesome nature of location-specific ownership advantages and context in OLI. First, the chapter explores the evolution of ownership advantages in IB research and outlines how resource-capability theories provide a robust conceptualisation of firm-specific ownership advantages. Dunning's OLI framework can assimilate resource-capability approaches to firm-specific advantages, but the concept of location-specific ownership advantages remains under-theorised. OLI's modelling of location remains problematic, with location reduced to little more than a list of contextual factors. The chapter then sets out the NIE framework, including its contribution for analysing international business problems.

Limits to OLI

There are two broad classes of OLI frameworks: Dunning's OLI framework (with reappraisals and personal perspectives) and other academics' interpretations and uses of OLI. Given the pervasive influence of OLI and the large number of disciplines required to understand the origin and growth of multinational companies (MNCs), it is unsurprising that the diverse interests of researchers have led to widely differing interpretations of OLI, perhaps giving the impression that OLI is a 'shopping list' (Dunning 2000b: 126). Dunning's own interpretation of the OLI framework has been broadened to take into consideration dynamics, through path-dependent trajectories and investment development paths (Dunning 1993; Dunning and Narula 1996), advantages of multinationality (Dunning 1995), strategic alliances (Dunning 1995, 1997), asset-acquiring FDI (Dunning 2000b), e-commerce (see Chapter 11) and relational assets (see Chapter 2).

The contribution of NIE for international business research can be demonstrated through an analysis of the troublesome nature of ownership advantages in OLI. To do this, it is important to return to the origins of the OLI paradigm. Originally, Dunning (1970, 1973) asked why there was international direct investment and production. After identifying the limitations of capital and trade theory, Dunning used location and industrial organisation models to identify advantages, some of them locational, which can be enjoyed by all firms, and others that were firm-specific (or not transferable between firms). More prescient, Dunning (1973) identified alternatives to FDI, including subcontracting and licensing, which were precursors to the full range of international forms of involvement, including alliances that Dunning would study two decades later (Dunning 1995). In his early work, Dunning was also prophetic on two other matters: that ownership advantages were not static, but dynamic (Dunning 1973: 328), and that ownership advantages could arise from the multinationality of a company (Dunning 1977: 401). While neither of these early insights was developed in his early work, both would form part of Dunning's (2000a, 2000b) reappraisals of OLI.

Outlining a research agenda at the end of his 1973 article, Dunning (1973: 325–6) stated that the question 'why international production' was now less interesting than 'why the particular geographical or industrial patterns of international production'. Dunning (1977: 401; 2000b: 119; see also 1979, 1995, 2000a, Chapter 2) has consistently drawn an industry–country link, arguing that OLI's 'focus of interest has always been directed to explaining the level and pattern of the foreign value-added activities of firms, and/or countries'. To understand the pattern of international production, Dunning (1977: 399) drew on different theories for each element in the OLI framework, relying on the theory of the firm for internalisation advantages, location theory for location advantages, and the theory of industrial organisation and market entry for ownership advantages (Dunning 1995, 2000a). Ownership advantages might be either firm-specific or location-specific ownership endowments, where the endowments were assets capable of generating future income streams (rents). Location-specific ownership advantages were used to help explain why the ownership advantages of 'Japanese iron and steel firms over South Korean will be very different from those of UK tobacco firms over Brazilian tobacco firms or U.S. computer firms over French computer firms' (Dunning 1977: 407). We believe that the industry–country ownership advantages are not rigorously developed or theoretically consistent, posing major problems for the applicability of OLI. This is also a criticism that can be levelled against other early, partial theories of international investment (Caves 1971; McManus 1972).

The dual meaning of ownership advantages in OLI required complex definitions of both ownership and location advantages. First, Dunning differentiated between location factors that gave rise to ownership advantages and those that were location advantages. According to Dunning (1977: 408) 'country or industry variables affecting ownership advantages are not the same as the location-specific endowments', where location-specific endowments were assets that could only be used by enterprises in the locations in which they were sited; were unavoidable or non-transferrable costs (such as taxes) or were costs of shipping products.

Second, Dunning differentiated between ownership advantages 'available to all firms' in a country or industry, and firm-specific ownership advantages, which 'stem from their exclusive possession and use' (ibid.: 399, 406). This duality of ownership advantages to cover both firm-specific and country-industry specific location ownership advantages was necessary if OLI was to explain both the internationalisation of the individual firm and the pattern of industry and country FDI. Linking the individual firm to industry–country patterns of overseas production through the duality of ownership advantages was at once obvious, but theoretically tricky.

Such a link was obvious because Dunning's articles were driven by empirical evidence from industries and countries with pervasive MNC activity, starting with his own work on whether American manufacturing subsidiaries in Britain performed differently from their home parent and host country counterparts (Dunning 1958). What stands out is the remarkable elasticity of OLI to incorporate different types of ownership advantages. In 1977, Dunning defined ownership advantages as: those that a branch plant could gain from the parent relative to an indigenous firm; those that any firm may have had over another firm producing in the same location (including monopoly power); and those that arose specifically from the nationality of a company (Dunning 1977: 401). These all appear to be firm-specific ownership advantages. Twenty-five years later, Dunning (2000a: 168–9) defined three categories of ownership advantages (two largely consistent with his earlier categories): those arising from a bundle of resources unique to one firm relative to others; those related to monopoly power; and those competencies that were management, rather than firm, specific. By 2001, Dunning (see Chapter 2) described ownership advantages as '[t]he (net) competitive advantages which firms of one nationality possess over those of another nationality . . . [arising] . . . either from the firm's privileged ownership of, or access to, a set of income-generating assets or from their ability to coordinate these assets'.

According to Dunning (2000a: 169), the test of whether a new conceptualisation or extension of ownership advantages can be incorporated into OLI is that the new examples do 'not undermine the basic tenets of the paradigm'. We believe that location-specific ownership advantages do not pass this test. In fact, the theory of ownership advantages poses a major challenge to the OLI paradigm's ability to explain the geographic and industrial pattern of MNC activity. Location-specific ownership advantages are under-'theorised', a failing common to all IB models treatment of context. Dunning (1998) has recognised this problem, most recently in his treatment of location as a 'neglected factor'. Over-emphasis on firm-specific ownership advantages has been to the detriment of correctly formulating location in IB models. The following sections trace the role of firm-specific ownership advantages in explaining the boundaries of the international firm; the theoretical problems of location-specific ownership advantages; and develop a NIE analytical framework for IB to integrate context into modelling of the international firm, competitive advantage and the geographic and industrial distribution of FDI.

Origins of ownership advantages and the distribution of FDI

Dunning (1973, 1977, 1979) has been a leading contributor and synthesiser of the theoretical underpinnings of ownership advantages in IB. The conceptualisation of ownership advantages began with Hymer's (1976: 32–54) definition of advantages specific to parents in imperfect markets, which led 'the possessor of the advantages to choose to supersede the market for his advantages' (ibid.: 49; see also Dunning and Rugman 1985). While Hymer introduced the idea of market failure, Johnson (1970) provided the public good explanation for firm-created know-how, and its implications for economic inefficiency. Drawing on Johnson's model, Caves (1971) expanded ownership advantages to include a range of firm-specific assets with public good characteristics, such as product differentiation know-how. With a similar industry–country focus as Dunning's, Caves (1971: 8) argued that casual inspection revealed a high correlation between product differentiation in an industry and the proportion of foreign subsidiaries.

McManus (1972) closed the circle, arguing that the coordination of resource owners, of which the international firm was only one coordinating organisation, arose due to transaction costs related to the transferability of firm-specific property rights (McManus 1972: 74). From a Coasean property rights perspective (Coase 1937, 1960), McManus explicitly introduced both transaction costs and comparative contracting into the theory of MNC, which laid the foundation for integrating both ownership and internalisation. Like Dunning, McManus (1972: 68) was interested in both 'why foreign direct investment occurs and why it is observed in some activities or industries and not others'. Dunning (1973: 314) contributed to these theoretical foundations of the firm, defining ownership advantages as 'enterprise-specific', 'internal to particular enterprises' and 'stem[ing] from their exclusive possession and use' (Dunning 1977: 401, 406). As we have seen, Dunning's (1973: 328) focus had already shifted from the individual firm to the industrial and national distribution of FDI.

By the mid-1970s, the concept of firm-specific ownership advantages, as tangible and intangible assets that yielded rents when exploited by international firms, provided the foundations for emerging theories of the MNC (Buckley and Casson 1976; Casson 1979; Magee 1977; Hennart 1982; Calvert 1981; Rugman 1981, 1982; Davidson and McFetridge 1984; Teece 1982, 1985, 1986). Internalisation took centre stage, whereby the international firm was one organisation for appropriating returns from unique assets, including information and knowledge. Acute as ever to the evolving literature in international business, Dunning (1979: 274) saw that this new focus involved a 'switch of attention from the act of foreign direct investment . . . to the institution making the investment'.

Dunning's promotion of OLI as a way of understanding the country–industry distribution of FDI contrasted with this narrower focus of IB research on the international firm. This focus on the institution making the investment saw organisational design (internal management) and entry mode choice become the dominant research issues. Statistical studies of entry mode and HQ-subsidiary relations took on a formulaic quality, with the theory listing firm-specific assets as determinates of the entry mode (Agarwal and Ramaswami 1992; and summaries by Caves 1996; Andersen 1997) and delineators of HQ-subsidiary types (Martinez and Jarillo 1989; Birkinshaw and Morrison 1995). The case study approach, such as the studies of Unilever, Kao, General Electric and NEC, identified internal resources, management capabilities and administrative heritage as the firm-specific factors creating multidomestic, global or transnational firms (Doz 1986; Hedlund 1986; Prahalad and Doz 1987; Bartlett and Ghoshal 1989; Kogut 1990). As Dunning (1977: 401) had predicted, such multinationality advantages arose when the firm gained experience from operating in different economic environments. Dunning (1995: 477) expanded OLI to embrace the advantages of multinationality.

Both the international management and entry mode literature relied on the new strategic management, resource-based and organisational capability paradigms that were revolutionising business policy and strategy. The analytical focus of the resource-capability paradigm was on the firm *qua* firm, with a sharp distinction between firm-specific and industry–country advantages. Not surprisingly, the thrust of the resource-capability approaches to understand 'why firms differ', presented a direct challenge to OLI, particularly its focus on the geographic and industrial distribution of FDI.

There was also a change in nomenclature, with ownership advantages relabelled core competencies, where core competencies were the roots of competitive advantage (Prahalad and Hamel 1990). By definition, core competencies applied across a wide variety of markets, made a significant contribution to customer benefits, and were difficult to imitate (Prahalad and Hamel 1990: 83–4). This approach to competitive advantage placed global growth strategies centre stage (Rumelt *et al.* 1994). The resource-capability approach to corporate growth strategies and competitive advantage depended on a rigorous theoretical definition of firm-specific resources and capabilities. While resources and core competencies were easily incorporated into OLI, as firm-specific ownership advantages, they challenged OLI's emphasis on location-specific ownership advantages.

Ownership advantages refined: resources and capabilities

Resource-capability research began by analysing the resource heterogeneity of firms. Wernerfelt (1984: 174) identified tangible and intangible resources 'tied semi-permanently to the firm' and Rumelt (1984) defined resource heterogeneity (or the firm's bundle of unique resources and relationships) by ambiguity that generated the heterogeneity, and non-imitability that maintained it. The strategic management literature recognised that non-imitability was linked to the process by which the unique (resources) assets were obtained (Barney 1986, 1991) or accumulated (Dierickx and Cool 1989). All authors sought isolating mechanisms that enabled rents from non-mobile 'unique factors' to be appropriated and competitive advantages to be sustained (Montgomery and Wernerfelt 1988; Peteraf 1993; Kogut and Zander 1992, 1993; Kogut 1993). More generally, in order for monopoly and Ricardian rents to be earned, *ex post* limits to competition, either through imperfect imitability or imperfect substitutability of the firm's capabilities, must be imposed. Imperfect imitability means that competitors cannot reproduce the firm's heterogeneous resources, which are linked to specialised (idiosyncratic) assets that earn quasi-rents. The model also assumed *ex ante* limits to competition, which ensured that competitive advantages are not immediately competed away. As Peteraf (1993: 187) noted a 'Nobel prize winning scientist may be a unique resource, but unless he has firm-specific ties, his perfect mobility makes him an unlikely source of competitive advantage'. In short, these limits to competition speak directly to the firm-specific nature of the firm's resources.

The culmination of the strategic management approach was the idea of dynamic capabilities, which explained the renewal of a firm's specific resources and competitive advantages (Rumelt *et al.* 1991; Nelson 1991; Winter 1993; Teece *et al.* 1997; Kogut and Zander 1992, 1993). Dynamic capabilities are the firm's ability to integrate, build and reconfigure its unique resources to create new competitive advantages. Drawing on business history (Chandler 1962, 1977, 1990), path dependency (David 1986, 1994; Cantwell 1989) and evolutionary economics (Nelson and Winter 1982; Winter 1987, 1993; Nelson 1991), Teece *et al.* (1997) described dynamic capabilities as coordination and integration attributes and inter-organisational learning processes. The key attribute defining dynamic capabilities was that they are 'a collection of routines, skills and complementary assets that are difficult to imitate' (Teece *et al.* 1997: 524; Nelson and Winter 1982). By rigorously defining the theoretical basis for firm-specific heterogeneity, the resource-capability paradigm's answer to 'why firms differ' was that, *ceteris paribus*, different competitive advantages (themselves the product of heterogeneous resources and capabilities) lead to different levels of behaviour and performance. Evidence on intra-industry profits confirmed the high importance of firm-specific differences and the relative unimportance of industry effects (Rumelt 1991).

The challenge of resource-capability models for OLI

While acknowledging that firms differ by nationality (Korean versus Japanese firms) and industry (computer versus steel) due to the context in which they operate, strategic management holds that differences between firms are mainly discretionary (Nelson 1991: 61). By asking 'why firms are different', the resource-capability approach directed attention away from the common properties of all firms to the phenomena that produce and sustain heterogeneity among firms (Rumelt *et al.* 1994). Clearly, the resource-capability approach is no better placed than OLI to explain how the distribution of FDI differs across countries and industries. The growing dominance of the resource-capability approach in international business has meant that location-specific ownership advantages have been largely ignored. This overwhelming focus on the individual firm has reduced the effect of location to a list of contextual factors, such as production costs, psychic distance, culture and government. Neither OLI nor resource-capability approaches provide a theoretically consistent integration of context and competitive advantage, although OLI correctly emphasised the importance of context.

Introduction to new institutional economics

NIE has a rich, cross-disciplinary lineage. Although the 'initial scaffolding' was erected by economists, most notably Nobel recipients Coase (1937, 1960) and North (1981, 1990a), and Williamson (1979, 1985),[1] contributions to the framework have emerged from a broad cross-section of social sciences. The cross-fertilisation of theoretical and methodological developments is reflected in the diversity of discliplinary backgrounds of leading contributors, such as economists Demsetz (1964), Arrow (1974), Joskow (1988, 1991), Cheung (1970, 1983); political economists and economic historians Olson (1965, 1982), Libecap (1989), Ostrom (1990), Greif (1989, 1993); sociologist Coleman (1990); and legal theorist Macneil (1974, 1981).[2]

As shown in Figure 3.1, the scaffolding of NIE has evolved into three core branches, anchored by an overarching research question of what explains the differential growth and development of societies across time and across geographic space. The framework cascades from a central theory of institutional change, or economic growth, that defines the key relationships linking individuals, organisations, institutions and the long-run economic evolution of societies. The three branches of analysis are delineated by theories to explain the main forms of exchange conducted by societies: political, economic and social.

Analysis of political exchange builds from the basic premise that complex institutions and organisations evolve from the willingness of individual members of society to cede freedom of action and choice in return for guarantees of freedom from violence. This branch of analysis incorporates theories from political science, political economy and transaction cost economics (Olson 1965, 1982; North 1990b; Buchanan 1980; Buchanan and Tullock 1962; Krueger 1974; Dixit 1996). The creation of the property rights system is a key focus of research, drawing a

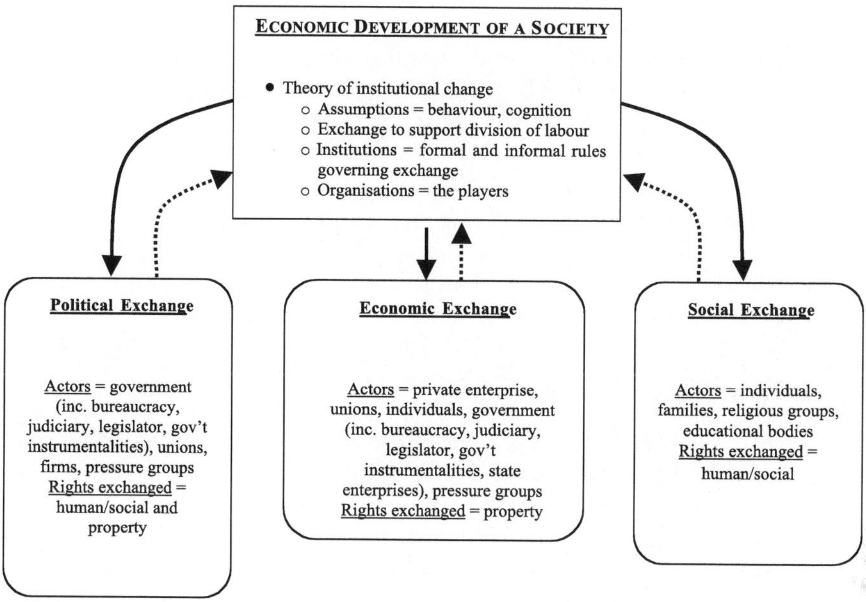

Figure 3.1 New institutional economics: three branches of analysis.

link between the political structure of a society and the system of rights to wealth-creating assets. Analysis encompasses models that consider, for example, representative government, military dictatorship, collective choice, corruption, voting systems, and social welfare (Weingast and Marshall 1988; Krueger 1996; Shleifer and Vishny 1993).

Economic exchange, in Figure 3.1, is the necessary counterpart to the specialisation of labour, including corresponding innovations in technology, which facilitates improvements in living standards. The property rights system provides the incentives for economic activity and the allocation of productive effort. For example, when an innovator cannot secure rights to new knowledge, due to weaknesses in the legal system, individuals and organisations shift their efforts away from innovation. Similarly, competition policy, deregulation, privatisation and the introduction of market-based pricing in socialist economies affect economic choices (see, for example, Joskow 1999; Oi and Walder 1999).

Finally, Figure 3.1 considers social exchange through organisations, such as marriage, religion and family. These organisations affect economic and political exchange through, for example, religious values on the rights of women to work; environmental activism on the use and ownership of resources, such as air, old-growth forests and rivers; and ideology on the rights of the state to tax and compel military service. Informal ordering within social groups may also substitute for costly and possibly ineffective enforcement of economic contracts through the courts. For example, Macmillan and Woodruff (1999) illustrated the reliance on social ordering by newly emerged Vietnamese private enterprises, rather than the erratic enforcement capabilities of the courts, to achieve reliable commercial transacting.[3]

The following sections set out NIE's core theory of institutional change and then focus specifically on economic organisation and the costs of comparative contracting. Finally, the chapter applies the framework to international business research questions, including those posed by Dunning (1973, 1977) in his original conceptualisation of OLI.

Analysing exchange

Mutually beneficial exchange relies on cooperation and coordination, which North (1997b: 149) argued are 'the key dilemma of societies past, present and future'. The need for exchange arises as the inevitable product of the division and specialisation of labour within a society, for it is only through exchange that rewards to specialisation can be distributed and sustained. As the level of specialisation increases, exchange becomes increasingly cyclical,[4] rather than simply bilateral (i.e. the exchange of labour for sustenance). Due to the costs of transacting, cyclical exchange becomes increasingly expensive and difficult for individuals to achieve. They must search for someone with whom to trade, negotiate the terms of trade, and ensure the party with whom they have chosen to interact fulfils their obligations.

Transaction costs are not only important in the new theory of the firm, but to the full study of comparative costs of organisation, whether political, social or economic. As North (1995: 259) noted, 'whereas Coase was concerned with the transaction cost that determined the existence of firms, I was concerned with the transaction cost that determined overall economic performance'. NIE argues these transaction costs arise because individuals have limited and differing cognitive abilities and integrity. This modification of the rationality and behavioural assumptions provides the unifying thread for the cross-disciplinary evolution of NIE as an analytical framework for the political, economic and social choices made by societies. For IB research, the power of NIE is its ability to understand the interaction between political, economic and social organisations, as well as the impact of context on economic outcomes.

New institutional economics: understanding context

As shown in Figure 3.1, NIE analyses the differential growth and development of societies across time and across geographic space. Clearly, OLI's aim to understand the industrial and geographic distribution of FDI forms part of NIE's larger research agenda. Although a fully specified theoretical understanding of the long-run evolution of societies has yet to emerge,[5] North (1997a: 5–6), delineated five key characteristics of institutional change:

1 The continuous interaction of institutions and organisations in the economic setting of scarcity and hence competition is the key to institutional change.
2 Competition forces organisations to continually invest in skills and knowledge to survive. The kinds of skills and knowledge individuals and their organisations acquire will shape evolving perceptions about opportunities and hence choices that will incrementally alter institutions.
3 The institutional framework dictates the kinds of skills and knowledge perceived to have the maximum pay-off.
4 Perceptions are derived from the mental constructs of the players.
5 The economies of scope, complementarities and network externalities of an institutional matrix make institutional change overwhelmingly incremental and path-dependent.

As a number of empirical studies have revealed, the direction of institutional development and change can be towards growth, stagnation or decline (Greif 1989, 1993; Greif *et al.* 1994; Eggertsson 1990; Landa 1994; Libecap 1996; North 1990a). As societies grow and become more complex in terms of the degree of interdependence, more complex institutions – rules – are required to coordinate production and exchange.

NIE posits that societies generate formal and informal institutions to provide certainty and structure to human interaction by inducing – or incentivising – cooperation and coordination of effort. The need for rules arises because individuals have limited and differing cognitive abilities and integrity. Relaxing the assumption of perfect rationality, the less extreme notion of bounded rationality recognises that individuals are '*intendedly* rational, but only *limited* so' (Simon 1976: xxviii; as quoted in Williamson 1985). The computational power of individuals is constrained by their capacity to receive, store, retrieve and process information (Williamson 1979, 1985). Individuals are unable to predict or know all future states of the world or to solve complex equations with multiple variables.

Coincident with these cognitive limitations is the ability of individuals to exercise choice or freedom of will in electing among a range of behaviours. Conscious choice of behaviour creates problems for exchange because of the possibility that individuals may engage in incomplete or distorted disclosure of information. They may intentionally mislead, disguise or cloud their intentions or motivations. The introduction of an assumption of opportunism does not presuppose that individuals seek to cheat others all of the time, it simply recognises that the possibility for such behaviour exists, requiring appropriate monitoring devices for successful exchange.[6] The acknowledgement of conscious choice of behaviour introduces not only opportunism as a variable but also tactics, strategy, honesty, openness, trust and reputation (Demsetz 1997).

The extent to which individuals are able to choose different types of behaviour is influenced by the context within which they make decisions. Behaviour may be conditioned by social structures, such as kinship ties, status or religious beliefs, or limited by coercive pressures, such as convict labour or slavery.[7] Each of these behavioural influences forms part of the wider institutional environment or rules of the game that structure all human interaction (North 1981, 1990a). Functioning either as informal or formal constraints, institutions limit the choice set of individuals. Formal institutions encompass political, judicial and economic rules, such as legislation, common law decisions, and government competition and welfare policies. Informal institutions are the conventions or codes of behaviour in a society, including religious rules governing food consumption, respect for the elderly and sharing of common resources among indigenous communities.

At any point in time, the institutional environment – the sets of formal and informal rules held and enforced by a society – defines the organisations that will be viable. Organisations are the players of the game, ranging from political bodies (such as the organs of government, regulatory agencies and political parties), economic groups (including firms and trade unions), to social bodies (such as churches, clubs and educational groups). As players, organisations form, shift and, sometimes, decay, in response to the opportunities created by institutions (North 1990a). Organisations in turn mould changes in the rules of the game in a complex process of institutional evolution. Change may be in the form of new legislation or gradual shifts in social attitudes towards, for example, foreign investment, racism or slavery, with the effectiveness of each new institution heavily dependent on its enforceability within a society.

Institutions are designed to decrease the costs of exchange. A common language or means of communication, standardised units of money and measurement, and a system of order inhibiting violence are basic institutions that decrease transaction costs. In developed economies, contract law is a crucial, formal institution facilitating economic exchange. It reduces transaction costs by supplying contracting parties with a set of standard terms that, in the absence of law, the parties would have to negotiate for themselves, as well as supplying information concerning unforeseen contingencies that may stymie exchange. Transaction costs are further economised by the state imposing sanctions for breach of contractual commitments through the courts. Formal and informal institutions are designed by society to decrease the costs of all types of exchange – political, social, and economic.

Property rights: institutions for economic growth

Historically, the state has facilitated cooperative exchange through its emergence as the leading creator and protector of property rights. At the heart of economic exchange is the transfer of property rights in an asset (Cheung 1970; Demsetz 1964). These rights encompass the right to earn income (rent) and to contract over the terms of asset use with other parties, the right to permanently transfer the asset, as well as user rights. The latter category defines the legitimate potential uses of an asset, such as the right to physically alter its composition or even destroy the asset (Eggertsson 1990).

Determining and enforcing property rights frequently requires extraordinarily complex rule structures, entailing costly diversion of resources into the production of property rights and maintaining a credible threat discouraging violation of rights (Libecap 1989). While transitional economies find it difficult to implement the rules structures of developed economies, developed economies themselves are struggling to introduce new rule structures defining and assigning ownership, for example, to genetic and stem cell material. Failures to comprehensively define property rights, whether through formal or informal institutions, may result in resources becoming *de facto* common property, with associated problems of free-riding and exploitation (Olson 1965; Libecap 1989). For individuals and organisations, uncertainty of ownership carries disincentives to investing in the acquisition or production of knowledge and skills, denying society the products developed from such investments in knowledge and skill.

By specifying who (or what) bears the rewards and costs associated with particular assets, property rights provide incentives for economic behaviour within a society. The property rights system, encompassing the rules and the degree to which they are enforced, is a crucial element of the insti tutional framework. The system shapes the types of economic organisations formed by a society and, ultimately, the level of labour specialisation and knowledge creation undertaken. The firm is one type of economic organisation. Given the overarching assumption of scarcity, competition for resources implies that the firm only survives when it can both protect its property rights in valuable assets and renew its asset base in sympathy with society's demand for particular goods and services.

For the individual firm, the institutional environment, therefore, shapes the choice of knowledge-creating strategies and the choice of contractual form to capture rents on its unique knowledge stock. The firm is a nexus of contracts, selected to maximise the rents on its distinctive resources and capabilities (Jensen and Meckling 1976). Hence, the next step in analysing the firm in its institutional context is to understand how it selects, across a range of forms for conducting exchange.

Economic organisations: choice of contracting

Economic organisations comprise a spectrum of contractual arrangements. As shown in Figure 3.2, markets, short-term forward contracts,[8] long-term intermediate contracts, equity joint ventures and firms are alternative forms for conducting exchange. The differences between forms depend on:

i the bundle or set of property rights transferred from one party to the other[9] (Cheung 1970);
ii the duration of the arrangements and, therefore, the potential for divergence of goals over time (Macneil 1974);
iii the degree of relations between the parties, to mediate the effect of the time element on the protection of property rights from incursion by the other party.

Representing one of the contractual poles, pure spot market transactions commence sharply, are short-lived and end sharply, either from clear performance or clear breach. Planning, such that there is, focuses on price and payment terms, and occurs almost instantaneously between autonomous parties. Property rights in the asset(s) are permanently transferred from the seller to the buyer, between whom there will be no future relations and whose identity is immaterial to the substance of the exchange.

Movement along the spectrum is distinguished by increasing interaction, or relations, between parties, increasing duration of the arrangement, and

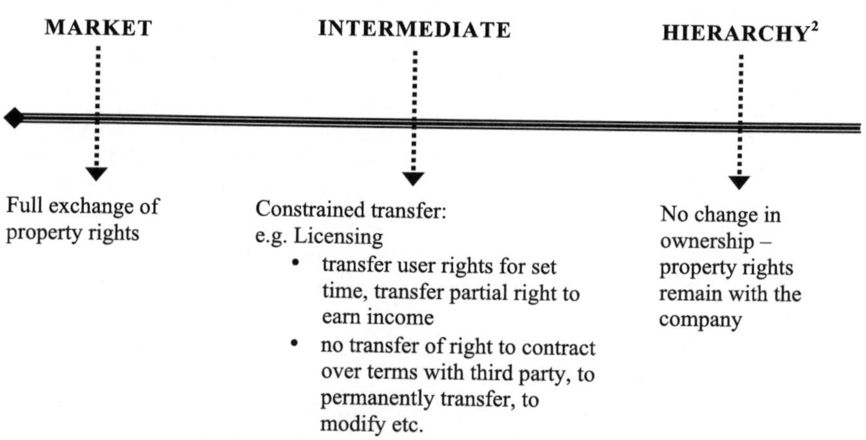

Figure 3.2 The contractual spectrum.[1]

[1] Adapted from ideas presented in Macneil (1981).
[2] As Macneil (1981) noted, the ultimate relational pattern of modern contracts in most respects is not the firm, small or large, but the nuclear family. Hence, the end of the spectrum is not delineated in the diagram.

the delegation, rather than transfer, of property rights.[10] For example, in licensing agreements, the licensor delegates only partial rights to use a patented technology and earn income from it, by withholding, for example, the rights to transfer the technology to third parties and to transform it (without transferring technical improvements back to the licensing company). All delegated rights are subject to a period of expiry (Mathewson and Winter 1984).

Intermediate and hierarchical arrangements, in Figure 3.3, rest on the concept of bilateral governance or self-enforcing contracts, whereby the parties to the agreement create an incentive structure for performance (Williamson 1979, 1985). Moving away from market exchange, the effect of the time element becomes critical. Once exchange is of an extended and potentially undefined duration, *ex ante* planning cannot be fully binding on the parties (Macneil 1974, 1981). Flexibility is required to enable adaptation at some future point to events that could not have been foreseen. The task is to write or create an incentive contract that inhibits potential opportunism, while recognising the constraints imposed by cognitive limitations and the excessive time and, hence, costs of specifying a contract covering all possible future contingencies (Kreps 1984). To operate successfully, the incentives need to be such that at any stage during the life of the contract the ongoing relationship makes the parties better off than does breaching.

When elaborate incentive contracts are necessary to protect against potential opportunism, the firm is the most efficient contractual choice. The hierarchical structure of the firm is a series of structured agency relations designed to constrain opportunism and economise on bounded rationality, through promoting incentives and information flows among members of the firm (Cheung 1970; Jensen and Meckling 1976; Williamson 1985). In place of the market's price mechanism, the firm relies on behavioural constraints (Hennart 1991). As shown in Figure 3.3, forms closer to hierarchy than pure-spot market exchange, rely on rich information flows, achieved through high degrees of interaction, to ensure co-operation (Hennart 1982; Williamson 1979, 1985).

Organisational choice: property rights over resources

Firms select contractual arrangements to maximise potential rents on their distinctive resources and capabilities. The choice of contractual form balances production cost (i.e. scale economies) and transaction cost considerations. The size of transaction costs associated with alternative forms for exploiting assets is a function of the ability to define rights to the assets; to separate (transfer) these rights from the original owner; and the owner's ability to appropriate the return on the rights (the rent).

Drawing on the resource-capability approach, asset mobility can be identified by the extent to which property rights can be defined, separated

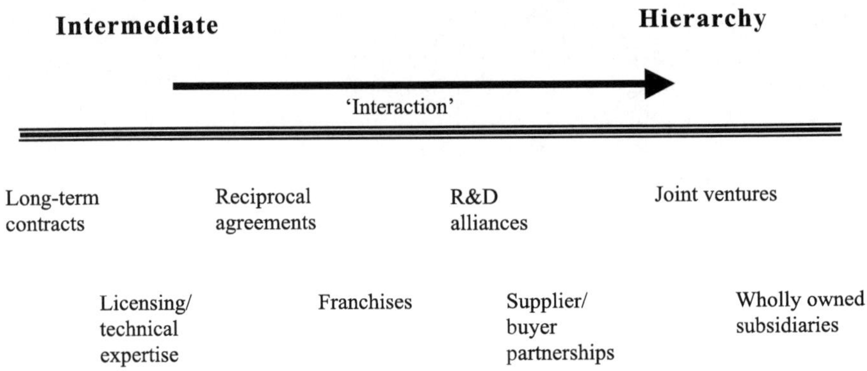

Figure 3.3 Relational contracting.

and appropriated. Perfectly immobile assets are those for which the potential rent on any alternative use by the firm or another party is zero. Imperfectly mobile assets are characterised by appropriable quasi-rents, or an excess value to the firm of an asset over its value to the second highest valuing use or user (Klein *et al.* 1978; Peteraf 1993). The degree of asset mobility is determined by the technical attributes of the asset – whether it is tangible or tacit, or co-specialised with other assets[11] – and the institutional setting.

The institutional context may prevent the separation of property rights in an asset from the firm, such as a government-granted monopoly prohibited by statute from transfer to a third party. Institutional weaknesses may render a firm-specific asset perfectly immobile simply because a patent or copyright is unenforceable through legal means. In transitional economies, the intellectual property rights regime is often underdeveloped (Maitland 2001), a deficiency exacerbated by an inexperienced, incompetent and, possibly corrupt, judiciary (Williamson 1993).

As part of the institutional setting, informal norms may effectively substitute for formal constraints. Private ordering within close-knit groups often spontaneously generates rules that promote cooperative outcomes among group members (Ellickson 1994; Landa 1994).[12] Non-legal sanctions within business communities or commercial networks are frequently complex and multidimensional, and include the desire to maintain reputation, profitable relationships and standing among peers (Macauley 1963; Charny 1990; Beale and Dugdale 1975; Arrighetti *et al.* 1997). By identifying and ostracising opportunistic agents (or identifying and rewarding trustworthy agents), such networks facilitate exchange by decreasing transaction costs for participants (Williamson 1993).

Asset mobility is also affected by its technical attributes. Highly tacit and/or idiosyncratic know-how creates distinct property rights problems for firms. When know-how is embedded in the firm's human capital, it is difficult to separate the firm's rights to the knowledge from the person or team of people in which it resides, inhibiting market sale as a method for maximising rents on the know-how. Similarly, firm-specific know-how may have strong public-good characteristics affecting disclosure. The more innovative the know-how, the greater the cost of verifying assertions regarding its characteristics and, hence, the greater uncertainty surrounding its application (Magee 1977; Davidson and McFetridge 1984). Under such conditions transfer from the firm to another party, through sale or licensing, may be prohibitively costly. Even where knowledge can be codified into a form that can be patented, trademarked or copyrighted, weaknesses in the intellectual property rights regime may render know-how perfectly immobile.

Exploiting firm-specific know-how also turns on consideration of co-specialisation of assets. As Nelson and Winter (1982) argued, codified knowledge may not convey all necessary know-how. Specialist skill may be required to interpret and implement codified knowledge and may not exist outside the firm, limiting transfer when divorced from the accompanying tacit skill of firm employees. Co-specialisation, or the presence of transaction-specific investment, also extends to physical assets (Williamson 1979, 1985; Masten *et al.* 1989; Stuckey 1983; Lyons 1996). In the presence of co-specialised assets, the firm faces a potential hold-up problem, in which, for example, suppliers or distributors opportunistically force ex post re-distributory negotiations (Goldberg 1976). To avoid hold-up, the firm can internalise the transaction or, in certain institutional settings, rely on relational norms, such as reputation, to constrain opportunism (see Masten *et al.* 1989).

The institutional context shapes not only the appropriability of rents, it also conditions the nature of the firm's knowledge base and innovative capacity. Learning is not only specific to the firm (or to its accumulated stock of knowledge) but also to the institutional context in which learning and innovation take place. Paths of development are generated both internally and externally, with external paths shaped by, for example, shifts in scientific knowledge, government competition policy and consumer preferences. This means that differentiation between firms also occurs because of different institutional environments, such that Australian firms are different from American, and computer firms are different from steel firms. Clearly, international expansion represents an opportunity for exploiting existing advantages and developing new capabilities because the institutional environments in which firms operate and grow are different.

International expansion

When the firm expands internationally, it makes two simultaneous and interdependent decisions: the form of international involvement and the location. Firms design international contractual arrangements to protect property rights in firm-specific and co-specialised assets. Designing contracts to inhibit breach and non-cooperative behaviour turns on the governance structures contained in the arrangement and the institutional context. In an unfamiliar institutional setting, the firm faces higher costs of search, stemming from unfamiliarity with host country commercial norms, such as reputation, and heightened costs of bonding and enforcement. The success of the growth strategy depends on institutional parameters, the dynamic capabilities of the firm to adapt its stock of advantages to a new environment, and production costs.

As Guisinger (see Chapter 6) observes, institutional variables can be divided into formal and informal constraints affecting the ability of the firm to protect property rights in its distinctive assets. Critical formal institutions include laws on foreign investment, competition, contract, corporate governance, industrial relations, consumer welfare, intellectual property, tariffs and non-tariff barriers, although the list is by no means exhaustive. For example, manipulation of property entitlements, under the aegis of the state, represents one of the greatest threats to rents on firm-specific assets. Subject to international agreements and the restraining influence of reputation, the principle of sovereignty endows the state with many degrees of freedom when determining the definition, allocation and enforcement of rights. A wide range of manipulations is available to the state to erode the competitiveness of foreign firms, including tax and tariff rate changes, blocked profit remittance, forced renegotiation of government contracts and concessions, and, at the extreme, expropriation. Manipulations of the investment environment also occur through less visible formal institutions, such as selective enforcement of environmental and workplace safety regulations, and delays in the issue of expatriate visas and import licences.

Economic activity is not only supported by formal and easily observable regulations, but also conducted within and shaped by a web of informal norms. For example, local norms may support public officials extorting bribes from MNCs, given their 'outsider' status. Equally, the MNC may not be able to utilise bribery as a competitive tool to match the actions of domestic rivals. This latter problem arises for two reasons. First, the costs of engaging in corruption are likely to be higher for the MNC relative to local firms. These costs include the potential loss of rent-earning opportunities in other countries, due to damage to the MNC's reputation. If the MNC's home government exerts extra-territoriality and fines its companies for engaging in corruption the costs of discovery for the MNC are further raised, relative to local firms. Second, the MNC may not possess the skills to solicit corruption, a condition that is likely to be strongest in the early post-entry period, when it also lacks knowledge of corruption patterns and networks in the host location. The MNC may be forced to watch from the sidelines, while host country companies engage in bribery to manipulate public tenders, induce public officials to nationalise, re-zone or restrict the use rights on the assets of rivals, and utilise technologies that erode public rights to common resources (Maitland 2001).

The influence of the institutional environment is also reflected in the constraints on the foreign firm's participation in local commercial networks. Foreign firms lack 'local' reputation and links with domestic players, and may be excluded by 'suspicious' local firms from informal networks (Schanze 1994). Precluded from reliance on informal institutions that decrease costs of searching for partners and enforcing breaches of contract, the foreign firm is heavily reliant on host country legal systems.

It is important to understand the central role of the institutional context of exchange because commodities and resources are not given by nature, but defined by society. Property rights systems are the product of the distribution of power within a society. The state may redefine or reallocate an MNC's property rights to fulfil its responsibilities to the greater good of society, or as the outcome of lobbying activities by select domestic interest. For the foreign firm, discerning and interpreting the influence of various organisations, such as unions, domestic companies and organised religions, is difficult. When a foreign firm is a 'new' player in the institutional setting, it is at a distinct disadvantage relative to domestic companies through its inferior knowledge of host country political, economic and social institutions and organisations. It faces a cost of learning 'how things are done' that carries implications for its competitiveness, relative to domestic competitors, and for its ability to appropriately select and design contractual arrangements.

Equity joint ventures and other forms of alliance, including licensing and acquiring existing companies, present methods for overcoming the intrinsic disadvantage of foreign firms in new environments. They provide a means for deciphering the complexity of different institutional environments. Linking up with a host country partner provides access to existing supplier, distribution and retail networks, diminishing the costs of establishing reputation and goodwill. The political and bureaucratic contacts of local partners are multi-faceted resources for the foreign firm. The political know-how of a local partner, which ranges from lobbying capabilities and knowledge of routinised processes for applying political pressure, to highly nuanced knowledge of when a 'word in the right ear' is appropriate, can be effectively exploited as a mechanism for protecting property rights.

Obviously, the institutional setting determines innovative activity, knowledge-creation and the ability of firms to create and sustain core competence, which varies across countries and industries. From a richer understanding of how location influences form selection, asset transfer and learning, the focus of analysis can be narrowed to look at specific issues of parent-subsidiary monitoring and control devices, or widened to consider the influence of the multilateral institutional framework on the design of the MNC's subsidiary network. By drawing on theories of path-dependency, location and industry structure, NIE provides an analytical framework for explaining the international and industrial distribution of FDI. There is no need for troublesome ideas, such as Dunning's location-specific ownership advantages. What is crucial is understanding the incentive set for economic activity, encompassing property rights regimes and the interaction of different organisations competing for access to value-creating assets within a society. The shopping-list of psychic distance, culture and political structure can be organised into a theoretically consistent and rigorous framework for IB research.

NIE and IB: the future

One of IB's pioneer theorists recently suggested 'that the international business research agenda is running out of steam after a period of vibrancy' (Buckley 2002: 365). Dunning's OLI paradigm was an early, prescient and influential force in the development of IB, periodically undergoing revision and adjustment to address new, empirically driven, research questions. Chapter 11 encapsulates this evolutionary path. Nevertheless, the very malleability of the paradigm has been a source of weakness, with each re-assembly of the constituent elements of O, L and I lacking theoretical consistency. We have argued that location-specific ownership advantages and the incorporation of location, more generally, are major problems affecting OLI's applicability. The resolution of this lacuna lies in harnessing the theoretical tools of NIE to IB research questions. These tools include the resource-capability theories of firm-specific advantages and, importantly, a theoretical basis for the inclusion of location and its interaction with competitive advantage into IB models.

New institutional economics provides a rigorous analytical framework, anchored by core assumptions of cognition and behaviour, to advance and reinvigorate the IB research agenda. Equally, IB researchers are well positioned to contribute to the ongoing construction of NIE. Addressing the annual meeting of the International Society for New Institutional Economics as president-elect, Williamson (2000) described NIE as 'a boiling cauldron of ideas . . . [but] its many accomplishments notwithstanding, there is a vast amount of unfinished business . . . the new institutional economics is the little engine that could. Its best days lie ahead. Who could ask for more?' We have outlined the rich, cross-disciplinary heritage of NIE, delineating its analytical strengths for understanding the evolution and operation of societies, their institutional environment and their organisations, including the multinational enterprise. Cross-country by orientation, IB researchers can bring unique insight to unlocking the influence of context on the patterns and design of economic activity. No one has been more attuned to these differences than John Dunning, nor more influential in posing questions of location to the discipline.

Notes

1 See also Coase (1992) and North (1994) for their respective Nobel Prize acceptance speeches, which summarise their contributions to the science of economics.

2 Equally, the intellectual influences on such theorists reflect a broad grouping of disciplines with Williamson (1993, 1999, 2000), for example, tracing the threads of his own intellectual development to organisational theorists (March and Simon 1958; Simon 1957), neoclassical economists (Hayek 1945; Barnard 1938; Stigle, 1961), business historians (most notably, Chandler 1962, 1977), and legal theorists (particularly Macneil 1974).

3 See Ekelund *et al.* (1996) for an interesting study of the growth of the Roman Catholic Church as a classic, hierarchical (and cross-border) economic organisation designed to capture rents in, for example, markets for marriage and the eternal soul.

4 In cyclical exchange 'no benefactor ever receives return benefits from the one upon whom he confers benefits; the first beneficiary may confer benefits on a different beneficiary, who in turn confers benefits on another, and so on until the original benefactor receives a 'return' from someone in the cycle, but not a return directly traceable to the original beneficiary' (Macneil 1974: 697).

5 North (1994) argued there probably never will be a theory of economic dynamics. See also Alston (1996) and Krueger (1996).

6 The criticism of transaction cost economics for what has been perceived as an overwhelmingly negative perception of human behaviour in the adherence to the concept of opportunism is misplaced (for a typical example, see Ghoshal and Moran 1996). As Williamson (1985: 64) clearly enunciated, opportunism is not an all-prevailing condition: 'I do not insist that every individual is continuously or even largely given to opportunism. To the contrary, I merely assume that some individuals are opportunistic some of the time and that differential trustworthiness is rarely transparent *ex ante*'.

7 While assigned set tasks and prevented from choosing the time and place of their labour, slaves and convict labour still retain some choice over their work intensity (Nicholas 1988).

8 Such as supply of a fixed quantity of a commodity once a week for eight weeks at a fixed price.

9 In technical terms, their implied residual property rights that have not been contracted for *ex ante*.

10 Contracts may or may not be formal documents.

11 The term co-specialised comes from Teece (1986) but is analogous to the terms idiosyncratic and transaction-specific investments favoured by Williamson (1979, 1985).

12 Ellickson (1994) identified relational signalling as the major instrument of social control in sharing groups. By choosing to cooperate, even where there are no explicit prescriptions to do so, the individual signals they care about their relations with others and would not jeopardise these relations: it is a signal of mutual social approval, thereby, facilitating incomplete contracting. See Schanze (1994) for an extension to inter-firm cooperation in advanced economies.

References

Agarwal, S., Ramaswami, S.N. (1992) 'The choice of foreign market entry mode: Impact of ownership, location and internationalisation factors', *Journal of International Business*, 23: 1–27.

Alston, L.J. (1996) 'Empirical work in institutional economics: An overview', in L.J. Alston, T. Eggertsson and D.C. North (eds), *Empirical Studies in Institutional Change*, Cambridge: Cambridge University Press.

Andersen, O. (1997) 'Internationalization and market entry mode: A review of theories and conceptual frameworks', *Management International Review*, 37: 27–42.

Arrighetti, A., Bachmann, R. and Deakin, S. (1997) 'Contract law, social norms and inter-firm cooperation', *Cambridge Journal of Economics*, 21: 171–95.

Arrow, K. (1974) *The Limits of Organization*, New York: W.W. Norton.

Barnard, C. (1938) *The Function of the Executive*, Cambridge, MA: Harvard University Press.

Barney, J.B. (1986) 'Strategic factor markets: Expectations, luck, and business strategy', *Management Science*, 32: 1231–41.

Barney, J.B. (1991) 'Firm resources and sustained competitive advantage', *Journal of Management*, 17: 99–120.

Bartlett, C.A. and Ghoshal, S. (1989) *Managing Across Borders: The Transnational Solution*, Boston: Harvard Business School Press.

Beale, H. and Dugdale, T. (1975) 'Contracts between businessmen: Planning and the use of contractual remedies', *British Journal of Law and Society*, 2: 45–60.

Birkinshaw, J.M. and Morrison, A.J. (1995) 'Configurations of strategy and structure in multinational subsidiaries', *Journal of International Business Studies*, 26: 729–94.

Buchanan, J.M. (1980) 'Rent seeking and profit seeking', in J.M. Buchanan, R.D. Tollison and G. Tullock (eds), *Toward a Theory of the Rent-seeking Society*, College Station: Texas A&M University Press.

Buchanan, J.M. (1995) 'James M. Buchanan', in W. Breit and R.W. Spencer (eds) *Lives of the Laureates: Thirteen Nobel Economists*, 3rd edn, Cambridge, MA: MIT Press.

Buchanan, J.M. and Tullock, G. (1962) *The Calculus of Consent*, Ann Arbor: University of Michigan Press.

Buckley, P.J. (2002) 'Is the international business research agenda running out of steam?', *Journal of International Business Studies*, 33: 365–73.

Buckley, P.J. and Casson, M. (1976) *The Future of Multinational Enterprise*, London: Macmillan.

Calvert, A.L. (1981) 'A synthesis of foreign direct investment theories and theories of the multinational firm', *Journal of International Business Studies*, 12: 43–60.

Cantwell, J.A. (1989) *Technological Innovation and Multinational Corporations*, Oxford: Basil Blackwell.

Casson, M. (1979) *Alternatives to the Multinational Enterprise*, London: Macmillan.

Caves, R.E. (1971) 'International corporations: The industrial economics and foreign investment', *Economica*, 38:1–27.

Caves, R.E. (1996) *Multinational Enterprise and Economic Analysis, Second Edition*, Cambridge: Cambridge University Press.

Chandler, A.D. (1962) *Strategy and Structure: Chapters in the History of the Industrial Enterprise*, Cambridge, MA: MIT Press.

Chandler, A.D. (1977) *The Visible Hand: The Managerial Revolution in American Business*, Cambridge, MA: Belknap Press of Harvard University Press.

Chandler, A.D. (1990) *Scale and Scope: The Dynamics of Industrial Capitalism*, Cambridge, MA: Belknap Press of Harvard University Press.

Charny, D. (1990) 'Nonlegal sanctions in commercial relationships', *Harvard Law Review*, 104(2): 373–467.

Cheung, S.N.S. (1970) 'The structure of a contract and the theory of a non-exclusive resource', *Journal of Law and Economics*, 13 (April): 49–70.

Cheung, S.N.S. (1983) 'The contractual nature of the firm', *Journal of Law and Economics*, 26 (April): 1–21.

Coase, R.H. (1937) 'The nature of the firm', *Economica*, 4: 386–405.

Coase, R.H. (1960) 'The problem of social cost', *Journal of Law and Economics*, 3: 1–44.

Coase, R.H. (1992) 'The institutional structure of production', *American Economic Review*, 82(4): 713–19.

Coleman, J.S. (1990) *Foundations of Social Theory*, Cambridge, MA: Harvard University Press.

David, P. (1986) 'Understanding the economics of QWERTY: The necessity of history', in W.N. Parker (ed.), *Economic History and the Modern Economist*, Oxford: Blackwell.

David, P.A. (1994) 'Why are institutions the 'carriers of history'?: Path dependence and the evolution of conventions, organizations and institutions', *Structural Change and Economic Dynamics*, 5: 205–20.

Davidson, W.H. and McFetridge, D.G. (1984) 'International technology transactions and the theory of the firm', *Journal of Industrial Economics*, 32: 255–64.

Demsetz, H. (1964) 'The exchange and enforcement of property rights', *Journal of Law and Economics*, 3 (October): 1–44.

Demsetz, H. (1997) *The Economics of the Business Firm*, Cambridge: Cambridge University Press.

Dierickx, I. and Cool, K. (1989) 'Asset stock accumulation and the sustainability of competitive advantage', *Management Science*, 35: 1504–11.

Dixit, A. (1996) *The Making of Economic Policy: A Transaction Cost Politics Perspective*, Cambridge: Cambridge University Press.

Doz, Y. (1986) *Strategic Management in Multinational Companies*, Oxford: Pergamon Press.

Dunning, J. (1958) *American Investment in British Manufacturing Industry*, London: George Allen and Unwin.

Dunning, J. (1970) *Studies in International Investment*, London: George Allen and Unwin.

Dunning, J. (1973) 'The determinants of international production', *Oxford Economic Papers*, 25: 289–336.

Dunning, J. (1977) 'Trade, location of economic activity and the MNE: A search for an eclectic approach', in B. Ohlin, P. Hesselborn and P.M. Wijkman (eds), *The International Allocation of Economic Activity: Proceedings of a Nobel Symposium held at Stockholm*, London: Macmillan.

Dunning, J. (1979) 'Explaining changing patterns of international production: In defence of the eclectic theory', *Oxford Bulletin of Economics and Statistics*, 41: 269–95.

Dunning, J. (1993) *The Globalization of Business*, London: Routledge.

Dunning, J. (1995) 'Reappraising the eclectic paradigm in an age of alliance capitalism', *Journal of International Business Studies*, 26: 461–93.

Dunning, J. (1997) *Alliance Capitalism and Global Business, Globalization of Business*, London: Routledge.

Dunning, J. (1998) 'Location and the multinational enterprise: A neglected factor?', *Journal of International Business Strategies*, 29: 45–66.

Dunning, J. (2000a) 'The eclectic paradigm as an envelope for economic and business theories of MNE activity', *International Business Review*, 9: 163–90.

Dunning, J. (2000b) 'The eclectic paradigm of international production: Personal perspective', in C.N. Pitelis, and R. Sugden (eds), *The Nature of the Transnational Firm*, London: Routledge: 119–39.

Dunning, J.H. (2001) 'The eclectic paradigm of international production: Past, present and future', *International Journal of the Economics and Business*, 8: 173–90.

Dunning, J.H. and Narula, R. (eds) (1996) *Foreign Direct Investment and Governments*, London: Routledge.

Dunning, J.H. and Rugman, A.M. (1985) 'The influence of Hymer's dissertation on the theory of foreign direct investment', *American Economic Review*, 75(2): 228–32.

Eggertsson, T. (1990) *Economic Behavior and Institutions*, Cambridge: Cambridge University Press.

Ekelund, R.B., Hebert, R.F., Tollison, R.D., Anderson, G.M. and Davidson, A.B. (1996) *Sacred Trust: The Medieval Church as an Economic Firm*, New York, Oxford: Oxford University Press.

Ellickson, R.C. (1994) 'The aim of order without law', *Journal of Institutional and Theoretical Economics*, 150(1): 97–100.

Ghoshal, S. and Moran, P. (1996) 'Bad for practice: A critique of the transaction cost theory', *Academy of Management Review*, 21(1): 13–47.

Goldberg, V. (1976) 'Regulation and administered contracts', *Bell Journal of Economics*, 7 (Autumn): 426–48.

Greif, A. (1989) 'Reputation and coalitions in medieval trade: Evidence on the maghribi traders', *Journal of Economic History*, 49: 857–82.

Greif, A. (1993) 'Contract enforcement and economic institutions in early trade: The maghribi traders coalition', *American Economic Review*, 83(3): 525–48.

Greif, A., Milgrom, P. and Weingast, B.R. (1994) 'Coordination, commitment and enforcement: The case of the merchant guilds', *Journal of Political Economy*, 102(4): 745–76.

Hayek, F. (1945) 'The use of knowledge in a society', *American Economic Review*, 35 (September): 519–30.

Hedlund, G. (1986). 'The hypermodern MNC – A heterarchy?', *Human Resource Management*, 25: 9–35.

Hennart, J.-F. (1982) *A Theory of Multinational Enterprise*, Ann Arbor: University of Michigan Press.

Hennart, J.-F. (1991) 'The transaction costs theory of joint ventures: An empirical study of Japanese subsidiaries in the United States', *Management Science*, 37(4): 483–97.

Hymer, S.H. (1976) *The International Operations of National Firms: A Study of Direct Foreign Investment*, Cambridge, MA: MIT Press.

Jensen, M.C. and Meckling, W.H. (1976) 'Theory of the firm: Managerial behavior, agency costs and ownership structure', *Journal of Financial Economics*, 3(4): 305–60.

Johnson, H.G. (1970) 'The efficiency and welfare implications of the international corporation', in C.P. Kindleberger (ed.), *The International Corporation*, Cambridge, MA: MIT Press.

Joskow, P.L. (1988) 'Asset specificity and the structure of vertical relationships: Empirical evidence', *Journal of Law, Economics and Organization*, 7: 95–117.

Joskow, P.L. (1991) 'The role of transaction cost economics in anti-trust and public utility regulation policies', *Journal of Law, Economics and Organization*, 7 (Special Issue): 53–83.

Joskow, P.L. (1999) 'Introducing competition into regulated network industries: From hierarchies to markets in electricity', in G.R. Caroll and D.J. Teece (eds), *Firms, Markets, and Hierarchies: The Transaction Cost Economics Perspective*, New York, Oxford: Oxford University Press.

Klein, B., Crawford, R.G. and Alchian, A.A. (1978) 'Vertical integration, appropriable rents, and the competitive contracting process', *Journal of Law and Economics*, 21(2): 297–326.

Kogut, B. (1990) 'International sequential advantages and network flexibility', in C.A. Bartlett, Y. Doz, and G. Hedlund (eds), *Managing the Global Firm*, London: Routledge.

Kogut, B. (1993) 'Learning, or the importance of being inert: Country imprinting and international competition', in S. Ghoshal and D.E. Westney (eds), *Organisation Theory and the Multinational Corporation*, New York: St. Martin's Press.

Kogut, B. and Zander, U. (1992) 'Knowledge of the firm, combinative capabilities, and the replication of technology', *Organization Science*, 3: 383–97.

Kogut, B. and Zander, U. (1993) 'Knowledge of the firm and the evolutionary theory of the multinational corporation', *Journal of International Business Studies*, 24: 625–46.

Kreps, D.M. (1984). 'Corporate culture and economic theory', Second Mitsubishi Bank Foundation Conference on Technology and Business Strategy.

Krueger, A.O. (1974) 'The political economy of the rent-seeking society', *American Economic Review*, 64(3): 291–303.

Krueger, A.O. (1996) 'The political economy of controls: American sugar', in L.J. Alston, T. Eggertsson and D.C. North (eds), *Empirical Studies in Institutional Change*, Cambridge: Cambridge University Press.

Landa, J. (1994) *Trust, Ethnicity and Identity: Beyond the New Institutional Economics of Ethnic Trading Networks, Contract Law, and Gift Exchange*, Ann Arbor: University of Michigan Press.

Libecap, G.D. (1989) *Contracting for Property Rights*, Cambridge: Cambridge University Press.

Libecap, G.D. (1996) 'Economic variables and the development of the law: The case of western mineral rights', in L.J. Alston, T. Eggertsson and D.C. North (eds), *Empirical Studies in Institutional Change*, Cambridge: Cambridge University Press.

Lyons, B.R. (1996). 'Empirical relevance of efficient contract theory: Inter-firm contracts', *Oxford Review of Economic Policy*, 12(4): 27–52.

Macauley, S. (1963) 'Non-contractual relations in business: A preliminary study', *American Sociological Review*, 45.

McManus, J. (1972) 'The theory of the multinational firm' in G. Paquet (ed.), *The Multinational Firm and the Nation State*, Ontario: Collier Macmillan.

Macmillan, J. and Woodruff, C. (1999) 'Dispute prevention without courts in Vietnam', *Journal of Law, Economics and Organization*, 15 (October): 637–58.

Macneil, I.R. (1974) 'The many futures of contract', *Southern California Law Review*, 47: 691–816.

Macneil, I.R. (1981) 'Economic analysis of contractual relations: Its shortfalls and the need for "rich classificatory apparatus"', *Northwestern University Law Review*, 75(6): 1017–63.

Magee, S.P. (1977) 'Information and the multinational corporation: An appropriability theory of foreign direct investment', in J.N. Bhagwati (ed.), *The New International Economic Order: The North-South Debate*, Cambridge, MA: MIT Press.

Maitland, E. (2001) 'Corruption and the outsider: multinational enterprises in the transitional economy of Vietnam', *Singapore Economic Review*, 46(1): 63–82.

March, J.G. and Simon, H. (1958) *Organizations*, New York: John Wiley & Sons.

Martinez, J.I. and Jarillo, J.C. (1989) 'The evolution of research on coordination mechanisms in multinational corporations', *Journal of International Business Studies*, 20: 489–514.

Masten, S., Meehan, J.W. and Snyder, E.A. (1989) 'Vertical integration in the U.S. auto industry: A note on the influence of transactions specific assets', *Journal of Economic Behavior and Organization*, 12: 265–73.

Mathewson, G.F. and Winter, R.A. (1984) 'An economic theory of vertical restraints', *Rand Journal of Economics*, 15 (Spring): 27–38.

Montgomery, C. and Wernerfelt, B. (1988) 'Diversification, Ricardian rents, and Tobin's q', *Rand Journal of Economics*, 19: 623–33.

Nelson, R.R. (1991) 'Why do firms differ, and how does it matter?', *Strategic Management Journal*, 12: 61–74.

Nelson, R.R. and Winter, S. (1982) *An Evolutionary Theory of Economic Change*, Cambridge, MA: The Belknap Press of Harvard University Press.

Nicholas, S. (ed.) (1988) *Convict Workers: Reinterpreting Australia's Past*. Cambridge and Melbourne: Cambridge University Press.

North, D.C. (1981) *Structure as Change in Economic History*, New York: Norton.

North, D.C. (1990a) *Institutions, Institutional Change and Economic Performance*, Cambridge: Cambridge University Press.

North, D.C. (1990b) 'A transaction cost theory of politics', *Journal of Theoretical Politics*, 2(4): 355–67.

North, D.C. (1994) 'Economic performance through time', *American Economic Review*, 84(3): 359–68.

North, D.C. (1995) 'Douglass C. North', in W. Breit and R.W. Spencer (eds), *Lives of the Laureates: Thirteen Nobel Economists*, Cambridge, MA: MIT Press: 251–67.

North, D.C. (1997a) 'Towards a theory of economic change', paper presented at the ISNIE Meeting, St Louis, September, 19–21.

North, D.C. (1997b) 'Transaction costs through time', in C. Ménard (ed.), *Transaction Cost Economics: Recent Developments*, Cheltenham: Edward Elgar.

Oi, J.C. and Walder, A.G. (1999) *Property Rights and Economic Reform in China*, Stanford: Stanford University Press.

Olson, M. (1965) *The Logic of Collective Action*, Cambridge, MA: Harvard University Press.

Olson, M. (1982) *The Rise and Decline of Nations*, New Haven: Yale University Press.

Ostrom, E. (1990) *Governing the Commons*, Cambridge: Cambridge University Press.

Peteraf, M.A. (1993) 'The cornerstones of competitive advantage: A resource-based view', *Strategic Management Journal*, 14: 179–91.

Prahalad, C.K. and Doz, Y. (1987) *The Multinational Mission: Balancing Local Demands and Global Vision*, New York: Free Press.

Prahalad, C.K. and Hamel, G. (1990) 'The core competence of the corporation', *Harvard Business Review*, 68: 79–91.

Rugman, A.M. (1981) *Inside the Multinationals: The Economics of Internal Markets*, London: Croom Helm.

Rugman, A.M. (1982) 'Internalisation and non-equity forms of international involvement', in A.M. Rugman (ed.), *New Theories of the Multinational Enterprise*, Kent: Croom Helm.

Rumelt, R.P. (1984) 'Towards a strategic theory of the firm', in R.B. Lamb (ed.), *Competitive Strategic Management*, Upper Saddle River, NJ: Prentice Hall.

Rumelt, R.P. (1991) 'How much does industry matter?, *Strategic Management Journal*, 12: 167–85.

Rumelt, R.P., Schendel, D.E. and Teece, D.J. (1991) 'Strategic management and economics', *Strategic Management Journal*, 12: 5–29.

Rumelt, R.P., Schendel, D.E. and Teece, D.J. (1994) *Fundamental Issues in Strategy*, Boston: Harvard Business School Press.

Schanze, E. (1994) 'Towards a realistic science of institutional design: Comment', *Journal of Institutional and Theoretical Economics*, 150: 117–22.

Shleifer, A. and Vishny, R.W. (1993) 'Corruption', *The Quarterly Journal of Economics*, 108 (August): 599–617.

Simon, H. (1957) *Models of Man*, New York: Wiley.

Stigler, G.J. (1961) 'The economics of information', *Journal of Political Economy*, 69: 213–25.

Stuckey, J. (1983) *Vertical Integration and Joint Ventures in the Aluminium Industry*, Cambridge, MA: Harvard University Press.

Teece, D.J. (1982) 'Towards an economic theory of multiproduct firm', *Journal of Economic Behavior and Organization*, 3: 39–63.

Teece, D.J. (1985), 'Multinational enterprise, internal governance, and industrial organization', *American Economic Review*, 75: 233–8.

Teece, D.J. (1986) 'Transaction cost economics and the multinational enterprise', *Journal of Economic Behavior and Organization*, 7: 21–45.

Teece, D.J., Pisano, G. and Shuen, A. (1997) 'Dynamic capabilities and strategic management, *Strategic Management Journal*, 18: 509–33.

Weingast, B.R. and Marshall, W.J. (1988) 'The industrial organization of Congress; or, why legislatures, like firms, are not organized as markets', *Journal of Political Economy*, 96(1): 132–63.

Wernerfelt, B. (1984) 'A resource-based view of the firm', *Strategic Management Journal*, 5: 171–80.

Williamson, O.E. (1979) 'Transaction-cost economics: The governance of contractual relations', *Journal of Law and Economics*, 22(2): 233–61.

Williamson, O.E. (1985) *The Economic Institutions of Capitalism: Firms, Markets, Relational Contracting*, New York: Free Press.

Williamson, O.E. (1993) 'The evolving science of organization', *Journal of Institutional and Theoretical Economics*, 149(1): 36–63.

Williamson, O.E. (1999) 'Strategy research: Governance and competence perspectives', *Strategic Management Journal*, 20: 1087–108.

Williamson, O.E. (2000) 'The new institutional economics: Taking stock, looking ahead', *Journal of Economic Literature*, 38 (September): 595–613.

Winter, S.G. (1987) 'Knowledge and competence as strategic assets', in D.J. Teece (ed.), *The Competitive Challenge*, Cambridge, MA: Ballinger Pub. Co.

Winter, S.G. (1993) 'On Coase, competence, and the corporation', in O.E. Williamson and S.G. Winter (eds), *The Nature of the Firm: Origins, Evolution and Development*, New York: Oxford University Press.

4 FDI and endogenous growth: IB perspectives

*Terutomo Ozawa and Sergio Castello**

Introduction

Endogenous growth theory (or also called the 'new' growth theory) (*inter alia*, Arrow 1962; Romer 1986; Lucas 1988; Grossman and Helpman 1991) has recently revitalised the field of economic development. This new growth theory treats a number of growth-inducing factors, such as 'learning by doing', 'human capital formation via education and training', 'R&D', 'public goods and infrastructure', and 'knowledge spillovers', as *endogenous* variables. All these growth factors are related to the generation and flows of knowledge. Knowledge is self-augmenting (cumulative and path-dependent) and its use is characterised by the law of increasing returns. As Alfred Marshall observed,

> the part of nature in production may show a tendency to diminishing returns, but the part of man shows a tendency to increasing returns . . . Knowledge is our most powerful engine of production: it enables us to subdue nature and satisfy our wants.
>
> (Marshall 1920, as cited in Meier 1999)

Knowledge thus helps us overcome the forces of diminishing returns. Hence, long-term vigorous growth becomes a possibility.

Multinational corporations (MNCs) are widely recognised as a crucial catalytic institution which creates and transfers knowledge across borders (both intentionally and unintentionally) – and are generally welcomed as such – in the developing host economies.[1] The literature on international business (IB) is full of fascinating ideas and conceptual analyses, as this chapter demonstrates, about the roles of MNCs and governments in knowledge creation and dissemination and their impact on local economic

* This chapter is a revised version of our article, 'Toward an "international business" paradigm of endogenous growth: multinationals and governments as co-endogenisers', *International Journal of the Economics of Business*, 8(2), 2001: 211–28.

development. Nevertheless, there has so far been no attempt to link these ideas to the notion of endogenous growth. On the other hand, the mainstream economic (neoclassical in origin) literature on endogenous growth has not yet drawn on any of these new exciting ideas introduced in the IB literature. For that matter, it has not yet explored the dynamic interactions between MNCs and government policies for economic growth as closely as the IB discipline has done. Indeed, the MNC–government nexus is a topic of central importance for IB scholars. True, some mainstream works on endogenous growth (e.g. Grossman and Helpman 1991; Eicher and Kalaitzidakis 1997) have explored this topic, but have done so only within the straightjacket (mathematical formalisation) of neoclassical analysis. In such a constrained approach, MNCs are not adequately portrayed as *an institutional agent* of dynamic growth and change, interacting with host governments' development policies.

The purpose of this chapter is to examine the MNC–government nexus and derive policy implications to the catch-up process of developing host countries. Hence, IB ideas are explored in comparison with mainstream theory. Here, John Dunning's contributions, especially the eclectic paradigm of international production (Dunning 1981) and his notion of macro-organisation (Dunning 1992, 1997), provide the necessary foundations to construct a conceptual framework. The following section first briefly reviews the 'mechanics' of endogenous growth introduced in mainstream economics and then identifies and discusses some relevant works in the IB field, the works that present the new 'mechanics' of enhanced growth. We will conceptualise MNC-cum-government-driven endogenous growth as a frame of reference and explore how major developmental gaps may be closed, thereby enabling developing host countries to achieve fast growth.

Conceptual exposition

Mainstream economic approach

As summarised in Figure 4.1(A), those growth factors that have so far been identified by the proponents of the theory of endogenous growth in the mainstream economic literature include: (a) learning-by-doing, (b) skills of workers, (c) human capital formation (education and training), (d) research and development (R&D), (e) knowledge spillovers, both at home and across borders, (f) infrastructure and public goods, and (g) trade liberalisation/deregulation.[2] Interestingly, the first five elements, (a) through (e), are directly related in one way or another to the knowledge-related activities of MNCs. Indeed, mainstream economics began to examine the link between FDI and knowledge spillovers in a traditional endogenous growth model (for example, De Mello 1997, 1999). As detailed below, investment in R&D and knowledge spillover ((d) and (e)) in particular are further explored in the IB literature.

A. Mainstream economic approach

- a Learning by doing (Arrow 1962)
- b Skill level of workers (Uzawa 1965)
- c Human capital formation (Lucas 1988)
- d Investment in R&D (Chipman 1970; Grossman and Helpman 1991)
- e Knowledge spillover, both at home and across borders (Romer 1986; Krugman 1979; Grossman and Helpman 1991)
- f Infrastructure and public goods (Barro 1990; Barro and Sala-i-Martin 1995; Turnovsky 1997)
- g Trade liberalisation and deregulation (Rivera-Batiz and Romer 1991)

B. International business-related approach

- a Technology transfer and spillovers by MNCs and their networks (Dunning 1958, 1981; Hymer 1960, 1971; Vernon 1966, 1979; Blomstrom 1989; Bende-Nabende 1999; Ernst 2000, 2001; Rugman and D'Cruz 1996, 1997)
- b Dynamic evolution of OLI configuration (Dunning 1993; Dunning and Narula 1996; Narula 1996; Narula and Dunning 2000; Tolentino 1993)
- c Cross-border technological development and sourcing (Lall 1979; Casson 1991; Cantwell 1989, 1995, 1999; Pearce 1997)
- d Cumulative causation in technological competitiveness (Cantwell 1987)
- e Locational agglomeration of innovative activities and production (Porter 1990; Dunning 1991, 1996, 1997, 2000; Gray and Dunning 2000; Nachum 1999, 2000; Enright 2000; Ernst 2000)
- f Pro-trade (as opposed to anti-trade) FDI (Kojima 1973, 1975; Kojima and Ozawa 1985)
- g Industrial restructuring and comparative advantage recycling (Ozawa 1992, 1993)

Figure 4.1 Basic mechanics of endogenous growth.

Note: A complete survey on conceptual contributions is not intended here. These two lists are meant to illustrate some representative ideas about the mechanics of endogenous growth.

Infrastructure and liberalisation ((f) and (g)) belong to the domain of government tasks and responsibilities. A large part of human capital formation (notably basic education and public health) is also carried out by the government. Thus, *both* MNCs and governments need to be recognised as *the crucial facilitators (institutions) of endogenous growth.* In fact, the recent popularity of the new growth theory as opposed to the 'old' growth theory (which predicts a path of growth convergence among countries) derives partly from the fact that because of wide disparities in growth rates among countries, 'peoples are interested in knowing the implications of different government policies on growth . . . The more practical implication of these models was that the government has *a role in economic growth*' (Long and Wong 1997: 64, emphasis added).[3]

Yet surprisingly – or rather expectedly from its narrow disciplinary orientation – the two key institutions, MNCs and governments, that can facilitate endogenous growth – and particularly their interactions and synergies in joint production of created assets – have so far not adequately been examined in mainstream economics; in fact, it leaves these two key institutions of endogenous growth unlinked and their strategic relations unexplored. In contrast, the IB discipline has recently witnessed the emergence of some interesting ideas about, and analyses of, the critical role of MNC–government relations in promoting rapid growth. The IB scholars are squarely focused on the role of MNCs as a business institution which serves as an organiser of globalised production networks.

International business-related approach

So, what are these ideas of the IB genre? We can identify seven major ones as shown in Figure 4.1(B): (a) technology transfer and spillovers by MNCs, (b) dynamic evolution of OLI configuration *pari passu* with economic growth, (c) cross-border technological development and sourcing, (d) cumulative causation in technological competitiveness, (e) locational agglomeration of innovative activities, (f) pro-trade (as opposed to anti-trade) FDI, and (g) industrial restructuring and comparative advantage recycling in a regionalised context. All these mechanics in IB facilitate a continuous expansion of business activities in the world economy by restraining the law of diminishing returns – hence, endogenously driven growth. What follows will highlight the nature of each of these IB-based sources of endogenous growth.

Technology transfer and spillovers by MNCs and their networks

One of the keys to endogenous growth is technology transfer and knowledge spillovers. Dunning (1958) examined both the direct knowledge transfers and spillovers of American multinationals in British manufacturing industry. In his seminal theory of MNCs' overseas investment, Hymer (1960) identified 'the flow of business techniques and skilled personnel' as a major feature of 'the international operations of firms'. Vernon (1966) introduced the product cycle theory of trade and investment, depicting the investment of American firms in less developed countries as the major vehicle of manufacturing technology (both product and process) transfer and as the cause of trade reversal (i.e. initial exports turning into imports in the mature stage of product cycle). Later on, Vernon (1979) explored the altered locus of innovations and their diffusions in light of globally dispersed R&D centres.

Focusing on the imperfect nature of the factor markets, Buckley and Casson (1976) theorised the efficacy of the MNC as a superior institution to organise production than markets (hence stimulating economic growth worldwide), an institution in which knowledge transfers are more efficiently executed. Blomstrom (1989) specifically analysed the technological spillovers associated with inward FDI (in the context of a host economy, Mexico).[4] Kokko (1992) looked at how FDI interacts with host country characteristics in producing spillovers. Kokko and Blomstrom (1995) studied the policies to encourage technology inflows through FDI. Lall (1997) similarly examined the host country strategies to promote the upgrading of technology advantages by intervening in MNC activities (in the four Asian NIEs). Bende-Nabende (1999) provided an econometric investigation for the ASEAN-5 economies of the impact of FDI and its spillover effects on economic growth.

Most recently, Ernst (2000, 2001) developed a concept of 'global production networks (GPN)', a cross-border hierarchical organisation which is established by flagship multinational corporations and which creates knowledge spillovers to their overseas affiliates and suppliers through global outsourcing activities. An empirical study of the interactions between such networks and the growth of Asia's electronics industry was made in a number of essays edited in Borrus *et al.* (2000). The GPN model is reminiscent of Hymer's (1971) 'corporate headquarters' model of the globalised and hierarchical corporate economy in which the benefits of such an economy 'trickle down' from 'the metropolis to the hinterland'. It also parallels with Rugman and D'Cruz's (1996, 1997) 'five-partners business network model' and 'theory of the flagship firm'. To the extent that the spread of the globalised corporate economy facilitates knowledge dissemination, global economic growth no doubt accelerates, so long as its efficiency-enhancing force outweighs its market-control power.

Dynamic evolution of OLI configuration pari passu *with economic growth*

Dunning's eclecticism of the OLI (ownership-, location-, and internalisation-specific advantages) framework has recently developed into a dynamic evolutionary theory of configurations of these three factors (Dunning 1993; Narula 1996; Dunning and Narula 1996; Narula and Dunning 2000; Tolentino 1993). Dunning (1991) explained how the OLI configuration changes over time:

> At a given moment of time, the pattern of international production represents a point on a set of trajectories towards (or, for that matter, away from) the internationalisation of production by firms. That trajectory itself is set by the continuous and iterative interaction between the OLI configuration over a succession of time periods and the strategy of firms in response to these configurations, which, in turn, will influence the OLI configuration in subsequent time periods.
>
> (Dunning 1991: 126)

Following Dunning's lead, Narula (1996) argued that a firm's O advantages at a given period of time depend on L factors in a previous period, while a country's L advantages are in turn augmented by its ability to capitalise on the O advantages of foreign MNCs in a previous period. In short, a virtuous circle emerges between O and L advantages through 'continuous and iterative interaction'.

Dunning and Narula (1996) and Narula and Dunning (2000) further explored the interactions among host governments' trade and FDI policies, foreign MNCs' activities, and the development of intellectual capital at large and created assets at firm level – and their growth-inducing impact on a host country's ability to 'catch up' and join the ranks of advanced countries (or otherwise a country's fate to 'fall behind'). Their analyses are based on the five stages of 'investment development path (IDP)' through which the configurations of O and L factors evolve in a mutually reinforcing manner, exhibiting the path of 'net outward investment (NOI)' position.

It should be noted that O and L advantages play different roles in endogenous growth because of their different characteristics; the former are firm-specific (i.e. of a private property nature) and mobile but are normally transferred on an exclusive basis (i.e. not necessarily available to all outsiders), whereas the latter are of a public good nature[5] (equally available to all firms in principle) but immobile. At low levels of development, L advantages are a more important determinant of *localised* endogenous growth (say, in a particular developing host country or region rather than anywhere else) than their O advantage counterparts. Yet as developing countries catch up and converge with the advanced world, their general L advantages grow similar at high levels of development, and their O advantages become less home country-specific, more industry-specific, and more firm-specific. Indeed, the O advantages are then increasingly derived from firm-specific multinationality.

Cross-border technological development and sourcing

As Cantwell (1989, 1995, 1999; Cantwell and Janne 1999) observed, through international business operations, MNCs in a high-tech/R&D-based industry are not merely involved in technology transfer and knowledge spillovers but more importantly are increasingly engaged in new knowledge creation via a cross-border network of R&D facilities. In fact, this trend is accelerating as more and more R&D-based industries are involved in overseas production of knowledge itself – in addition to overseas production of products. International business thus serves as a catalyst for knowledge creation – hence as an engine of endogenous growth. In other words, the more globalised and interconnected the world economy becomes (instead of subdivided and segmented), the greater the chances of border-straddling knowledge production – hence the more rapid growth throughout the world.

Cumulative causation in technological competitiveness

The cumulative causation (both *virtuous* and *vicious* circles) between the technological activities of MNCs and the international competitiveness of an open economy may be intensified (Cantwell 1987). A virtuous circle occurs because inward FDI is likely to be attracted into innovative industries caught up in a virtuous circle in the first place, industries with local R&D facilities and a rising indigenous technological capacity, and because newly established foreign affiliates increase technological dissemination to suppliers and customers and spur local rivals to a higher rate of innovation. This virtuous circle is nothing but a powerful engine of MNC-driven endogenous growth. Here, the size of the firm and the size of the market attainable by way of expanding multinationality of the firm also give strong micro-economic incentives to innovate, because the firm's enlarged operations across international markets facilitate transfer of intangible assets at nominal marginal costs.

On the other hand, a vicious circle may also occur as a result of MNCs' market power, which may drive out local competitors and hinder the technology creation of local suppliers. As Cantwell (1987) summarises:

> The overall conclusion to be drawn from the basic model is that, as a long run process, an internationally trading industry will gradually become increasingly divided into some dynamic and some stagnant production locations. The former will be characterized by a high proportion of research intensive activity and a relatively steep technological progress function, the latter by a low proportion of research linked production and a comparatively shallow technical progress function.
>
> (Cantwell 1987: 134)

Locational agglomeration of innovative activities

Cantwell's analysis thus directly leads to the notion of locational agglomeration of innovative activities. This notion has only recently begun to be explored by IB scholars (*inter alia*, Porter 1990; Dunning 1991, 1996, 1997; Nachum 1999, 2000), although conventional economics has long been familiar with it ever since Alfred Marshall (1920) conceptualised an 'industrial district', and a revival of interest in it has occurred with a work of Krugman (1991). The IB literature has lately begun to zero in on this important topic. For example, an idea of cluster-based development strategies was discussed in Enright (2000). The creation of competitive advantages by MNCs through geographical agglomeration of innovation with all its multifarious aspects is explored in a conference volume (Dunning 2000) from IB perspectives. For example, the volume contains Gray and Dunning (2000) which initiates 'a theory of regional policy' by exploring how a particular macro-region interacts with micro-regions in generating virtuous or vicious circles in industrial and service agglomeration through their policies and economic activities. Ernst (2000) also introduces the 'flagship model of concentrated dispersion', in which rapid cross-border dispersion coexists with agglomeration. IB-related research is expected to step up on the dynamics of cluster-based L advantages, which is no doubt a main wellspring of endogenous growth.

One should remember, however, that forces of agglomeration work in both ways, positively and negatively. Diseconomies occur in term of deteriorations in the environment (pollution and ecological destruction) and social infrastructure (congestion, crimes and rising housing costs, etc.), and these need to be tackled at policy level to ensure endogenous growth. Diseconomies of this type weighs more heavily as a growth policy constraint in advanced countries than in developing countries.

Pro-trade (as opposed to anti-trade) FDI

Focusing on the trade implications of FDI, a distinction can be made between pro-trade and anti-trade types of FDI from a host country's perspective and in terms of the Ricardian doctrine of comparative advantage (Kojima 1973, 1975; Kojima and Ozawa 1984). When trade induces strong growth, the pro-trade type of FDI is a powerful promoter of comparative advantage. The origination of this idea lies in the recognition that MNCs as organisers of FDI are business institutions (as opposed to markets) that can create this macro-economic effect as a result of their multinational business operations. There is no place for MNCs to exist in the traditional trade theories which assume atomistic competition ('firms don't matter') and no cross-border factor movement (including FDI and knowledge diffusion). Therefore, the pro-trade/anti-trade FDI model, though based on the neoclassical Heckscher–Ohlin framework, is classified here as an IB approach because of its focus on MNCs as a market-transcending institution.

The 'maximum' growth-inducing effect of trade is based on three key propositions (Kojima and Ozawa 1985: 135–9):

Proposition I: Countries gain from trade and maximise their economic welfare when they export comparatively advantaged goods and import comparatively disadvantaged goods.

Proposition II: Countries gain even more from expanded trade when superior entrepreneurial assets are transferred through FDI (or through non-equity types of transactions) from the home countries' comparatively disadvantaged industries or segments in such a way as to improve the efficiency of comparatively advantaged (existing as well as potential) industries or segments in the host countries.[6]

Proposition III: The process of transferring comparative-advantage-augmenting assets is facilitated when the home countries are capable of generating new goods or industries in which they can continuously renew comparative advantages and retain full employment at home, particularly employment of those resources released from comparatively disadvantaged (hence contracting) industries.

These propositions represent a country's *triple pro-trade orientation,* which can magnify the growth-enhancing effect of trade. The first proposition is Ricardo's pro-trade specialisation which is comparative (not absolute) advantage-induced; the second is pro-trade asset transfers; and the third is pro-trade structural upgrading (analogous to Schumpeter's 'creative destruction' at the macro level). The third proposition, in particular, is now important, since the cost of globalisation on domestic employment needs to be minimised; here the government is expected to play the role of a facilitator of MNC-triggered structural change via human resource development and technology policies (this directly connects to the IB theme of the interactive roles of governments and MNCs) (Ozawa 1997a).

On the other hand, anti-trade FDI (replacement of exports) is most likely to occur, for example, when the host countries pursue import-substituting development policies,[7] or when innovating firms decide to specialise in knowledge creation (R&D) rather than production and exporting by quickly transplanting manufacturing overseas, especially in rapidly growing foreign markets.

Pro-trade FDI also generates scale and learning economies – in addition to allocative efficiency through trade-induced specialisation in comparatively advantaged host industries. FDI itself (even if it is of the neutral type) brings with it the superior technologies and organisational skills hitherto unavailable in the host economies. How the host countries will benefit from inward MNC activities, however, depends ultimately on their 'social capability' (Abramovitz 1986) and 'national technological capabilities' (Lall 1992). But the host country's capability itself may or may not be enhanced by foreign MNCs' participation in the domestic industries for a variety of reasons (Lall 1999; UNCTAD 1999). But when the host countries are able to capitalise on MNCs' pro-trade investment in the interest of structural upgrading, economic growth will be all the more accelerated and prolonged – that is, more endogenously driven than otherwise.

Industrial restructuring and comparative advantage recycling

Closely related to the idea of pro-trade FDI is the use of FDI, both inward and outward, as agent of industrial restructuring and upgrading along the lines of dynamic comparative advantage (Ozawa 1992, 1993). Supergrowth (a success case of endogenous growth in its ultimate form) thus becomes a possibility, especially for outward-looking, export-oriented economies with MNC-friendly development policies.[8] Moreover, when a regionally clustered group of countries which are at staggered stages of economic development adopt the similarly outward-focused strategies of development *simultaneously,* such a region becomes a dynamo of economic development, as has been evidenced in East Asia.[9] A proper alignment of countries along the different stages of industrialisation, along with stability in foreign exchange rates, creates an ideal set of conditions for comparative advantage recycling

(Ozawa 1993). For example, labour-intensive manufacturing (e.g. apparel, toys and standardised electronics goods such as radios, TV sets, microwave ovens and keyboards) has been quickly shifted first from Japan to the NIEs, then from the NIEs to ASEAN-4, and most recently from these Asian countries to China, creating the job-creating, wage-boosting and growth-inducing effects of FDI in *each* successive round of comparative advantage recycling and generating a 'tandem development multiplier' effect.

This analysis actually constitutes an extension (restatement) of the so-called 'flying-geese' theory of economic development (Akamatsu 1935, 1961; Kojima 1958) in terms of the role of multinationals as the augmenters and recyclers of trade advantages (Ozawa 1993, 1996, 1997b, 2000).[10] Comparative advantage recycling is thus a powerful mechanism of endogenous growth.

Focus on MNC–government interactions

It is now clear that the IB literature has produced some analytically fascinating ideas which can constitute the critical 'mechanics' of endogenous growth and which are equal in explanatory power and supplementary to those identified in the mainstream economic literature. One major strength of IB, compared with the traditional branch of economics, is that it is focused squarely on the relationships between MNCs and the host governments charged with the task of promoting domestic economic growth. And in fact, this focus makes IB distinctly different from conventional economics (especially, traditional international economics). In the words of Grosse and Behrman (emphases added):

> International business has existed as a distinct field of study for the past three decades, but it does not have a widely accepted explanatory theory on which to base its uniqueness as a discipline . . . . Since international business is the study of business activities that cross national borders and, therefore, is *fundamentally concerned with the firms that undertake that business and the national Governments that regulate them,* a theory that is unique to such business must explain the responses of businesses to government policies and the policy-making of Governments themselves towards international firms.
>
> (Grosse and Behrman 1992: 93)

They insist that 'any theory of *international* business must be a theory of policies and activities of business *and* Governments, in conflict and co-operation' and that 'a theory of international business should explain how the issues of government concerned with TNC activities are defined, how they are negotiated, what trade-offs are involved, how differences are resolved, what adjustments are made over time and why' (ibid.: 97). Consequently, Grosse and Behrman introduced a bargaining theory which explains how MNCs and host governments come to terms by trading off their 'relative bargaining resources' and 'relative stakes' within the range of 'similarity of interests'.

What this chapter is concerned with is, however, not the bargaining process *per se*, nor how the two parties arrive at an agreement or end up sharing the spoils of MNCs' operations. Our primary focus is beyond the bargaining and settlement stage: it is on the possible macro-organisational (*à la* Dunning) developmental impact of both MNCs' operations and the host governments' policies toward trade and FDI on the process of local economic growth and change.

Dunning's eclectic paradigm has the well-known threefold conditions that a firm must meet if it is to invest and produce abroad: (1) the possession of net O advantages *vis-à-vis* firms of other nationalities in serving particular markets; (2) it should be more profitable to use O advantages within its hierarchy (i.e. to 'internalise'); and (3) it must be in the interests of the firm to utilise these advantages in conjunction with at least some factor inputs outside its home country (Dunning 1988: 25–6). And it is this third condition that directly leads to the arena of interactions through international business operations between MNCs as seekers of foreign factor inputs and markets and host governments as managers of host-specific location factors. Here, O advantages need to be *both efficiently matched and effectively linked up with* some L advantages or assets of the host economy. There are, however, a number of political economy issues in the 'new diplomacy' of bargaining between foreign firms and host governments (Stopford and Strange 1991). Both benefits and costs need to be thrashed out and managed when developing host countries intervene in investment promotion, domestic-content requirements, export performance requirements and technology transfers for local economic development (Moran 1998).

It is this focus on the MNC–government interface that makes the IB approach distinct from mainstream economics and places the former in a more advanced state of understanding about the dynamics of endogenous growth and its policy implications for catch-up development. In addition to this key difference, the latter is devoted to 'formal theorising', while the former is oriented to 'appreciative theorising' – to borrow Nelson and Winter's (1982) classification of theorising. Furthermore, mainstream economics is somewhat 'reductionist' and 'deeply specialised', while IB is more 'holistic' and interdisciplinary in analysis – and more directly policy relevant. Needless to say, however, these two approaches are complementary with each other, although the IB literature unfortunately remains unorthodox to – and largely neglected by – mainstream economics.

MNC-cum-government-driven endogenous growth

By combining all the ideas of the IB genre reviewed above into a unitised frame of reference, we can thus develop what may be called 'MNC-cum-government-driven endogenous growth'. This framework should become the hallmark of IB theory of endogenous growth.

As stipulated in Dunning's eclectic paradigm, O advantages or firm-specific intangible assets (such as product and process technologies, and managerial and marketing skills) are possessed and controlled by MNCs, while L factors – inputs (such as labour, natural resources, and industrial infrastructure), local markets (demand conditions), and the general economic ambience – are all under the purview, and in the bailiwick, of the host government.

It is, therefore, worth stressing that an act of 'internalisation' (setting up and running a subsidiary) – and especially *how efficiently and how effectively (hence how profitably) internalisation is carried out* – is actually not a one-sided act on the part of MNCs alone. Internalisation efficiency is obviously affected by the host government's management and control of L factors, as well as by the quality of participation of local interests in MNCs' ventures as partners (in the case of joint ventures, which are often strongly promoted by the host country as a measure to build its national industrial capacity). In other words, a process of internalisation needs to be examined from both MNCs' and host countries' points of view, involving 'MNC-side internalisation' and 'host-side internalisation' (Ozawa 1997c). By host-side internalisation is meant that the host government endeavours to *maximise and retain within its own economy* as much as possible all the *direct* and *indirect* benefits of foreign MNC activities which accrue from their effective combinations with existing L factors.[11] To this end, the government must create the 'right' type of L advantages suitable for the 'right' type of foreign MNCs and their activities at the 'right' stage of economic development (Narula and Dunning 2000), the L advantages appropriate for the prevailing factor endowment and technological conditions. Thus, the government's ability to do so basically determines national competitiveness and economic growth. It should be noted, however, that the present trend toward globalisation has tilted the balance of power in favour of MNCs (Dunning 1997).

This two-sided process of internalisation is likely to contribute to FDI-led locational agglomerations – another cause of endogenous growth. Since the host governments' abilities to capitalise on and make use of MNC activities differ, divergent growth paths and rates will result.

Removal of growth constraints and endogenous growth

At the macro level, moreover, MNC–government collaboration can address the problems of developmental gaps or bottlenecks that have long hindered the developing countries' catch-up efforts and even trapped them in vicious circles. The developing countries are confronted with a variety of growth constraints. Four major constraints are typically listed in the standard literature on economic development: (1) a savings gap (due to the tendency of investment exceeding saving at low income levels); (2) a foreign exchange gap (due to the tendency of payments for imports exceeding revenue from exports and inflow of foreign capital at low income levels); (3) a lag in human resources development (due to the tendency of demand for skilled workers to exceed the supply in growing low-income countries); and (4) institutional/macro-organisational inadequacies. We will briefly discuss how each of the gaps may be closed in terms of inward MNC activities.

Savings gap

Any fast-growing economy encounters a shortage of savings due to a high rate of investment. There are two basic solutions to this gap. One is a 'central-bank-based' approach and the second is a 'current-account-deficit-based' approach (Ozawa 1999). The former uses the credit-creating capacity of a central bank to provide loanable funds to domestic capital formation, an approach which represents a self-reliant *dirigiste* financial strategy to avoid dependence on foreign capital – but which is accompanied by the danger of inflation unless managed judiciously by the central bank.[12] On the other hand, the current-account-deficit approach fills in a savings gap through foreign capital inflows by running a current-account deficit. The second approach, which can be identified as 'borrowed growth' (Ozawa 2001), is more oriented to the open-economy but exposes the country to the danger of a sudden reversal of capital flows.

Along with multinational manufacturers, other types of MNCs, notably in banking and finance (e.g. banks and investment funds), are ever more active in bringing capital (bank loans and portfolio capital) to rapidly developing countries, which in turn deliberately seek such investment to remove the savings constraint. Aliber (1993) stressed national differences in capitalisation rates as a determinant of FDI and the flows of portfolio capital from the low-rate-of-return mature economies to the higher-rate-of-return, rapidly growing economies. Rapidly catching-up countries are likely to attract both FDI and foreign portfolio investment (FPI) simultaneously. Dunning and Dilyard (1999) and Wilkins (1999) examined whether FDI and FPI are complements, substitutes or independent (unrelated). Good market conditions in the host economies normally induce the inflows of both types at the same time. The government, as well as culture and mores, can play a crucial role in promoting domestic savings. For example, the high rates of savings in East Asia owe much to these factors.

Foreign exchange gap

The stages theory of balance-of-payments (Kindleberger 1963) posits that any developing country which opts for an open-economy industrialisation path initially experiences a growing current-account deficit as it draws on capital goods and industrial knowledge from the advanced world. As the economy succeeds in industrialisation, its current account is expected to show continual improvements and eventually turns to a surplus. This phenomenon is also emphasised in the inter-temporal trade theory of external balance (*inter alia,* Sachs 1981). But until the economy reaches a robust current-account phase, it must finance the deficit by capital inflows, thereby securing the necessary foreign exchange reserves when it operates under fixed exchange rates. In this respect, MNCs can cover the both sides of the ledger of international payments, since they can bring in capital goods, technology and access to export markets (on the current-account side) and provide all necessary finance, including working capital in foreign exchange (on the capital-account side).

Being thus a critical resource, foreign exchange is normally allocated only to important uses under government control in the early developmental stages.[13] Developing countries usually promote the trade-supporting type of outward FDI, which augments their capacity to earn foreign exchange from exports – and accelerating trade-led endogenous growth. This process illustrates a transition of development policy from one-sided dependence on inward FDI as a remover of growth constraints to the active use of outward FDI as an asset acquirer by integrating into the more advanced economies which are the key export markets for developing countries.

Human resource gap

Catch-up development is basically a learning process, and people are the receptacles and assimilators of knowledge. A country's absorptive capacity in learning thus depends on the quality of its people and their capacity for entrepreneurship (innovation).[14] Here, governments can provide basic/universal education and specialised occupational training at the national level, while foreign MNCs can give on-the-job training and skill formation at the company level and stimulate economic growth through various avenues.[15] If it is to achieve endogenous growth a country with scarce natural resources, especially if it is small, needs to develop its only and ultimate asset, i.e. human resources, as the source of international competitiveness.[16]

Macro-organisational/institutional inadequacies

Dunning's concept of macro-organisation is an important and analytically useful one in the IB discipline. For example, in order for even a market economy to function properly it requires adequate macro-organisational setups. The free enterprise system is based on property rights and the rule of law. The global capital (equities and bonds) markets can function efficiently only when open information is available to market participants. This in turn requires the internationally acceptable and unified standards of accounting and disclosure if each country's financial sector is integrated into the global economy as it is increasingly the case nowadays.

The recent financial crisis in East Asia occurred despite the effective macro-economic policies (low budgetary deficits, stable prices, high employment and high growth) pursued in those crisis-hit countries. But the trouble was that they had macro-organisational weaknesses, especially in their heavily protected and controlled financial sectors (which began to open up to foreign competition in earnest only after the crises).[17] In this respect, inward FDI can serve as a catalyst to restructuring and reforms; it can even free the economy from the shackles of over-regulation and unnecessary bureaucracy, thereby improving its capacity to adapt to changing external conditions and enhancing its trade competitiveness.[18]

In short, MNC-cum-host-government cooperation can lead to vibrant growth in the local host economy as it can fill the developmental gaps that would otherwise remain hindrances to economic development. There are, however, some costs of this type of outward-dependent process of growth. One cost is the inevitable weakening of 'nationhood' (Gray 1999), and some may even resurrect the ghost of the 'dependency' theory of development. But here lies the key role of government in maximising the benefits of MNC-based growth and minimising its costs. In this respect, the capability of government (national governance) itself increasingly becomes a key locational asset (an endogenous factor) in its own right, given the current trend of globalisation.[19]

Summing up

In this chapter, some new ideas developed in the IB-related field are linked to the notion of endogenous growth, contrasting with, and supplementing, the mainstream economic analyses. And what may be called 'MNC-cum-government driven endogenous growth' is stressed as the core of the IB-related approach. MNCs and governments play the critical roles in facilitating the efficient matching of firm-specific assets with location-specific advantages so as to produce a vigorous and enduring business expansion in the local economy. They (in collaboration with local industry) are *the chief co-drivers* of endogenous growth by turning development constraints (L factors) into both new business opportunities (opportunities for O advantages to be appropriated and reinforced) and growth augmentation (in which L advantages are appropriated as part of GDP), which can counteract the law of diminishing returns – all within Dunning's eclectic paradigm. Moreover, endogenous growth can be enhanced even further within an intra-regional hierarchy of economies where a new form of dynamic industrial agglomeration comes into existence via MNC-networks of knowledge generation and commercialisation, via cross-border networks of production and procurement, and via inter-development-stage recycling of dynamic comparative (and competitive) advantages. The 'international business' paradigm of endogenous growth presented in this chapter draws heavily on the *macro-organisational* model of MNC–government interactions (i.e. the interface between the O and L factors) as the vital factor in continuous and rapid economic development. Development and growth in the age of globalisation are increasingly becoming IB-driven endogenous growth within the key framework for interactions ('joint internalisation') between MNCs (possessors of O advantages) and host governments (managers of L advantages) – that is, co-endogenisation of advantages for vigorous business expansion and economic growth.

Notes

1 Just to cite a few notable examples of policy studies that specifically treat MNCs as an instrument of development. United Nations Conference on Trade and Development (UNCTAD) has been actively studying the role of MNCs in economic development. See its annual World Investment Reports (especially 1992, 1995 and 1999). The impact of globalisation with stepped-up MNC activities on economic development is the focus of a conference volume (Hood and Young 2000). How Central Europe is making use of inward FDI in its transition to a capitalist-path of development and growth is detailed, for example, in a collection of papers (Hunya 2000).

2 This list is not meant to be exhaustive. There are many ancillary ideas such as horizontal and vertical innovations.

3 For the origins of the new growth theory, see Romer (1994). Barro and Sala-i-Martin (1995) emphasised the role of government in promoting viable long-term growth: 'if we can learn about government policy options that have even small effects on the long-run growth rate, then we can contribute much more to improvements in standards of living than has been provided by the entire history of macroeconomic analysis of counter-cyclical policy and fine-tuning.'

4 Spillovers by outward FDI on domestic employment, exports, and research in Swede – and in general terms – are explored by Blomstrom and Kokko (1994, 1997).

5 The 'publicness' of L advantages is, of course, stressed as a key factor of endogenous growth in the mainstream approach, as mentioned as 'infrastructure and public goods'.

6 This important second proposition is lacking in Ricardo's trade model, since he intentionally assumed away cross-border factor movement. Ricardo reasoned that international factor movement (including technology) would destroy the basis for trade (i.e. productivity differentials) between countries: '. . . under [the circumstances of higher labor productivity in Portugal] . . . the wine and the cloth should both be made in Portugal . . . therefore the capital and labor of England . . . should be removed to Portugal.' Thus, Ricardo did not see how England's comparative advantage in cloth would be even enhanced if the Portuguese secrets of higher productivity in cloth are transferred to England's comparatively advantaged industry. See Ozawa (1997b).

7 If import substitution occurs as part of the long-term process of building up local industrial capacity which eventually develops into exporting, the anti-trade nature of FDI is a temporary (short-run) phenomenon. In fact, FDI is generally trade-creating in the long run.

8 A developing country's strong outward orientation is certainly not without drawbacks, as has been seen in the recent Asian crisis, especially when its financial sector is opened to foreign banks, portfolio investors and currency speculators.

9 Indeed, one key question raised in the New Growth Theory is: 'Why have countries, or groups of countries, been able to grow for decades in succession with no apparent tendency to slow down, despite rising capital–labour ratios?' (Boltho and Holtham 1992). Most recently, such successful growth did occur in East Asia, which was, in fact, once identified as 'the East Asian miracle' (World Bank 1993).

10 Both Dunning's IDP theory and the reformulated 'flying-geese' paradigm are the stages models of FDI-driven economic development. For discussions on their interrelatedness, see Ozawa (1996) and van Hoesel (1997).

11 Here, government active involvement toward inward FDI is postulated. If the host government adopts a hands-off policy leaving economic coordination to market forces, no internalisation efforts are made by the government, as is usually the case with advanced countries (notably the United States, although its state and local governments are engaged in capturing the benefits of inward FDI with incentives).

12 This approach was, for example, adopted most effectively in Japan and Korea.

13 For example, Altomonte *et al.* (2000) point out how deliberately the East Asian governments allocated scarce foreign exchange for their firms to acquire real assets abroad, especially via outward FDI, despite 'the apparent severity of their home countries' foreign exchange constraint'. In fact, they encouraged their own local firms in purchasing technology via licensing and setting up shop overseas, especially in warehousing, distribution and information gathering.

14 Korean experience shows that local firms early on often served – and learned – in the capacity of 'apprentices' to foreign MNCs which came to operate in Korea (Amsden 1989). Now that Korea has attained a more advanced status of industrialisation, many of its firms are using outward FDI in the advanced countries as a vehicle of learning.

15 For an econometric analysis of the relations between FDI inflows from advanced countries and economic growth rates in the developing host countries (69 countries altogether), see Borensztein *et al.* (1995).

16 See, for example, Castello and Ozawa (1999). In the context of East Asian experiences, Japan is a classic example. Some other resource-indigent small economies (e.g. the NIEs) also have quickly attained a higher level of development than resource-abundant large countries (e.g. Indonesia, the Philippines and Malaysia) because of the former's focus on human resource development.

17 Indeed, these Asian countries are now attracting inward FDI, mostly in the form of mergers and acquisitions (M&As), which, for example, accounted for over 80 per cent of total FDI inflows in 1999. These M&As play a critical role in introducing new skills of management, notably internationally acceptable standards of accounting, disclosure and corporate oversight, thereby enforcing corporate governance reform and industrial restructuring (Zhan and Ozawa 2000).

18 This is well illustrated by the 1984–94 experience of New Zealand (Castello 1998; Castello and Ozawa 1999).

19 True, the mainstream approach also stresses human capital formation, investment in R&D, and infrastructure and public goods. But these are not framed in terms of MNC–government interactions.

References

Abramovitz, M. (1986) 'Catching up, forging ahead, and falling behind', *Journal of Economic History*, June, XLVI (2): 385–406.

Akamatsu, K. (1935) 'Waga kuni yomo kogyohin no boeki susei [Trend of Japanese trade in woollen goods]', *Shogyo Keizai Ronso,* volume unknown, Nagoya Higher Commercial School.

Akamatsu, K. (1961) 'A theory of unbalanced growth in the world economy', *Weltwirtschaftliches Archiv*, 86(2): 196–215.

Aliber, R. (1993) *The Multinational Paradigm*, Cambridge, MA: MIT Press.

Almonte, C., Bolwijn, R. and Gray, P. (2000) 'Open industrialization as a development strategy: The example of East Asia', in K. Fatemi (ed.), *The New World Order*, Amsterdam: Pergamon.

Amsden, A. (1989) *Asia's New Giant: South Korea and Late Industrialization*, New York: Oxford University Press.

Arrow, K. (1962) 'The economic implications of learning by doing', *Review of Economic Studies*, 29: 155–76.

Barro, R. (1990) 'Government spending in a simple model of endogenous growth', *Journal of Political Economy*, 98: 103–25.

Barro, R. and Sala-i-Martin, X. (1995) *Economic Growth*, New York: McGraw Hill.

Bende-Nabende, A. (1999) *FDI, Regionalism, Government Policy and Endogenous Growth: A Comparative Study of the ASEAN-5 Economies, with Development Policy Implications for the Least Developed Countries*, Aldershot, UK, Brookfield, Vt. and Sydney: Ashgate.

Blomstrom, M. (1989) *Foreign Investment and Spillovers: A Study of Technology Transfer to Mexico,* London and New York: Routledge.

Blomstrom, M. and Kokko, A. (1994) 'Home country effects of foreign direct investment: Sweden', in S. Globerman (ed.), *Canadian-Based Multinationals*, Calgary: University of Calgary Press.

Blomstrom, M. and Kokko, A. (1997) 'How foreign investment affects host countries', Policy Research Working Paper No. 1745, Washington, DC: World Bank, International Economics Department.

Boltho, A. and Hotham, G. (1992) 'The assessment: new approaches to economic growth', *Oxford Review of Economic Policy*, 8 (Winter): 1–14.

Borensztein, E, De Gregorio, J. and Lee, J. (1995) 'How does foreign direct investment affect economic growth?', NBER Working Paper No. 5057.

Borrus, M., Ernst, D. and Haggard, S. (eds) (2000) *International Production Networks in Asia: Rivalry or Riches?*, London: Routledge.

Buckley, P. and Casson, M. (1976) *The Future of the Multinational Enterprise*, London: Macmillan.

Cantwell, J. (1987) 'The reorganization of European industries after integration: selected evidence of the role of multinational enterprise activities', *Journal of Common Market Studies*, 36: 127–51.

Cantwell, J. (1989) *Technological Innovation and Multinational Corporations*, Oxford: Basil Blackwell.

Cantwell, J. (1995) 'The globalization of technology: What remains of the product cycle model?' *Cambridge Journal of Economics*, 19: 155–74.

Cantwell, J. (1999) 'From the early internationalization of corporate technology to global technology sourcing', *Transnational Corporations*, 8 (August): 71–92.

Cantwell, J. and Janne, O. (1999) 'Technological globalisation and the innovative centres: the role of corporate technological leadership and locational hierarchy', *Research Policy*, 28(2–3): 119–44.

Casson, M. (1991) *Global Research Strategy and International Competitiveness*, Oxford: Basil Blackwell.

Castello, S. (1998) 'Outward direct investment orientation, structural adaptability and economic revival: a case study of New Zealand', in J. Dunning (ed.), *Globalization, Trade and Foreign Direct Investment*, Amsterdam: Elsevier.

Castello, S. and Ozawa, T. (1999) *Globalization of Small Economies as a Strategic Behavior in International Business*, New York: Garland Publishing.

Chipman, J. (1970) 'Induced technical change and patterns of international trade', in R. Vernon (ed.), *The Technology Factor in International Trade*, New York: National Bureau of Economic Research.

De Mello, L., Jr. (1997) 'Foreign direct investment in developing countries and growth: A selective survey', *Journal of Development Studies* 34: 1–34.

De Mello, L., Jr. (1999) 'Foreign direct investment-led growth: evidence from time series and panel data', *Oxford Economic Papers,* 51: 133–51.

Dunning, J.H. (1958) *American Investment in British Manufacturing Industry*, London: Allen & Unwin.

Dunning, J.H. (1981) *International Production and the Multinational Enterprise*, London: Allen & Unwin.

Dunning, J.H. (1988) *Explaining International Production*, London: Unwin Hyman.

Dunning, J.H. (1991) 'The eclectic paradigm of international production: a personal perspective', in C. Pitelis and R. Sugden (eds), *The Nature of the Transnational Firm*, London: Routledge.

Dunning, J.H. (1992) 'The global economy, domestic governance, strategies and transnational corporations: interactions and policy implications', *Transnational Corporations,* 1(3): 7–45.

Dunning, J.H. (1993) *Multinational Enterprises and the Global Economy*, Reading, MA: Addison-Wesley.

Dunning, J.H. (1996) 'The geographical sources of competitiveness of firms: the results of a new survey,' *Transnational Corporations,* 5(3): 1–30.

Dunning, J.H. (1997) 'A business analytic approach to governments and globalization', in J. Dunning (ed.), *Governments, Globalization, and International Business*, Oxford: Oxford University Press.

Dunning, J.H. (2000) *Regions, Globalization and the Knowledge Economy: The Issues Stated,* Oxford: Oxford University Press.

Dunning, J.H. and Dilyard, J. (1999) 'Toward a general paradigm of foreign direct and foreign portfolio investment', *Transnational Corporations,* 8: 1–54.

Dunning, J.H. and Narula, R. (1996) *Foreign Direct Investment and Governments: Catalysts for Economic Restructuring*, London and New York: Routledge.

Eicher, T. and Kalaitzidakis, P. (1997) 'The human capital dimension to foreign direct investment: training, address selection, and firm location', in B. Jensen and K. Wong (eds), *Dynamics, Economic Growth, and International Trade*, Ann Arbor: University of Michigan Press.

Enright, M. (2000) 'The globalization of competition and the localization of competitive advantage: policies towards regional clustering', in N. Hood and S. Young (eds), *The Globalization of Multinational Enterprise Activity and Economic Development,* London: Macmillan/New York: St. Martin's Press.

Ernst, D. (2000) 'Global production networks and the changing geography of innovation systems: implications for developing countries', East-West Center Working Paper, Economic Series, November, No. 9, *Journal of the Economics of Innovation and New Technologies.*

Ernst, D. (2001) 'The internet's effect on business organization: bane or boon for developing Asia?' *Asia Pacific Issues*, January, No. 48.

Gray, P. (1999) *Global Economic Involvement*, Copenhagen: Copenhagen Business School Press.

Gray, H. and Dunning, J. (2000) 'Toward a theory of regional policy', in J. Dunning (ed.), *Regions, Globalization, and the Knowledge-Based Economy*, Oxford: Oxford University Press.

Grosse, R. and Behrman, J. (1992) 'Theory of international business', *Transnational Corporations*, 1(1): 93–126.

Grossman, G. and Helpman, E. (1991) *Innovation and Growth in the Global Economy*, Cambridge, MA: MIT Press.

Hoesel, R. van (1997) *Beyond Export-led Growth: The Emergence of New Multinational Enterprises from Korea and Taiwan*, No. 174 of the Tibergen Institute Research Series, Amsterdam: Thesis Publishers.

Hood, N. and Young, S. (eds) (2000) *The Globalization of Multinational Enterprise Activity and Economic Devleopment*, London: Macmillan; New York: St. Martin's.

Hunya, G. (ed.) (2000) *Integration Through Foreign Direct Investment: Making Central European Industries Competitive*, Cheltenham, UK: Edward Elgar.

Hymer, S. (1960) *The International Operations of National Firms: A Study of Foreign Direct Investment*, Ph.D. dissertation, Cambridge: Massachusetts Institute of Technology (later published by MIT Press in 1976).

Hymer, S. (1971) 'The multinational corporation and the law of uneven development', in J. Bhagwati (ed.), *Economics and World Order*, New York: Macmillan.

Kindleberger, C. (1963) *International Economics*, New York: Irwin.

Kojima, K. (1958) 'Nihon keizai no gankokeitaiteki hatten to boeki no yakuwari', in K. Kojima (ed.), *Nihon Boeki no Kozo to Hatten [The Structure and Growth of Japanese Trade]*, Tokyo: Shiseido.

Kojima, K. (1973) 'A macroeconomic approach to foreign direct investment', *Hitotsubashi Journal of Economics*, 14: 11–21.

Kojima, K. (1975) 'International trade and foreign investment: substitutes or complements', *Hitotsubashi Journal of Economics*, 16: 1–12.

Kojima, K. and Ozawa, T. (1984) 'Micro- and macro-economic models of direct foreign investment: toward a synthesis', *Hitotsubashi Journal of Economics*, 25: 1–20.

Kojima, K. and Ozawa, T. (1985) 'Toward a theory of industrial restructuring and dynamic comparative advantage', *Hitotsubashi Journal of Economics*, 26: 135–45.

Kokko, A. (1992) *Foreign Direct Investment, Host Country Characteristics, and Spillovers*, Stockholm: Economic Research Institute.

Kokko, A. and Blomstrom, M. (1995) 'Policies to encourage inflows of technology through foreign multinationals', *World Development March*, 23: 459–68.

Krugman, P. (1979) 'A model of innovation, technology transfer, and the world distribution of income', *Journal of Political Economy*, 87: 253–65.

Krugman, P. (1991) *Geography and Trade*, Cambridge, MA: MIT Press.

Lall, S. (1979) 'The international allocation of research activity by U.S. multinationals', *Oxford Bulletin of Economics and Statistics*, 41: 187–96.

Lall, S. (1992) 'Technological capabilities and industrialization', *World Development*, 20: 165–86.

Lall, S. (1997) 'East Asia', in J. Dunning (ed.), *Governments, Globalization, and International Business*, Oxford: Oxford University Press.

Lall, S. (1999) 'Technological change and industrialization in the Asian NIEs', in L. Kim, and R. Nelson (eds), *Technological Learning and Economic Development: The Experience of the Asian NIEs*, Cambridge: Cambridge University Press.

Long, N. and Wong, K. (1997) 'Endogenous growth and international trade: a survey', in B. Jensen and K. Wong (eds), *Dynamics, Economic Growth, and International Trade*, Ann Arbor: University of Michigan Press.

Lucas, R. (1988) 'On the mechanics of economic development', *Journal of Monetary Economics*, 22: 3–42.

Marshall, A. (1920) *Principles of Economics*, 8th edition, London: Macmillan.

Meier, G. (1999) *The International Environment of Business*, New York: Oxford University Press.

Moran, T. (1998) *Foreign Direct Investment and Development: The New Policy Agenda for Developing Countries and Economies in Transition*, Washington, DC: Institute for International Economics.

Nachum, L. (1999) *The Origins of the International Competitiveness of Firms: The Impact of Location and Ownership in the Professional Service Industries*, Cheltenham, UK: Edward Elgar.

Nachum, L. (2000) 'Economic geography and the location of TNCs: financial and professional service FDI to the USA', *Journal of International Business Studies*, 31: 367–85.

Narula, R. (1996) *Multinational Investment and Economic Structure: Globalisation and Competitiveness*, London: Routledge.

Narula, R. and Dunning, J. (2000) 'Industrial development, globalisation and multinational enterprises: new realities for developing countries', *Oxford Development Studies*, 28: 141–67.

Nelson, R. and Winter, S. (1982) *An Evolutionary Theory of Economic Change*, Cambridge, MA: Harvard University Press.

Ozawa, T. (1992) 'Foreign direct investment and economic development', *Transnational Corporations*, 1 (February): 27–54.

Ozawa, T. (1993) 'Foreign direct investment and structural transformation: Japan as a recycler of market and industry', *Business & the Contemporary World*, 5 (Spring): 129–49.

Ozawa, T. (1996) 'Japan: the macro-IDP, meso-IDPs and the technology development path (TDP)', in J. Dunning and R. Narula (eds), *Foreign Direct Investment and Governments: Catalysts for Economic Restructuring*, London: Routledge.

Ozawa, T. (1997a) 'Japan', in J. Dunning (ed.), *Governments, Globalization, and International Business*, Oxford: Oxford University Press.

Ozawa, T. (1997b) 'MNCs and the flying-geese paradigm of tandem growth in the Asia-Pacific', in N. Kobayashi (ed.), *Management: A Global Perspective*, Tokyo: Japan Times.

Ozawa, T. (1997c) 'Excessive FDI and dilemmas of technology transfer' (in Japanese), *Developing Engineering*, 3: 27–42.

Ozawa, T. (1999) 'Bank loan capitalism and financial crises: Japanese and Korean experiences', in A. Rugman and G. Boyd (eds), *Deepening Integration in the Pacific Economies*, Cheltenham: Edward Elgar.

Ozawa, T. (2000) 'The 'flying-geese' paradigm: toward a co-evolutionary theory of MNC-assisted growth', in K. Fatemi (ed.), *The New World Order: Internationalism and Multinational Corporations*, 209–23, Amsterdam: Pergamon

Ozawa, T. (2001) 'Borrowed growth: current-account deficit-based development finance', in K. Fatemi (ed.), *The New World Order: Internationalism and Multinational Corporations*, Amsterdam: Pergamon.

Pearce, R. (1997) 'Decentralized R&D and strategic competitiveness: globalized approaches to generation and use of technology in MNEs', *Research Policy,* 28: 157–78.

Porter, M. (1990) *The Competitive Advantage of Nations*, New York: Free Press.

Rivera-Batiz, L. and Romere, P. (1991) 'International trade with endogenous technical change', *European Economic Review*, 35: 971–1004.

Romer, P. (1994) 'The origins of endogenous growth', *Journal of Economic Perspectives*, 8: 23–44.

Romer, P. (1986) 'Increasing returns and long-run growth', *Journal of Political Economy*, 94: 1002–37.

Rugman, A. and D'Cruz, J. (1996) 'Partners across borders: the five partners business network model', *International Management,* 1: 15–26.

Rugman, A. and D'Cruz, J. (1997) 'The theory of the flagship firm', *European Management Journal,* 15: 403–12.

Sachs, J. (1981) 'The current account and macroeconomic adjustment in the 1970s', *Brookings Papers on Economic Activity*, 1: 201–68.

Stopford, J. and Strange, S. (1991) *Rival States, Rival Firms: Competitions for World Market Shares,* Cambridge: Cambridge University Press.

Tolentino, P. (1993) *Technological Innovation and Third World Multinationals*, London: Routledge.

Turnovsky, S. (1997) 'Public and private capital in an endogenously growing open economy', in B. Jensen and K. Wong (eds), *Dynamics, Economic Growth, and International Trade*, Ann Arbor: University of Michigan Press.

UNCTAD (1992, 1995 and 1999) *World Investment Report,* New York: UN Publications.

Uzawa, H. (1965) 'Optimal technical change in an aggregative model of economic growth', *International Economic Review*, 6: 18–31.

Vernon, R. (1966) 'International investment and international trade in the product cycle,' *Quarterly Journal of Economics*, 80(2): 190–207.

Vernon, R. (1979) 'The product cycle hypothesis in a new international environment', *Oxford Bulletin of Economics and Statistics*, 41: 255–67.

Wilkins, M. (1999) 'Two literatures, two story-lines: Is a general paradigm of foreign portfolio and foreign direct investment possible?', *Transnational Corporations*, 8: 53–116.

World Bank (1993) *The East Asian Miracle: Economic Growth and Public Policy,* New York: Oxford University Press.

Zhan, J. and Ozawa, T. (2000) *Business Restructuring in Asia: Cross-Border Mergers and Acquisitions in the Crisis Period*, Copenhagen: Copenhagen Business School Press.

5 The co-evolutional advantage

Strategic management theory and the eclectic paradigm

Anoop Madhok and Anupama Phene

Introduction

It has been about a quarter century since the eclectic paradigm (Dunning 1977) was originally introduced as a framework for examining the international production and foreign direct investment decisions of firms. The eclectic paradigm offered a comprehensive approach hypothesising that foreign direct investment would occur when three conditions are satisfied. First, firms possessed a competitive advantage (the ownership or O advantage). This advantage was relative to firms from the host country and assumed to be necessary in order to offset the disadvantage of not being a local firm. In other words, the O advantage offsets the liability of foreignness. Second, to benefit fully from the potential created by the O advantage, there must be some need to combine them with some location-specific (L) advantages of host countries. Third, to complete the triangle, given the O advantage, there must be some benefit to undertaking the relevant set of activities internally (the internalisation or I advantage) within the firm rather than through more arm's-length contractual mechanisms like licensing or market mechanisms such as exports.

The following can be noted about the original OLI framework: (1) It is an efficiency argument. The decision to internalise or not is considered on the grounds of efficiency. (2) The central focus of interest in this regard has been the market for knowledge. Certain characteristics of knowledge, such as uncertainty in valuation, create difficulties both in its pricing and in its transfer through the contractual process (Buckley and Casson 1976; Hennart 1991). This undermines the efficiency of contractual mechanisms and is a driving force for internalisation. (3) As a logical extension of the above, alternative governance forms such as foreign direct investment and licensing serve to substitute for one another. (4) The primary role of the host country is to enable the multinational corporation (MNC) to exploit its home country-based assets through the provision of some complementary assets. Since the O advantage is relative to host country firms, the source of advantage lies at the level of the home country, i.e. the headquarters.

The eclectic paradigm is regarded by many as a leading explanation of the growth of multinational activity. Undoubtedly, the OLI framework was useful in the context within which it was introduced, when firms were beginning to embark on internationalisation of their activity. It answered the question, 'How does the multinational firm arise?' However, the question arises as to how the paradigm is relevant today, (a) when many firms have established a multinational presence, face global competitors, are undertaking repeat investments and/or are moving to higher-value creation activities such as innovation in overseas locations (Cantwell 1995; Casson 1991; Pearce 1989), and (b) when the basis of advantage is shifting towards a firm's ability to create and manage a knowledge portfolio. Does the OLI framework, even the later reformulated version (Dunning 1988), have adequate meaning in the current international business context?

In this chapter, we bring in a strategic management perspective to understanding multinational activity today. We turn to concepts from strategy, such as macro and micro co-evolution, to delineate differences across firms and, using the resource/capability-based view of the firm, extend the eclectic paradigm beyond the institutional dimension to incorporate the firm-specific level of analysis in terms of their capabilities. Given the increasing emphasis in the strategic management literature on a knowledge-based view of organisation, we focus on the management of knowledge.[1] Our position is that the OLI framework continues to maintain its relevance but only after making some significant modifications. The expanded framework extends the original question to 'Why do multinational firms continue to exist?' We suggest that multinational firms continue to exist (and even prosper) because of the unique possibilities they possess in the management of knowledge creation and transfer. Macro and micro co-evolutionary forces are critical in this regard and lend dynamism to this process.

The chapter is organised as follows. The next section discusses the OLI framework in the current business context and shows why the OLI arguments need to be expanded and modified to better explain multinational activity today. The following section discusses the contribution of strategic management theory in this regard, and emphasises in particular the role of co-evolution. The final section sums up the argument and some of its implications.

The OLI framework and multinational firm advantage

The 1980s can be considered the decade characterised by the onset of globalisation, with individual firms globalising their production and transactions (Dunning 1992). This was then followed by the internationalisation of research and development activity through the 1990s (Cantwell 1995; Dunning 1992). At the turn of the millennium, the emphasis is increasingly on researching the evolution of cross-border innovatory activity and understanding how it forms the basis of firm competitive advantage (Zander and Solvell 2000). The more noticeable trends in the international business environment of today include the increasing concentration of cross-border activity in the industrialised countries, especially those of the Triad (UN World Investment Report 1999). With the post-war maturing of the European and Japanese economies, flows of investment and technology among the Triad countries have over time become more evenly spread out and more participatory and reciprocal in nature. Other trends include: greater technological complexity and knowledge intensity of production; increasing incidence of and trend towards cross-border mergers/acquisitions and strategic alliances, including among competitor firms (UN World Investment Report 2000); and the shift in importance of location from the national to the subnational level, i.e. local clusters of innovative activity (Porter 1998; Saxenian 1990). With globalisation, there is increasing concentration in many industries and concomitant global interdependence of competitors. Also, a greater portion of trade is being conducted within the boundaries of multinational enterprises, in other words, increasing international integration of production within the sub-economy of the MNC.

The above trends pose some interesting challenges to the eclectic paradigm in terms of its ability to address them in its explanation of multinational activity. Recent developments in the strategic management field can be applied to the OLI framework and help understand the modern multinational more fully. In a modified version of the OLI framework, one could argue the following: the source of a firm's advantage lies potentially at the level of headquarters or subsidiary or the systemic relations between them. Management of a firm's global stocks and flows of knowledge has become a critical issue and perhaps the ultimate source of competitive (O) advantage. Besides enabling the MNC to exploit some home country-based advantage, the host country (L) potentially provides complementary assets to also enhance such advantage and build system-wide assets, which in turn enables the firm to become more robust in its competition with global or local rivals. Here, the motive is the exploitation of local assets to enhance the global capabilities of the firm.

In essence, rather than foreign firms operating in local economies, an MNC is better approached as a sub-economy in its own right. OLI-related decisions are made more and more in light of how they support the internal sub-economy of the firm. As a result of a firm's strategy, there is increasing intra-firm specialisation throughout its global configuration of operations in order to match affiliate capabilities to their respective environments, and affiliate charters to capabilities, many of which may be at least partially location-bound. This also manifests itself through increased levels of intra-firm trade. Moreover, there is a shift in emphasis from mainly the multinational firm's interface with the local environment to also address competition between multinational firms in local environments. Finally, FDI and other forms are not substitutes but complements.

Note that many of the above issues are concerned more with managing the multinational firm itself rather than just why, where and the means by which to engage in overseas activity. This is important because many of the trends noted above have resulted in a greater convergence among multinational firms from different home countries in terms of their global configuration of activities. This raises the question 'How do multinational enterprises differ from each other?', a question that falls squarely within the domain of strategic management theory.

The OLI framework is not adequately equipped to deal with this question of performance differences across firms and with the related issues. To understand why, one needs to separate out the institutional level argument from the firm level one. From an institutional approach, the question 'Why do multinational firms exist?' or, alternatively, 'Why does a firm become a multinational?' is a direct geographical extension of Coase's (1937) original question 'Why do firms exist?' If Coase's main purpose was to explain why economic activity was organised within firms, the purpose of the I aspect of the OLI triangle was to explain why international economic activity was organised within *the* multinational firm. However, there are two principal questions with respect to the organisation of economic activity. The first is 'Why is an activity organized within firms and not through the market mechanism?' The other, equally important, is 'Why is an activity organised within a particular firm (or firms) and not any other?' The main difference between the two questions is that the first operates at the institutional level, i.e. it is accordingly interested in the institution of firms and markets as a whole, while the latter operates at the firm level, and is accordingly interested in specific firms.

On closer examination, it is evident that the OLI argument operates at the institutional level. The O advantage is available to the home-country firm relative to that of the host-country firm. The argument is not about a specific firm but *the* home-country firm as an institution. To the extent that home location is a source of advantage (Vernon 1966; Porter 1990), it operates at the collective level, i.e. all the home firms from that location. In principle then, the framework is not interested in why a particular home-country firm may have a unique competitive advantage *vis-à-vis* other (global) competitors but almost exclusively over location-bound local competitors. Similarly, if the complementary assets offered by a host country enable superior exploitation of the home-country advantage, this is at least in principle available to all the foreign firms participating in that location.

An institutional level approach begs the question of why a particular firm has an advantage over other firms. This latter question has been the central focus of the strategic management field over the past decade or so, under the general rubric of the resource/capability-based view of the firm. Strategic management arguments operate at both the institutional and firm-specific level. At the institutional level, a growing body of work in strategic management has been arguing that the reason why an activity is conducted within the firm is not Coasean market failure (i.e. the cost of transacting through the market) but rather firm success, in that the firm as an institution enjoys an 'organizational advantage' which enables it to organise economic activity in a manner that markets simply cannot (e.g. Teece *et al.* 1997; Kogut and Zander 1992, 1996; Ghoshal and Moran 1996). Rather than pure price coordination and efficient contracting, it is the more qualitative coordination that underlines 'the distinctive ways that things are accomplished within the enterprise' (Teece *et al.* 1997: 528).

In the international domain, the evolutionary theory of the multinational corporation (Kogut and Zander 1993) posits that the reason multinational firms exist is their distinctive ability to disseminate, receive and absorb knowledge within the boundaries of the firm. From such a standpoint of firms as repositories of knowledge, the multinational firm can be viewed as a vehicle for creating, integrating and applying productive knowledge across its different locations. In this view, proposed as an alternative explanation to Coasean arguments (Kogut and Zander 1993; Madhok 1997), the multinational firm arises not because of market failure in transactions involving knowledge but because of its superior ability to transfer knowledge and knowledge-related processes internally. This defines the ownership advantages of firms. Differently put, the decision to maintain the transaction within the firm could be due to superior organisation, and relatedly, a superior capacity to generate and extract rents from its know-how.

Note that the multinational-firm advantage is available in principle to all multinationals with global configurations. Given the convergence of multinational firms' configurations in many industries, the argument of multinational-firm advantage still operates at the institutional level, that of *the* multinational corporation. Although the eclectic paradigm in its later (Mark II) version (Dunning 1988) recognised the MNC's superior ability to manage knowledge as a result of its global configuration, this advantage, referred to as economies of common governance, is relative to the local firm that does not and cannot enjoy the benefit of such economies. There is no advantage relative to other MNCs who also potentially enjoy the benefits of common governance. The argument therefore does not distinguish between the institutional level of the MNC and that at the level of a specific multinational firm. This leaves us with the question 'What accounts for performance differences across global competitor firms?'

The resource/capability-based view and the OLI framework: the co-evolutional advantage of the multinational firm

Strategic management theory is centrally concerned with the question of how firms differ and would argue, in particular, that the source of such differences lies in firm capabilities, i.e. the firm's ability to manage resources. As mentioned, a key capability that is crucial in the context of the technological complexity and intense competition that characterises modern commerce is a firm's ability to create, transfer and absorb knowledge. This recognition of knowledge as a particularly pivotal asset of firms in securing and sustaining competitive advantage (Bresnan *et al.* 1999; Nonaka and Takeuchi 1995) has resulted in an explosion of research in the strategic management literature on knowledge flows within and across firm boundaries, both domestic and international. However, much of the earlier research, especially the economics-based research, emphasises the costs incurred in the transfer of knowledge (Teece 1976; Davidson and McFetridge 1984). This tends to under-emphasise the long-term potential for value creation through knowledge transfer, which has been more recently emphasised by strategic management research (Szulanski 1996; Gupta and Govindarajan 2000). The emphasis on long-term value shifts the focus towards a more dynamic perspective of knowledge management by multinational firms, one which can accommodate concepts like firm evolution and changes in its capabilities (Nohria 1992). Rugman and Verbeke (2001) examine the different kinds of processes that lead to the development and diffusion of subsidiary capabilities within the multinational network. We also examine this dynamic, although through a different lens – that of co-evolutionary theory.

Research on the evolution of organisations has typically focused on adaptations in response to contingencies (Pettigrew 1985) or the role of the external environment in driving changes within organisations (Tushman and Anderson 1986) and the characteristics of the new organisational forms that emerge as a result of managerial adaptation or environmental selection (Volberda 1998; Miles *et al.* 1997). Co-evolutionary theory suggests, however, that change is the joint outcome of both managerial adaptation and environmental selection and not the outcome of one or the other (Lewin and Volberda 1999). It also suggests that co-evolution occurs at two different levels: macro co-evolution, representing co-evolution between the firm and its environment, and micro co-evolution, which is co-evolution within the firm among its constituent parts (McKelvey 1997). Scholars have tended to mainly examine co-evolution with the environment (Baum and Korn 1999; March 1994). Yet, the evolutionary trajectory of the MNC is driven by the simultaneous role of micro co-evolution and macro co-evolution of the MNC and the resulting influence on its knowledge generation, transfer and absorption capabilities. In the context of the OLI framework, the co-evolutionary process can lead to firm-level differences in the development of MNC capabilities.

Macro co-evolution

The concepts of macro co-evolution and micro co-evolution have a special relevance for the multinational corporation because of the diversity of environments in which the firm operates and the particular role each subsidiary occupies in the multinational firm's locational configuration. Within the MNC, there are two kinds of co-evolutionary processes unfolding simultaneously: macro co-evolution, or the co-evolution between the different parts (i.e. subsidiaries and HQ) of the MNC and their respective niches, and micro co-evolution, or the co-evolution among the constituent parts within the MNC which occurs through the interaction of subsidiaries with each other and with headquarters.

Each subsidiary occupies a particular niche in the multinational firm, for instance a geographic or product specialisation (or both). As the niche decays or expands due to processes operating at the environment level, a subsidiary may find itself becoming more or less important to the firm as it co-evolves with the niche. Macro co-evolution is reflected by changes in subsidiaries in response to changes in their niche and by changes in their charters in response to such changes in both their environment and their capabilities. For instance, in the IT sector, if liberalisation policies of the Government of India, both generally and with respect to IT, coupled with other factors such as a ready supply of IT-qualified engineers at reasonable wages, make India a much more pivotal player in the global IT industry, then the Indian affiliate of IT multinationals would become more critical to the firm and would play a larger role in the firm's IT strategy and operations worldwide.

The question of performance differences across global competitor firms is partly answered by the dynamics of the macro co-evolutionary process. In the context of OLI, macro co-evolution would be determined by the location-specific advantages (L) in a dynamic way. Macro evolutionary processes reflecting environmental selection pressures and the initiatives of local subsidiaries influence the resources available to the subsidiary and the capabilities of the subsidiary occupying that niche. MNCs that are better able to direct the macro co-evolutionary process, both through anticipation of the forces of environmental selection as well as managerial adaptation, will tend to create organisational structures and processes that allow subsidiaries to target and fit into expanding niches in the environment.

Micro co-evolution

Micro co-evolution reflects the co-evolution of intra-firm resources, capabilities and competencies (Lewin and Volberda 1999) within the MNC. Part of this co-evolution occurs through managerial adaptation as a result of interaction among the constituent parts of the firm, especially with respect to knowledge assets and routines, which are created, diffused, absorbed, augmented and transformed over time. The other part is driven by the natural selection of certain routines that enhance the knowledge processes above, and perhaps also due to their inimitability by competitors.

In the evolutionary theory of the MNC, the critical capability of the MNC is that of the management of knowledge on a global scale. Knowledge by virtue of some of its characteristics – stickiness within firm and location boundaries, potential for leverage, and increasing returns – becomes the ultimate source of firm competitive advantage. The quest for a knowledge-based advantage has made MNCs less hierarchical in that where earlier O advantages were considered to be rooted in the home country, today it is increasingly recognised that they could be more dispersed and located at different points or nodes in the multinational system as a result of macro co-evolutionary forces. To the extent that those closest to the source of knowledge understand it best and know best how to manage it, they should have decision rights over how the knowledge is accessed, developed and applied (Grant 1996).

Rugman and Verbeke (1992) have already challenged the assumption that an MNC's ownership advantage originates in the parent company and point to the role of the subsidiary in the creation of firm advantage. As a reflection of MNCs' increasing recognition of this logic, the locus of decision-making is evolving in many multinational firms as they seek to align the charters of various units with their capabilities and location (Birkinshaw and Hood 1998). Accordingly, the subsidiary becomes a repository of specific advantages, rather than the firm (i.e. HQ), and would tend to have specialised capabilities in knowledge creation and application in its particular domain of product-market activity.

Over the years, a number of researchers (White and Poynter 1984; Bartlett and Ghoshal 1986; Birkinshaw and Morrison 1995; Holm and Pedersen 2000) have emphasised a more dynamic view of the multinational corporation as an integrated network of subsidiaries with differing roles and responsibilities that are continually evolving. These differences in roles assumed by subsidiaries over time are determined by various drivers such as local environment (Gupta and Govindarajan 1991), entrepreneurial capacity of the subsidiary (Birkinshaw 1997) or membership in a leading edge industry cluster (Birkinshaw and Hood 2000).

Firm advantage therefore comprises two levels: HQ and subsidiary. The former is the traditional argument where firm advantage arises in the headquarters and is then transferred to its subsidiaries. Equally important, if not more so, is subsidiary-specific advantage (Rugman and Verbeke 2001). Much, though not all, of this is sticky at the subsidiary level since it alone is present in the particular niche and has an intimate knowledge of that niche. Subsidiary-specific advantage is important but clearly not sufficient. It is the ability of the multinational firm to tap into location-bound knowledge and make it non-location-bound, and thus available to the rest of the firm, through various organisational mechanisms, which is critical. This alone allows the firm to leverage its knowledge assets and extract the full rent potential, and could be the ultimate font of advantage for the multinational firm. There is therefore a third level of firm advantage, the network level (Rugman and Verbeke 2001), which is created through the interaction among the HQ and the various subsidiary-specific advantages. The management, through its direction of this micro co-evolutionary process, creates a dynamic O advantage for the multinational corporation.

The argument places a premium on the ability of a firm to marshal and manage the skills and knowledge necessary to be at the competitive forefront. If firm advantage is created by capabilities at the subsidiary level and the firm's ability to manage knowledge flows across this dynamic network of subsidiaries, then the role of headquarters becomes in effect an enabler and catalyst, a facilitator and coordinator of knowledge stocks and flows. This requires management (a) to become more knowledgeable and informed about the activities of a particular affiliate in order to understand and appreciate its needs and contributions, (b) to become, in a general sense, both more knowledgeable and globally informed about the multiple kinds of activities being performed within the various parts of the organisation, and outside, and (c) to harness, leverage and channel the more pertinent information and knowledge to the relevant units and subunits in a productive manner.

Firm advantage – a joint outcome of macro and micro co-evolution

As the multinational corporation is transformed by both macro and micro co-evolutionary processes, these processes drive the developmental trajectory of the firm's capabilities. Macro co-evolution, driven by environmental selection of specific technologies and geographic niches as well as by managerial adaptation through change in subsidiary charters and roles, dictates the configuration and scope of the multinational firm. Micro co-evolution, due to natural selection of routines that enhance knowledge transfer and due to managerial adaptation of organisational structure and processes, determines the firm's ability to manage the configuration and scope. Thus, the evolution of firm capabilities as a whole is a joint outcome of the macro and micro co-evolutionary processes. The firm's ability to manage subsidiaries in a macro co-evolutionary way so that they ebb and flow with respect to their respective niches, cemented with the ability to manage them in a micro co-evolutionary way so that the entire firm benefits from various subsidiary initiatives, can be a distinctive feature underlining the O advantage.

The co-evolutionary framework answers the question, posed earlier, 'What accounts for performance differences across global competitor firms?' There are differences among firms in how they manage the co-evolutionary process. The O advantage could also be viewed as the organisational advantage that is 'owned' by the firm. To put it in the context of the resource-based argument of strategic management theory, the ability to manage knowledge flows globally can be characterised by certain key properties, which result in it becoming the key source of firm-specific rents. Being embedded in firm routines, this ability is tacit, socially complex, causally ambiguous and thus difficult to imitate (Barney 1991). As a result, not only is it a valuable asset but also potentially the source of sustainable competitive advantage and resultant performance differences among multinational competitors.

The O and I decisions are interlinked with the co-evolution of the MNC network. The evolution and success of the MNC network is a result of the O and I advantages it has created with respect to knowledge creation and transfer. With its emphasis on the institutional level, the OLI framework evaluates the capabilities of the multinational firm at a point in time relative to the market. The co-evolutionary argument shifts the emphasis from the more static efficiency comparison of the FDI decision in local markets towards the multinational firm's prowess in managing knowledge *vis-à-vis* global competitors. Moreover, from the co-evolutionary argument, the yardstick for comparison of these evolving capabilities also incorporates other MNCs. MNCs that are able to direct the capability development, transfer and exploitation process relatively smoothly can create a superior competitive advantage *vis-à-vis* other multinational corporations. This, in turn, influences the internalisation decision, since a highly co-evolved network with significant capabilities in knowledge generation and transfer can leverage such capabilities in a superior manner relative to alternative governance forms.

In this regard, the relative extent and relative speed of knowledge transfer are important when examining firm-specific capabilities. The relative extent of knowledge transfer reflects firm success in leveraging and building upon large amounts of dispersed knowledge within the organisation relative to other firms. The relative speed of knowledge transfer is crucial in dynamic industries, where a firm's distinctive ability is not merely whether it can leverage and transfer knowledge but whether it can do this quickly, before the knowledge becomes obsolete.

Absorptive capacity, an important notion in the strategic management literature (Cohen and Levinthal 1989), can therefore have a direct impact on the internalisation decision. Where the constituent parts of the firm already possess important related knowledge and where requisite routines are mainly present through prior activities, this being a direct outcome of the co-evolutionary processes, a superior ability to earn rents due to faster and more complete knowledge transfer may motivate firms to internalise a particular activity.

On the other hand, in today's environment characterised by rapid change and increasing technological complexity as well as fragmentation and dispersion of knowledge across firms and in sub-national clusters, it is unlikely that a firm's internal knowledge base would suffice in and of itself to remain competitive under all circumstances. In such an environment, collaboration with external actors who possess complementary skills can be attractive. Besides complementarities such as access to needed knowledge, such collaborations provide an opportunity to learn

(Hamel *et al.* 1989; Inkpen 1998) and thus enable a more diverse and robust platform for developing new internal capabilities (Cohen and Levinthal 1989).

However, the extent of learning depends on the existing knowledge base within the organisation. A multinational firm with effective knowledge transfer capabilities internally may very well be able to extend those and learn from other firms. While alliances expand a firm's absorptive capacity, which enables it to further assimilate external knowledge, such collaborations often need to be complemented with investments in in-house capabilities in order to best benefit from them. Moreover, the development of a firm's in-house knowledge base is akin to a currency for subsequent exchange with other firms, which then makes it a more attractive candidate for collaboration. Thus, an astute balance between in-house capability development (internalisation) and cooperation with other firms can result in a virtuous cycle whereby internal and external competencies and opportunities become mutually enforcing.

In accordance with the above logic, governance forms are not mutually exclusive but function as complements (Cohen and Levinthal 1989; Madhok and Osegowitsch 2000), creating synergies that enhance the O advantages of the firm. The argument that foreign modes of entry or governance forms are complements is supported by Benito and Welch (1994) who suggest that the economics and process approaches to understanding mode of entry have failed to deal with the reality that different modes of entry are often used in combination with each other. The internalisation decision needs to be understood in this context, and also explains why, in many countries, firms have both in-house investments as well as alliances. In contrast, under the OLI framework, the internalisation decision views different governance forms such as own subsidiary, alliance, licensing or market as substitutes.

Summary, implications and concluding remarks

In today's environment, the imperative for firms to continually be at the competitive forefront is ever more pressing. We began by noting some of the trends that characterise the international business arena today. We then examined some of the developments in the strategic management field, emphasising in particular the central role of knowledge management in the multinational firm. To sum up the argument, we would like to make the following points, which are interrelated to one another.

First, firms are not just efficient governance structures but also institutions for learning (Kogut and Zander 1992; Ghoshal and Moran 1996). In this regard, governance structures serve to not only align transaction and governance characteristics, but also serve as a vehicle to generate as well as transfer knowledge, both embodied and otherwise. Moreover, to learn, firms need to broaden the focus away from just the macro-structural features, such as organisational form, to also address micro-structural features such as specific organisation mechanisms within organisational forms.

Second, with regard to O, the critical issue is increasingly the management of knowledge throughout the multinational system. This is what lies at the heart of firm-specific ownership advantage and differentiates the firm from other firms. Better management of the O advantage requires firms to recognise, shape and respond to both macro and micro co-evolutionary forces. Unlike the globalisation of the value chain, which leads to economies of location, scale and scope that can be imitated, the management of intangible assets, especially knowledge, offers a potentially inimitable advantage.

Third, the O advantages are increasingly dispersed throughout the firm. That is, they increasingly extend to the subsidiary level. The increasing dispersion of innovatory activity to overseas locations (Zander and Solvell 2000; Zander 1999; Dunning 1992) indicates that knowledge is being created in different subsidiaries of the firm. The importance of L lies in how it contributes in the leveraging and transformation of a firm's global capabilities. Here, both the density and quality of linkages to the local milieu as well as to the rest of the firm are of equal importance.

Fourth and importantly, following from the above, the firm should be viewed as a sub-economy in its own right. The multinational dynamic has changed from the ownership of a HQ-based advantage to an advantage embedded in its internal network. Therefore, I-related decisions need to be understood in this context since the co-evolutionary process may well result in a preference to organise the activity internally. If the preference is otherwise, then rather than FDI and other forms like licensing or alliances behaving as substitutes driven by efficiency considerations, as represented by the original OLI framework, different organisational forms in the more modern MNC may behave as complements in the overall knowledge management process. As elaborated in the previous section, the potential learning benefits through cooperation can be captured more fully when the firm itself has prior knowledge through internalised activities, thus reinforcing absorptive capacity.

Broadly speaking, the intensity of global competition, the increased complexity and global dispersion of knowledge and the accelerated rate of technological change is daunting to even the largest multinational firms. In this regard, the ability to manage the co-evolutionary process, i.e. to globally harness, coordinate and leverage knowledge-related resources and capabilities in a value-added manner, is a key source of advantage (O). Multinational firms' efforts today are increasingly geared towards developing and exploiting the capacity for such behaviour.

The ability to work well with other entities could be a critical asset in another different way. Given increasing concentration and interdependence of competitors in most industries, and given that a few global firms are competing feverishly both to develop and exploit a unique advantage as well as to prevent each other from doing so, the question arises as to why a potential partner in a particular location would prefer one firm versus another, i.e. what unique benefit can a particular global firm provide to a potential partner firm. In this regard, the ability to efficiently and effectively create and manage network-level knowledge-sharing processes at least partially explains the relative productivity advantages enjoyed by a firm over its competitors. This ability evolves as a simultaneous result of the macro and micro co-evolutionary processes operating on the MNC. Such an ability potentially promises greater benefits to partner firms. To the extent that its affiliates learn and benefit more rapidly from participating in the multinational firm's knowledge-sharing network, the firm becomes a more attractive partner, which in turn further reinforces its competitive advantage by providing it with opportunities to access and absorb external resources.

In conclusion, with the advent of the knowledge economy, the key advantage of firms has moved towards the management of global knowledge flows. Moreover, the primary arena of competition has moved away from the global – local interface to that of global competitors in multiple locations. In order to develop theoretically, the OLI framework, to remain robust, will benefit from examining and incorporating current strategic management research in a more discerning manner. Whereas economists focus on efficiency and the institution of firms, strategic management scholars tend to emphasise the role of competitive advantage. Whereas the former is directed mainly towards the I advantage, the latter is directed mainly towards the O advantage. The argument in this chapter, however, underlines the need to integrate the O and I advantages of firms. Whereas the former looks at why MNCs exist, the latter additionally poses the question of performance differences among firms, and examines much more carefully the role of knowledge and its management in this regard. Whereas the former argues that the yardstick for comparison is the market, the latter uses the yardstick of other firms. We believe that the critical issue is effectiveness and efficiency of knowledge creation, coordination, transfer, and leveraging within the firm relative to the market and relative to other firms. The eclectic paradigm provides an envelope for bringing together and incorporating both the economic and strategic management views, although much more work needs to be done in this direction. This chapter is intended to contribute in this regard.

Note

1 While researchers have also examined the use of collaborative arrangements to access knowledge (Hamel *et al.* 1989; Mowery *et al.* 1996; Inkpen 1998), we limit our scope to how firms manage knowledge within their boundaries, and only touch on collaborations at the end.

References

Barney, J.B. (1991) 'Firm resources and sustained competitive advantage', *Journal of Management*, 17: 99–120.

Bartlett, C.A. and Ghoshal, S. (1986) 'Tap your subsidiaries for global reach', *Harvard Business Review*, 64(6): 87–94.

Baum, J. and Korn, H. (1999) 'Dynamics of dyadic competitive interaction', *Strategic Management Journal*, 20(3): 251–78.

Benito, G.R.G. and Welch, L.S. (1994) 'Foreign market servicing: beyond choice of entry mode', *Journal of International Marketing*, 2(2): 7–27.

Birkinshaw, J. (1997) 'Entrepreneurship in multinational corporations: the characteristics of subsidiary initiatives', *Strategic Management Journal*, 18(3): 207–29.

Birkinshaw, J. and Hood, N. (1998) 'Multinational subsidiary evolution: capability and charter change in foreign owned subsidiary companies', *Academy of Management Review*, 23(4): 773–95.

Birkinshaw, J. and Hood, N. (2000) 'Characteristics of foreign subsidiaries in industry clusters', *Journal of International Business Studies*, 31(1): 141–54.

Birkinshaw, J. and Morrison, A. (1995) 'Configurations of strategy and structure in subsidiaries of multinational corporations', *Journal of International Business Studies*, 26(4): 729–54.

Bresnan, H., Birkinshaw, J. and Nobel, R. (1999) 'Knowledge transfer in international acquisitions', *Journal of International Business Studies*, 30(3): 439–50.

Buckley, P. and Casson, M. (1976) *The Future of the Multinational Enterprise*, London: Macmillan.

Cantwell, J. (1995) 'The globalization of technology: what remains of the product life cycle model?', *Cambridge Journal of Economics*, 19(1): 155–75.

Casson, M. (1991) 'International comparative advantage and the location of R&D', in M. Casson (ed.), *Global Research Strategy and International Competitiveness*, Oxford: Basil Blackwell.

Coase, R.H. (1937) 'The nature of the firm', *Economica*, 4: 386–405.

Cohen, W. and Levinthal, D. (1989) 'Innovation and learning: two faces of R&D', *Economic Journal*, 99: 569–96.

Davidson, W. and McFetridge, D. (1984) 'International technology transactions and the theory of the firm', *Journal of Industrial Economics*, 32: 253–64.

Dunning, J.H. (1977) 'Trade, location of economic activity and the multinational enterprise: the search for an eclectic approach', in B. Ohlin, P.O. Hessleborn and P.M. Wijkman (eds), *The International Allocation of Economic Activity*, London: Macmillan.

Dunning, J.H (1988) 'The eclectic paradigm of international production: a restatement and possible extensions', *Journal of International Business Studies*, 19: 1–32.

Dunning, J.H. (1992) 'Multinational enterprises and the globalisation of innovatory capacity', in O. Granstrand, L. Hakanson, and S. Sjolander (eds), *Technology Management and International Business, Internationalisation of R&D and Technology*, Chichester, UK: John Wiley.

Ghoshal, S. and Moran, P. (1996) 'Bad for practice: a critique of the transaction cost theory', *Academy of Management Review*, 21: 13–43.

Grant, R.M. (1996) 'Prospering in dynamically-competitive environments: organizational capability as knowledge integration', *Organization Science*, 7: 375–87.

Gupta, A. and Govindarajan, V. (1991) 'Knowledge flows and the structure of control within multinational corporations', *Academy of Management Review*, 16(4): 768–92.

Gupta, A. and Govindarajan, V. (2000) 'Knowledge flows within multinational corporations', *Strategic Management Journal*, 21: 473–96.

Hamel, G., Doz, Y. and Prahalad, C.K. (1989) 'Collaborate with your competitors – and win', *Harvard Business Review*, 67(1): 133–9.

Hennart, J. (1991) 'The transaction costs theory of joint ventures: an empirical study of Japanese subsidiaries in the United States', *Management Science*, 37(4): 483–98.

Holm, U. and Pedersen, T. (eds), (2000) *The Emergence and Impact of MNC Centers of Excellence: A Subsidiary Perspective*, London: Macmillan.

Inkpen, A. (1998) 'Learning and knowledge acquisition through international strategic alliances', *Academy of Management Executive*, 12(4): 69–80.

Kogut, B. and Zander, U. (1992) 'Knowledge of the firm, combinative capabilities and the replication of technology', *Organization Science*, 3: 383–96.

Kogut, B. and Zander, U. (1993) 'Knowledge of the firm and the evolutionary theory of the multinational corporation', *Journal of International Business Studies*, 29(3): 625–45.

Kogut, B. and Zander, U. (1996) 'What firms do? Coordination, identity and learning', *Organization Science*, 7: 502–18.

Lewin, A. and Volberda, H. (1999) 'Prolegomena on co-evolution: a framework for research on strategy and new organizational forms', *Organization Science*, 10: 519–34.

McKelvey, B. (1997) 'Quasi-natural organization science', *Organization Science*, 8(4): 352–80.

Madhok, A. (1997) 'Cost, value and foreign market entry mode: the transaction and the firm', *Strategic Management Journal*, 18: 39–61.

Madhok, A. and Osegowitsch, T. (2000) 'The international biotechnology industry: a dynamic capabilities perspective', *Journal of International Business Studies*, 31: 325–36.

March, J. (1994) 'The evolution of evolution', in J. Baum and J. Singh (eds), *Evolutionary Dynamics of Organizations*, New York: Oxford University Press.

Miles, R., Mathews, J., Miles, G. and Coleman Jr., H. (1997) 'Organizing in the knowledge age: anticipating the cellular form', *Academy of Management Executive*, 11(4): 7–20.

Mowery, D., Oxley, J. and Silverman, B. (1996) 'Strategic alliances and inter-firm knowledge transfer', *Strategic Management Journal*, Special Issue, 17: 77–91.

Nohria, N. (1992) 'Introduction: Is a network perspective a useful way of studying organizations?', in N. Nohria and R.G. Eccles (eds), *Networks and Organizations*, Boston, MA: Harvard Business School Press.

Nonaka, I. and Takeuchi H. (1995) *The Knowledge-Creating Company*, New York: Oxford University Press.

Pearce, R.D. (1989) *The Internationalisation of Research and Development by Multinational Enterprises*, London: Macmillan.

Pettigrew, A. (1985) *The Awakening Giant: Continuity and Change in Imperial Chemical Industries*, Oxford: Basil Blackwell.

Porter, M. (1990) *The Competitive Advantage of Nations,* New York: Free Press.

Porter, M. (1998) 'Clusters and the new economics of competition', *Harvard Business Review*, 76(6): 77–90.

Rugman, A. and Verbeke, A. (1992) 'A note on the trans-national solution and the transaction cost theory of multinational strategic management', *Journal of International Business Studies*, 23(4): 761–77.

Rugman, A. and Verbeke, A. (2001) 'Subsidiary specific advantages in multinational enterprises', *Strategic Management Journal*, 22(3): 237–50.

Saxenian, A. (1990) 'Regional networks and the resurgence of Silicon Valley', *California Management Review*, Fall, 39–112.

Szulanski, G. (1996) 'Exploring internal stickiness: impediments to the transfer of best practice within the firm', *Strategic Management Journal* (Winter special issue) 17: 27–44.

Teece, D. (1976) *The Multinational Corporation and the Resource Cost of International Technology Transfer*, Cambridge, MA: Ballinger.

Teece, D., Pisano, G. and Shuen, A. (1997) 'Dynamic capabilities and strategic management', *Strategic Management Journal*, 18: 509–33.

Tushman, M. and Anderson, P. (1986) 'Technological discontinuities and organizational environments', *Administrative Science Quarterly*, 31(3): 439–65.

United Nations World Investment Report (1999) Geneva: United Nations.

United Nations World Investment Report (2000) Geneva: United Nations.

Vernon, R. (1966) 'International investment and international trade in the product cycle', *Quarterly Journal of Economics*, 80: 190–207.

Volberda, H. (1998) *Building the Flexible Firm: How to Remain Competitive?* New York: Oxford University Press.

White, R.E. and Poynter, T.A. (1984) 'Strategies for foreign owned subsidiaries in Canada', *Business Quarterly*, Summer, 59–69.

Zander, I. and Solvell, O. (2000) 'Cross border innovation in the multinational corporation: a research agenda', *International Studies of Management and Organization*, 30(2): 44–67.

Zander, I. (1999) 'How do you mean global? An empirical investigation of innovation networks in the multinational corporation', *Research Policy*, 28: 195–213.

6 From OLI to OLMA

Incorporating higher levels of environmental and structural complexity into the eclectic paradigm

Stephen Guisinger

Introduction

If the principal criterion of a paradigm's value is fertility, then the eclectic paradigm has passed its test with flying colours. The vast number of articles spawned by John Dunning's seminal article in 1978 validates his vision of a unified set of propositions about ownership, location and internalisation.

That paradigms emerge in science in response to the conditions of their times is well established. In the 1960s and 1970s, economists such as Dunning were grappling with problems of multinational firm activity: where do firms locate and with what mode of entry? The opportunities for multinational firm expansion seemed unbounded: the only question was when, where and how. If a firm's performance was unsatisfactory, it was generally assumed that its managers had chosen the wrong location or entry mode. The OLI trilogy responded precisely and effectively to the major concerns of that period.

In recent years, there has been a gradual shift away from a pre-occupation with multinational firm *activity*, as captured by location and mode of entry, toward multinational firm *profitability*. This is not to say that location and mode are no longer of interest, but there is a realisation that multinational firm performance depends on more than just location and entry mode. Multinational firms in the same industry that have located in the same market, such as China, using the same entry modes, can have strikingly different rates of profitability. These variations in profitability are often due to operational decisions independent of location and entry mode. Multinational firms that fail to use 'best practices' in their business processes suffer by comparison to competitors who do. In a recent interview, Michael Porter emphasised the importance of operational improvement, placing it on an equal footing with strategic positioning (Hodgetts 1999). The next wave of research on the multinational firm is likely to focus on operational issues, even more so than in the past. Is the eclectic paradigm prepared to meet this challenge?

The short answer is yes, but only with a substantial modification and extension. To put the issue in its most basic terms: the eclectic paradigm captures much but not all of the variations in profitability observed among multinational firms. But with an extension to cover the operational adjustments that multinational firms must go through in order to survive and prosper in unfamiliar international environments, an enhanced eclectic paradigm could prove to be a powerful, if not the dominant, theoretical approach to the study of the multinational firm.

To understand the need for modification and extension of the eclectic paradigm, it is necessary to review briefly the evolution of two bodies of literature: management theory, principally organisational theory, and economic theory, both as they apply to the multinational firm.

The state of management theory's approach to the multinational firm in the early 1990s was best captured by Yves Doz and C.K. Prahalad, who stated that, 'very little attention has been paid to the conceptual and theoretical frameworks used to analyze DMNCs [diversified multinational corporations] and their management' (Doz and Prahalad 1991: 145). In their article, Doz and Prahalad propose a managerial-based paradigm that addresses several important features of the multinational firm left unexplored by previous international business research. Any paradigm applicable to research on the DMNC must, in their view 'incorporate a *differentiated approach to businesses, countries and functions* [authors' italics], and provide enough flexibility for different trade-offs between multiple dimensions to be made' (ibid.: 146). They argue that organisational theorists have been too narrowly focused on constructs that lend themselves to easy theoretical manipulation. At the same time, multinational firm scholars have 'often been engrossed in the complexity of what they studied, and failed to develop, or borrow, a sufficiently powerful conceptual framework to shed light on the observed phenomenon. As a result the bridge between the MNC phenomenon and organisation theory was not built.' Doz and Prahalad list the criteria for a paradigm – which they do not name but can be termed the environmental adaptation approach – providing the differentiated approach to businesses, countries and functions that they deem necessary for research on the multinational firm (MNC).[1] Since Doz and Prahalad wrote their article, a great deal of research has been produced on multinationals' approach to businesses and functions, but very little interest has been shown by management researchers in defining and integrating country variations into the work on businesses and functions.

Economic theory, in contrast, has spent considerable effort in defining the international business environment (the differentiated approach to countries in Doz and Prahalad's terms) and have attempted to 'build a bridge to the MNC phenomenon' by examining how firms have accommodated themselves to variations in this environment. The eclectic paradigm is perhaps the epitome of environmental accommodation research, but there are important other contributors as well.

In effect, these two bodies of theory have been building bridges toward the 'MNC phenomenon' but from two different directions: management theory outwards from the firm toward the environment; and economic theory from the environment back to the firm. Do these two bodies of theory meet in the middle, so to speak, and what is the nature of multinational theory at this intersection? Can the comparative advantages of both bodies of literature be leveraged to produce a new approach to explanations of multinational firm behaviour and performance that draw simultaneously on both theories?

Structural and environmental complexity

Limitations of existing theory

A hallmark of recent developments in organisational theory has been their ability to incorporate higher levels of structural complexity into their models. Structural complexity refers to the numbers of businesses, functions and products that the firm's managers must control. A high degree of structural complexity is one of the defining features of large, diversified MNCs. A recent example is Nitin Nohria and Sumantra Ghoshal's study of the difficulties that MNC managers have in coordinating all the facets of structurally complex firms (Nohria and Ghoshal 1997).

Richard Scott has written:

> Every organisation exists in a specific physical, technological, cultural and social environment to which it must adapt. No organisation is self-sufficient: all depend for survival on the types of relations they establish with the larger systems of which they are a part. Early analysts of organisations . . . tended to overlook or underestimate the importance of organisations-environmental linkages, but recent work places great emphasis on these connections. And, indeed, the number and variety of these connections are impressive.
>
> (Scott 1998: 21)

However, organisational theorists' conceptualisations of the environmental influences relevant for MNCs are either too coarse – the environment is described in non-operational terms – or too selective – only certain differentiated country effects are included with no rationale for the exclusion of other country effects. For example, research shows that foreign exchange risk affects an MNC's decisions regarding market entry, mode of entry, production levels, product mix, preferred suppliers, finance, legal form as well as the structure of its business processes that are affected by currency fluctuations. Foreign exchange risk is an environmental variable with substantial consequences for the internal organisation of MNCs but one that rarely finds its way into organisational theory. And foreign exchange is but one element in a rich variety of country effects – political risk, tariffs, physical distance from major markets, for instance, that studies have shown to have potentially significant and differential effects on the structure and operations of MNCs. The environmental adaptation approach is only partially realised because of its limited view of the environment

Economists have suffered from just the reverse shortcomings in their development of theories related to the MNC. Using a rich and operational definition of the international business environment, economics-based MNC research has produced useful theories about the ways in which firms accommodate their location, product mix, output levels and entry mode to differences in the environment. Yet, the environmental accommodation approach has so far stopped short of providing useful insights into the organisation's internal operations.

Structural and environmental complexity graphically depicted

The inadequacies of both the organisational and economic approaches to the MNC can be shown in a figure with structural and environmental complexity as axes. Figure 6.1 represents a hypothetical starting point for a firm's adjustment to the two types of complexity it faces.

A firm located at the origin has the simplest structural form (e.g. a sole proprietorship) and operates in the least demanding environment conceivable (minimal government intervention, homogeneous social culture, rudi-

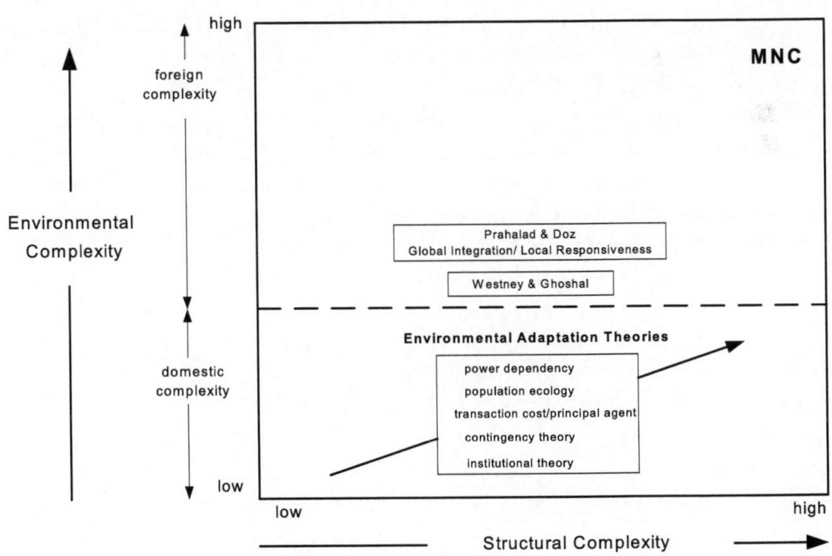

Figure 6.1 Environmental adaptation.

mentary technologies, etc.). The further along the structural complexity axis a firm finds itself, the greater is the diversity and difficulty of the managerial tasks performed. These include, but are not limited to, the variety of goods produced, types and sophistication of management functions performed, and numbers of employees to be coordinated. The further along the environmental complexity axis a point is from the origin, the greater the attention managers have to pay to each element in the environment and its impact on the firm's operations. These include, but are not limited to, numbers of countries in which the firm operates, cultural distances separating managers, employees and customers, differences in legal systems encountered and degrees of risk assumed through currency exposures.

The concept of structural complexity has been a core feature of organisational theory since its inception.[2] Structural complexity captures the varieties of organisational forms that a firm may assume. In Figure 6.1, structural complexity appears as an ordinal variable.

The concept of environmental complexity is also a familiar feature in the organisational theory literature. Several recent studies have used the concept in explaining some of the choices available to MNCs. Anant Sundaram and Stuart Black have applied environmental complexity to the study of market entry (Sundaram and Black 1992). Sumantra Ghoshal and Nitin Nohria have used it in analysing organisational choices among MNCs (Ghoshal and Nohria 1993). Tatiana Kostova and Srilata Zaheer have used the term in examining organisational legitimacy. For example, they state, 'the complexity of the MNE environment is reflected in the multiple domains of the institutional environment and in the multiplicity of institutional environments faced by MNEs' (Kostova and Zaheer 1999: 70).

Economists have used the term as well. Douglass C. North, who pioneered the field of institutional economics (North 1990), is perhaps the best-known user of the term. North further made the important distinction in the environment between institutions and organisations (competitors, suppliers and customers). North, however, did not focus on differences between international and domestic environments.

A common feature of environmental complexity in both bodies of theory has been a high level of abstraction. For any concept with the term 'complexity' in it, some ambiguity is to be expected. Yet rarely do authors even provide taxonomies of environmental elements and as yet have never described environmental complexity in precise terms that could lead to its operationalisation.

For the study of multinational firms, it is essential to divide the environment into home and foreign components in environmental complexity. An axiom of multinational studies is that from the perspective of the parent company the foreign environment is more complex than the domestic environment, shown in Figure 6.1 by a vertical axis divided into two parts: a 'domestic' section adjacent to the origin, representing the home-country environment, and a 'foreign' section further along the axis, representing the international business environment. A firm operating in the foreign section, by definition, takes on both domestic and foreign complexity, while a firm in the domestic portion faces only the less complex (from the perspective of domestic companies) home environment.

Points off these axes and in the interior of the grid represent various combinations of structural and environmental complexity. The fully adjusted MNC, one embracing the maximum amounts of structural and environmental complexity, is represented by a point in the upper right-hand corner of the grid.

The impact of structural and environmental complexity on environmental adaptations

Developments in organisational theory are portrayed in Figure 6.1, using the familiar shorthand labels given to these theory clusters. In varying degrees, each of these groups of theories have explored how the structural components of firms are shaped by their environment. In some cases, population ecology for example, firm structures are simple (individual firms) while others explore more complex structural forms. However, despite the environment being a principal force affecting the organisation, the nature of this environment has either been at a very high level of abstraction or has been limited to generic (domestic) business environments (Meyer and Rowan 1977).

As Doz and Prahalad imply in their survey of organisational theory developments, organisational theorists have steered away from the upper regions of environmental complexity in part, it would appear, because structural complexity provides by itself such a varied array of issues requiring explanation. Venturing too far into environmental complexity compounds the already difficult task of dealing with multiple managerial functions, lines of business and organisational forms. There are two notable exceptions to this trend. Doz and Prahalad developed a local-responsiveness/global-integration construct to integrate environmental issues (local responsiveness to host-country environments) into decisions managers make in coordinating multiple businesses and managerial functions (Prahalad and Doz 1987). Eleanor Westney and Sumantra Ghoshal organised a group of essays that further explored the relevance and adaptability of organisational theories to the MNC. These essays lacked, unfortunately, a clear and cogent analysis of environmental complexity.

The impact of structural and environmental complexity on environmental accommodation

While organisational theorists have explored the MNC principally through the lens of structural complexity, economic theorists have approached

MNC theory from the perspective of the firm's interaction with environmental complexity. Significant contributions to research on the MNC over the past three decades, especially those conducted under the eclectic paradigm, have provided a clearer explanation of the MNC's production location, output composition and mode of market entry. At the same time, another group of researchers has pioneered the development of new ways to define and quantify elements of the international business environment. A selection of these research studies is shown in Figure 6.2.

The positioning of these studies suggests an almost mirror image in the pattern of theory development originating from the side of environmental complexity. Studies that have focused primarily on quantifying dimensions of the environment (shown in Figure 6.2 with principal author's name in italics) have understandably made little headway in incorporating structural complexity into their analyses. Both John Dunning and Michael Porter have made brief forays into explanations of the environmental influences on the internal choices of MNCs but their main focus remains the external attributes of location, entry mode and product mix.

Environmental influences on MNCs

Studies that attempt to explain how the environment has influenced multinational firms go back to the origin of the field of international business more than four decades ago. Much of the early work was descriptive and

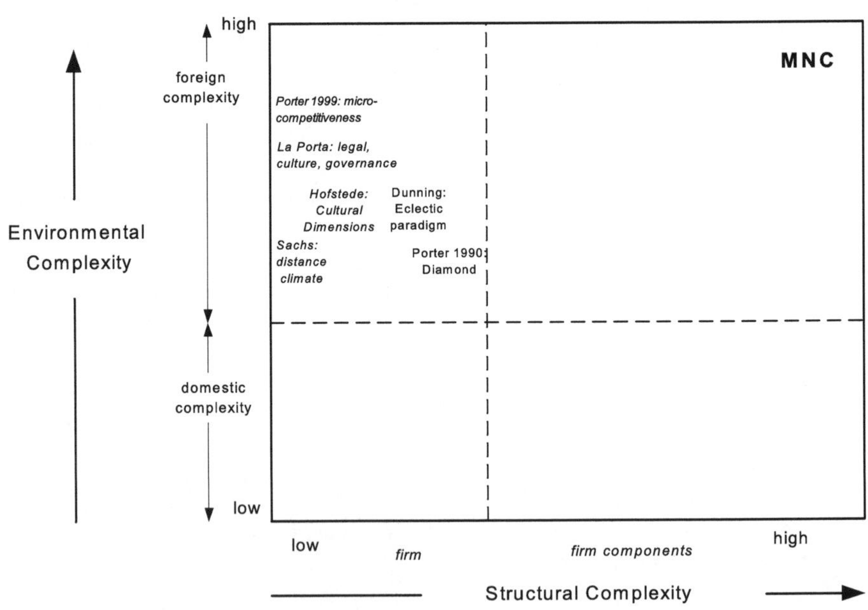

Figure 6.2 Environmental accommodation.

anecdotal and therefore not included in Figure 6.2. The early writers on the environment, such as John Fayerweather (1960), Richard Robinson (1964) and Raymond Vernon (1973) deserve considerable credit for both legitimising work on the multinational firm and identifying concepts that were later incorporated into formal models of multinational behaviour. Two of the most prominent attempts at formal modelling of multinational firm behaviour are included in Figure 6.2.

Perhaps the most important and comprehensive work on international business environments and their influence on the firm has been done by John Dunning (retrospectively surveyed in Dunning 2000). The purpose of the eclectic paradigm is to explain the 'extent, geography and industrial composition of foreign production undertaken by MN[C]s' (Dunning 2000). The unit of analysis is the firm itself and not its businesses or managerial functions.

Michael Porter's name appears twice in Figure 6.2, once for a contribution made to measuring aspects of the international business environment itself (shown in italics) and once for a contribution that showed the relation of the MNC's environment to its strategic choices. This last contribution is perhaps most frequently associated with Porter's *The Competitive Advantage of Nations* (Porter 1990). In this study he provides a conceptual framework for assessing a country's strengths and weaknesses, which captures many of the features familiarly associated with the environmental complexity facing the MNC. Porter focuses on mechanisms by which a firm's customers, competitors, suppliers and factors of production influence a firm's strategic decisions. He does not examine the implications of alternative strategies for the structure and operations of MNCs. In another study (italicised in Figure 6.2) Porter (1999) extends his work on national competitive factors by creating an index of national micro-competitiveness. This index provides empirical support for his contention that nations differ in their ability to support dynamic industrial and service sectors.

Quantitative measures of environmental complexity

A number of researchers have produced important new empirical measures that contribute to our understanding of environmental complexity. Geert Hofstede (1983), for example, has explored the nature of national cultures and the ways in which cultures affect individual decisions. Hofstede's principal interest has been focused less on the influence of cultures on the structure and operations of multinational firms and more on the ways in which individual managers should incorporate knowledge of cultures into their decision making. Rafael La Porta and associates have examined the influence of a variety of environmental attributes, such as legal systems, trust and governance, on economic growth (La Porta *et al.* 1997, 1998). Sachs and Warner have studied the influence of a country's physical features – temperate vs. tropical climate, amount of shoreline, distance from markets, etc. – and traced these influences back to the pace and character of a country's economic development (Sachs and Warner 1997).

Convergence of the environmental accommodation and environmental adaptation approaches

Figures 6.1 and 6.2 represent two research paths aimed at explaining the MNC but originating from different premises and drawing on different disciplinary methods. In the case of the studies in Figure 6.2, the principal motivating force for theorising was the presence of differentiated elements external to the firm that require MNCs to adapt their location, level of production or the composition of their output. In the case of the studies in Figure 6.1, the underlying force behind theory development was the difficulty the firm had in coordinating increased numbers of elements internal to the firm (businesses, divisions, managerial functions) but which additionally had to be accommodated to the firm's environment. These two forces are shown in Figure 6.3.

Figure 6.3 suggests that the two streams of research have the potential to converge toward a unified explanation of the MNC. Each research tradition contributes an important part toward solving the puzzle of MNC behaviour but neither is sufficient on its own. A central question is: can the eclectic paradigm be extended and enhanced to provide this marriage of the two disciplines?

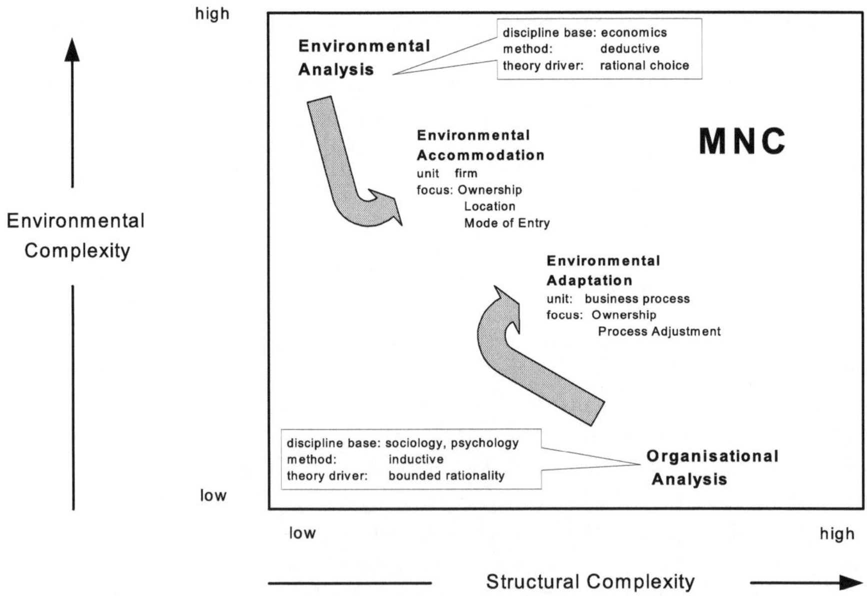

Figure 6.3 Synthesising environmental adaptation and accommodation.

Toward an evolved eclectic paradigm

Two modifications of the eclectic paradigm appear necessary to bring it closer to the ideal of merging the environmental accommodation and environmental adaptation approaches to the study of the multinational firm. The first is minor, the second more substantial.

The minor modification involves replacing the I (internalisation) in the traditional OLI framework with M for mode of entry. Internalisation focuses only on one entry mode – control of a subsidiary – when many more forms of international involvement are possible. Replacing I in OLI with M seems a logical step toward recognition of the variety of methods by which a firm can profit from foreign activity. It is fair to say that replacing I with M is more form than substance. The essence of the concept of internalisation is the option available to the firm's managers to choose new forms of service and product delivery.

The major modification involves adding a fourth domain of analysis to the familiar tripartite OLI (or OLM as suggested above). The fourth domain is the adaptation of the firm's operations – i.e. its business processes – to the international business environment. The addition of A (for adaptation) to OLM would effectively demarcate the four principal areas that determine multinational firm performance: (1) ownership of intellectual property, routines, etc. that create the basis for competitive advantage; (2) location of the firm's activities; (3) mode of entry selected; and (4) adaptation of business processes to the international business environment.

Toward a better definition of the international business environment

Despite a great deal of interesting exploratory work, our understanding of environmental complexity is not nearly as evolved as our knowledge of structural complexity (Boddewyn 1999). Authors have defined the international business environment (IBE) to suit their purposes and a common accepted definition remains elusive. A better definition of the IBE could provide a catalyst to closer integration of organisational and economic theories of the multinational firm.

The starting point for an improved definition of the IBE is institutional theory, which among the sub-fields of organisational theory has developed the most thorough treatment of the firm's environment. According to Westney (1993), there are a variety of contending concepts to represent the environment but a consensus appears to be building around 'organizational field'.

According to DiMaggio and Powell, the organisational field consists of 'those organisations that, in the aggregate, constitute a recognised area of institutional life: key suppliers, resource and product consumers, regulatory agencies, and other organisations that produce similar services or products (DiMaggio and Powell 1983: 151).

Two types of partitioning of this organisational field are possible. As noted previously, the first is axiomatic for international business theorists. For the field of international business to have any meaning at all, there must be a compelling distinction between domestic and foreign components of the environment. The second partition is less obvious but equally critical. Following the lead of Douglass North, the environment can be subdivided into organisations – called here 'interactors' – and institutions, called hereafter the 'geovalent component'. Interactors comprise the organisations that interact directly with the firm – suppliers, customers, and competitors. Interactors have other important properties: they can acquire other members (or be acquired), form alliances or simply cease to exist (liquidation/bankruptcy).

The geovalent component comprises all other environmental forces that impact on the firm but are not themselves organisations: institutional rules, regulations, cultures and exchange rates, for example. The geovalent elements, unlike organisations, are not mobile, but are fixed in geographic space, usually but not always, following national boundaries. For the purposes of developing usable theory, the elements of the geovalent component have two other necessary properties. First, they have the potential for directly and significantly affecting the performance of firms. Second, they are to some extent quantifiable, permitting measures of how they vary over time and space (physical attributes of countries are one exception).

What are the relevant elements that fill this geovalent component? This chapter can only sketch out an approach to defining the geovalent component. Both deductive and inductive reasoning will be required to provide greater clarity about the IBE.

For example, deductive logic would suggest that differences in legal systems must have a bearing on the performance of MNCs. According to the modern theory of the firm, characteristics of the legal system and the quality of legal enforcement determine firms' choices between market transactions and internal hierarchies. This finding is supported through inductive logic: empirical research in international business, economic development and other social science fields have found that differences in legal systems (code versus common law, for example, or the quality of judicial enforcement) are significant independent variables, explaining patterns of firm and country growth.

There is no widely shared definition of the geovalent part of the IBE in the same way that organisational theorists have developed naming conventions for categorising the principal structural forms of the firm or that economists have about factors of production such as land, labour and capital. Table 6.1 displays one possible taxonomy for geovalent components.

These eight broad categories of geovalent components have been selected on the basis of both deductive and inductive approaches, especially the frequency of their recurrence in literature on international business.

Table 6.1 Illustrative geovalent elements

Element	*Examples of attributes defining each element*	*Examples of empirical measurement of element*
Econography	Climate Proximity to major markets Physical size Infrastructure	Sachs & Warner 1997 Peterson & Malhotra 1997 Porter 1999 Sachs 1999
Culture	Values Attitudes Beliefs	Hofstede 1983 Kogut & Singh 1988
Legal systems	Common Civil Religious law	La Porta *et al.* 1998 Posner 1998
Income profile	GNP per capita Growth of GNP Income inequality	Golub 1995 Sachs 1999
Political risk	Government instability Corruption Bureaucratic instability Quality of government	La Porta *et al.* 1998 Transparency International 1999 Porter 1999 Howell & Chaddick 1994
Tax systems	Effective tax rate for multinational firms	Loree & Guisinger 1995
Exchange rates	Exchange rate variability Exchange rate overvaluation/ undervaluation	Miller & Reuer 1998
Government restrictions	Tariffs Quotas Investment controls	Laird & Yeats 1988 UNCTAD 1995 Porter 1999

Each category consists of more finely grained elements. These categories are exhaustive, mutually exclusive, quantifiable (to some degree) and reproducible by researchers in any country. International institutions have not been included as a geovalent component because their effect on MNCs is exercised indirectly through their influence on national policies, such as exchange rates, tariffs and tax systems, which are included in Table 6.1.

Seven of the eight categories in Table 6.1 are self-explanatory. The eighth, econography, is a portmanteau word, joining economic geography with demography to describe the physical and human assets possessed by countries. Some authors use the term 'physical environment', but this leads to confusion with 'green' issues relating to pollution, omits the country's human capital and, most importantly, risks further confusion with 'the firm's environment', as used in the preceding discussion. The international business environment relevant for any particular multinational firm comprises an aggregate of the geovalent profiles for the countries where the firm operates (or intends to operate in the case of prospective analysis).

Blending environmental and organisational analysis

A convergence of the environmental accommodation and environmental adaptation approaches to create an evolved eclectic paradigm requires new constructs and methods that are not currently available. The grid in Figure 6.4 shows one possible approach to this convergence. This figure blends environmental and organisational analysis by including key constructs from each of the approaches, geovalent components from environmental analysis and business processes from organisational theory.

The grid in Figure 6.4 contains eight rows representing broad process categories. The business process has become the focal point of much recent theorising in organisational theory. Thomas W. Malone *et al.* (1999) have written:

> . . . as the boundaries between functions and companies crumble, the old organizational chart loses its usefulness as a management tool. In tomorrow's companies, executives will likely depend on richly detailed process maps to guide their managerial and strategic decision making.
>
> (Thomas W. Malone *et al.* 1999: 427)

When domestic firms expand beyond their borders, they take on additional structural complexity as the scope of their operations increases (bigger size, more affiliates, etc.). Increased complexity requires firms to add new pro-

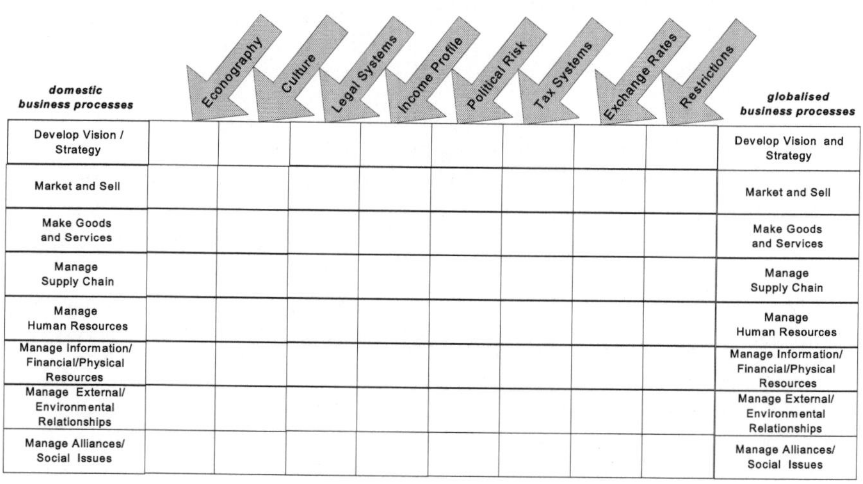

Figure 6.4 Geovalent adjustment of business processes.

cesses and modify existing ones. As Malone *et al.*'s (1999) research suggests, increased complexity requires greater concentration on managing processes compared to attention to managerial functions or organisational hierarchies.

The eight columns in the centre of the grid represent the eight categories of geovalent components. The far left-hand column represents the business processes of the domestic firm and the far right-hand column represents the business processes of the globalised firm. Venturing abroad requires domestic firms to take on additional environmental complexity, most immediately in the form of the geovalent components. In much the same way they cope with structural complexity, firms must add new processes and modify existing ones to manage environmental complexity. Researchers, such as David Ricks (1993), have collected evidence showing that firms' performance suffers when they fail to adapt to foreign environments.

If a purely domestic firm begins with its business processes at 'best practice' levels, two types of modifications are necessary to move to status as a global best practices firm once it decides to enter the international marketplace. First, it must adapt its processes to deal with the scope and scale required by its presence in a larger market (structural complexity). Second, it must adjust its processes to deal with the requirements of the international business environment (environmental complexity). If the grid in Figure 6.4 were completely filled with guidelines on the required adaptations, it would be relatively simple for domestic best practice firms to make the transition to being global best practice MNCs. This can be termed the geovalent adjustment of business processes. Figure 6.5 indicates how each of the two major streams of multinational firm theory contribute to geovalent adjustment.

Adding geovalent adjustment to the eclectic paradigm will permit more detailed work on the structure and operations of the MNC. Geovalent adjustment will require a robust approach: firms differ in so many ways that affect the selection of appropriate process adjustments: size; industry affiliation; degree of vertical integration; mode of market entry; numbers of countries where they operate; environmental differences among the countries where they operate, etc. For example, under 'Market & Sell', firms planning to market their products abroad must create new routines, even new organisational structures, to deal with different cultures they will encounter. A firm considering expansion into just one foreign market may alter its advertising message only slightly, depending on the degree of cultural distance between the home and foreign market. For this modest increase in geovalent complexity, a single local marketing consultant on a contract basis may be sufficient. By contrast, a firm making a long-term commitment to enter many markets with diverse cultures may need to make more fundamental adjustments, such as creating a new international advertising unit staffed with personnel equipped for multi-cultural advertising campaigns.

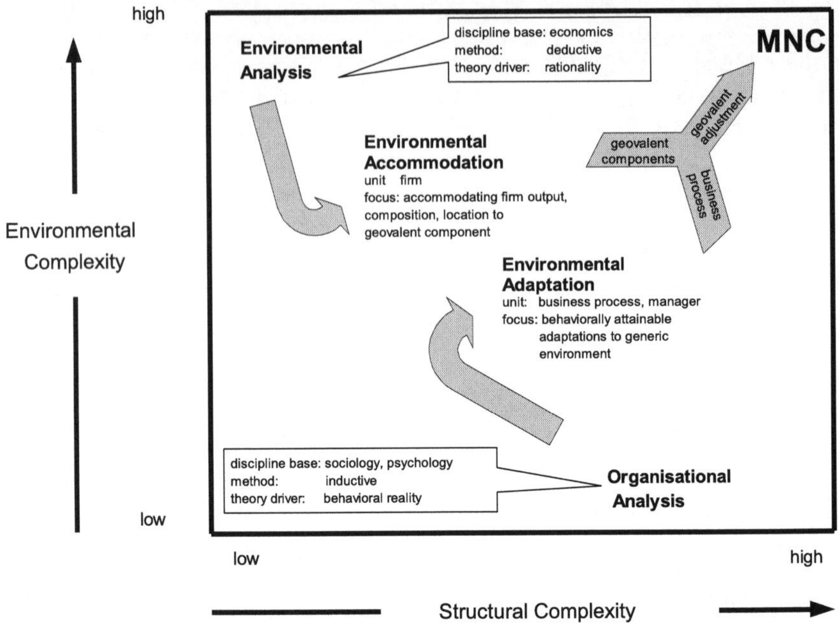

Figure 6.5 Geovalent synthesis of adaptation and accommodation.

Conclusion

This study has argued that economic and organisational theory can contribute importantly to the evolution of the eclectic paradigm. The strength of economic theory is its ability to show how firms have accommodated themselves to different international environments. The strength of organisational theory has been its ability to show that complex organisational forms are often necessary to cope with environmental variation.

This study has proposed a synthesis of organisational and economic theory through an analysis of how multinational firms' business processes are adjusted to deal with the geovalent elements of the international business environment. Expansion of the eclectic paradigm to incorporate geovalent adjustment results in an expanded eclectic paradigm, called OLMA, where ownership (O), location (L), mode of entry (M) and geovalent adjustment (A) explain the principal determinants of multinational firm performance.

Notes

1 Doz and Prahalad prefer the term 'diversified multinational firm' or DMNC to emphasise the fact that most multinational firms manage multiple businesses

and diversified product lines. This study uses the more familiar term 'multinational firm' or MNC but with the same meaning as Doz and Prahalad give to the DMNC.

2 Structural complexity differs from Doz and Prahalad's organisational complexity in an important way. Structural complexity describes the degree of a firm's structural diversity, including varieties of products, divisions and managerial functions. Organisational complexity is the result of adapting these structural forms to the environment. Organisational complexity describes all of the points on in the structural–environmental grid in Figure 6.1, whereas structural complexity is represented by points on the horizontal axis.

References

Boddewyn, J. (1999) 'The domain of international management,' *Journal of International Management*, 5: 3–14.

DiMaggio, P. and Powell, W. (1983) 'The iron cage revisited: institutional isomorphism and collective rationality', *American Sociological Review*, 48: 147–60.

Doz, Y. and Prahalad, C. (1991) 'Managing DMNCs: a search for a new paradigm', *Strategic Management Journal*, Special Issue, 12: 145–64.

Dunning, J.H. (2000) 'The eclectic paradigm as an envelope for economic and business theories of MNC activity', *International Business Research*, 9(2): 163–90.

Fayerweather, J. (1960) *Management of International Operations*, New York: McGraw-Hill.

Ghoshal, S. and Nohria, N. (1993) 'Horses for courses: organizational forms for multinational corporations', *Sloan Management Review*, 34(2): 23–35.

Golub, S. (1995) 'Comparative and absolute advantage in the Asia-Pacific region', Federal Reserve Bank of San Francisco Working Paper.

Hodgetts, R.M. (1999) 'A conversation with Michael E. Porter: a "significant extension" toward operational improvement and positioning', *Organizational Dynamics*, 28(1) (Summer): 24–33.

Hofstede, G. (1983) 'The cultural relativity of organizational practices and theories', *Journal of International Business Studies*, 14 (Fall): 75–89.

Howell, L.D. and Chaddick, B. (1994) 'Models of political risk for foreign investment and trade', *Columbia Journal of World Business*, 29(3) (Fall): 70–92.

Kogut, B. and Singh, H. (1988) 'The effect of national culture on the choice of entry mode', *Journal of International Business Studies*, 19(3) (Fall): 411–32.

Kostova, T. and Zaheer, S. (1999) 'Organizational legitimacy under conditions of complexity: the case of the multinational enterprise', *Academy of Management Review*, 24(1): 64–81.

La Porta, R., Lopez-de-Silanes, F. and Vishny, R. (1997) 'Trust in large organizations', *American Economic Review*, Papers and Proceedings, May, 333–42.

La Porta, R., Lopez-de-Silanes, F., Shleifer, A. and Vishny, R. (1998) 'The quality of government', NBER Working Paper, 6727.

Laird, S. and Yeats, A. (1990) *Quantitative Methods of Trade Barrier Analysis*, New York: New York University Press.

Loree, D. and Guisinger, S. (1995) 'Policy and non-policy determinants of U.S. foreign direct investment', *Journal of International Business Studies*, 26(2) Second Quarter.

Malone, T.W., Crowston K., Lee, J., Pentland B., Dellarocas, C., Wyner, G., Quimby, J., Osborn, C.S., Bernstein, A., Herman, G., Klein, M. and O'Donnell, E. (1999)

'Tools for inventing organizations: toward a handbook of organizational processes', *Management Science*, 45(3): 425–43.

Meyer, J. and Rowan, B. (1977) 'Institutional organizations: formal structure as myth and ceremony', *American Journal of Sociology*, 83(2).

Miller, K.D. and Reuer, J.J. (1998) 'Firm strategy and economic exposure to foreign exchange rate movements', *Journal of International Business Studies*, 29(3) Third Quarter.

Nohria, N. and Ghoshal, S. (1997) *The Differentiated Network*, San Francisco: Jossey-Bass Publishers.

North, D.C. (1990) *Institutions, Institutional Change and Economic Performance*, Cambridge: Cambridge University Press.

Peterson, M. and Malhotra, N.K. (1997) 'Comparative marketing measures of societal quality of life: substantive dimensions on 186 countries', *Journal of Macromarketing*, Spring: 25–38.

Porter, M.E. (1990) *The Competitive Advantage of Nations*, New York City: Free Press.

Porter, M.E. (1999) 'Microeconomic competitiveness: findings from the 1999 executive survey', in World Economic Forum (ed.), *World Competitiveness Report*, Davos: World Economic Forum.

Posner, R. (1998) 'Creating a legal framework for economic development', *World Bank Research Observer*, 13(1) (February): 1–11.

Prahalad, C. and Doz, Y.L. (1987) *The Multinational Mission*, New York: Free Press.

Ricks, D.A. (1985) 'International business research: past, present and future', *Journal of International Business Studies*, 16(2): 1–4.

Robinson, R. (1964) *International Business Policy*, New York: Holt, Rinehart and Winston.

Sachs, J. (1999) 'The year in review', in World Economic Forum (ed.), *World Competitiveness Report 1999*, Davos: World Economic Forum.

Sachs, J. and Warner, A. (1997) 'Fundamental sources of long term growth', *American Economic Review*, Papers and Proceedings, 184–7.

Scott, R. (1998) *Organization*, New Jersey: Prentice-Hall.

Sundaram, A.K. and Black, J.S. (1992) 'The environment and internal organization of multinational enterprises', *Academy of Management Review*, October, 17(4): 729–57.

Transparency International, '1999 Bribe Payers Index; 1999 Corruption Perceptions Index', Available: http://www.transparency.de/documents/cpi/index.html (Accessed 20/11/99).

UNCTAD (1995) *Incentives And Foreign Direct Investment*, Geneva: UNCTAD.

Vernon, R. (1973) *Sovereignty at Bay*, Harmondsworth: Penguin.

Westney, E. (1993) 'Institutional theory and the multinational corporation' in E. Westney and S. Ghoshal (eds), *Organizational Theory and The Multinational Firm*, New York: St. Martin's Press.

Westney, E. and Ghoshal, S. (1993) *Organizational Theory and The Multinational Firm*, New York: St. Martin's Press.

7 From a theory to a paradigm

Examining the eclectic paradigm as a framework in international economics

Paz Estrella Tolentino

Introduction

This chapter has the following objectives. The first objective is to examine the historical and theoretical forces that led to the emergence of John Dunning's eclectic theory in 1976, and its evolution into a paradigm since the mid-1980s. It is argued that both the history of the eclectic paradigm and its evolution over time is a product of dynamic forces of interaction between forces external and internal to the economics discipline. In studying the dynamics of the eclectic paradigm, the study also hopes to examine its relevance as a general framework in explaining changing patterns of international economic transactions, and to suggest some refinements to further enhance the broad appeal of the paradigm to various different theoretical approaches in the discipline of international economics.

The emergence of the eclectic theory

The seeds of the eclectic theory of international economic involvement were first sown by John Dunning in 1976 in a presentation to a Nobel Symposium in Stockholm on The International Allocation of Economic Activity, the proceedings of which were published by Ohlin *et al.* (1977). The theory was an intellectual response to the growing role of international production and the multinational corporation (MNC) in the world economy, and to developments within theory itself that led a scholar to refine the theoretical structures of economic analysis. In this respect, both the emergence of the eclectic theory and its evolution as a paradigm represented a close intertwining of the relativist and absolutist approaches in the formulation of economic theory.[1]

In explaining the growth and composition of international production and the MNC, the emergence of the eclectic theory in 1976 was a by-product in the evolution of at least six main branches of economic theory: the macroeconomic theories of trade, international capital movements and location; the mesoeconomic theories of industrial organisation and innovation; and the microeconomic theories of the firm (Dunning *et al.* 1986). The synthesis of the ownership, internalisation and location factors in the eclectic theory drew on macroeconomic theories in the elaboration of the concept of location advantages, and on mesoeconomic and microeconomic theories in the development of the concepts of ownership and internalisation advantages.[2] It is partly because it draws upon a variety of theoretical approaches in economics and partly because it explains a number of possible channels of international economic involvement, each of which is determined by a number of factors, that the theory (or the paradigm as it came to be known since the mid-1980s) is known as eclectic (Dunning 1977, 1981a, 1988b). The elaboration of the eclectic theory emerged from a required shift in the emphasis of neoclassical economic theory on location factors towards newer theories incorporating factors associated with the ownership and organisation of economic activity in an attempt to address the changing patterns of international economic transactions after the Second World War. The discussion below focuses on the theoretical antecedents that led to the emergence of the eclectic theory.

At the time of the emergence of the eclectic theory in 1976, neoclassical economics formed the backbone of mainstream orthodox theories – a trend evident since the 1890s (Landreth 1976).[3] Nevertheless, criticisms directed against theoretical neoclassical economics have given rise to the elaboration of a plethora of more heterodox theories which have either provided seminal ideas that helped to advance orthodox theories or have become part of accepted theoretical structures over time. The development of theories outside of conventional neoclassical economics that affected the emergence of the eclectic theory spanned the microeconomic theories of the firm; the mesoeconomic theories of market structures and industrial organisation; and the macroeconomic theories of trade.

Among the most notable contributors in the elaboration of microeconomic theories of the firm other than the neoclassical model was the criticism of Coase (1937) directed against neoclassical economics that led to the further development of institutional economics based on theories of economic organisation and theories of international production based on internalisation. The theory of Coase (1937) explaining the role of transaction costs in superseding the market mechanism provided a general analytical framework for the existence of the firm and its size and scope.[4] Notwithstanding the criticism directed by Coase (1937) against neoclassical economics in its regard for markets as the sole determinant of the allocation of resources, his framework of analysis based on the transactions costs of a two-party exchange remained in the neoclassical approach (see Cantwell 1984; Hébert and Link 1988). Various theories of the firm or the growth of the firm outside the neoclassical mould were also developed that sought to overcome the perceived lack of descriptive realism of the profit-maximising neoclassical model of the firm. Among these were the static and dynamic optimising models advanced by Baumol (1958, 1967), Penrose

(1959), Williamson (1963, 1964), Marris (1964) and Williamson (1966) which built on the seminal work of Berle and Means (1932). There were also the behavioural or satisficing models of the organisational theorists evident in Simon (1947, 1955, 1959), Cyert and March (1963), Cohen and Cyert (1965), Alchian and Demsetz (1972) and Williamson (1975, 1985, 1986) that helped to promote the theoretical economics of institutions and organisations. There had also been the elaboration of a set of theories that shed most light on the emergence of the large, modern business firm as the main organisational form of capital accumulation in the era of managerial capitalism, and their impact on social and economic structures (see Tolentino 2000a). This included the theories of Veblen (1904) and Hilferding (1910) which along with Liefmann (1897, 1903, 1905) and Levy (1911) have served to advance further the concepts of concentration and centralisation of capital in large corporations first elaborated by Marx. The work of Berle and Means (1932) also demonstrated that the modern corporation was not simply a larger version of the owner-managed firm but represented the divorce of ownership and control.

There had also been the development of formal theoretical models of market structures based on monopolistic competition and oligopoly initiated by Joan Robinson and Edward Chamberlin in the 1930s in response to a growing discontent in the empirical validity of the theoretical model of competitive markets based on atomistic competition (see Veblen 1919; Knight 1921; Clapham 1922; and Sraffa 1926). The implications of market power elaborated in the development of formal theoretical models of market structures based on monopolistic competition and oligopoly formed the basis of the work of Bain on barriers to entry to new competition (Bain 1954, 1956) and on industrial economics more generally (Bain 1959).

Evolutions in the theory of international trade in the late 1950s and 1960s included the elaboration of newer theories that sought to provide more realism into the neoclassical Heckscher–Ohlin–Stolper–Samuelson comparative advantage model of international trade. This involved the elaboration of neo-factor trade theories or the introduction of more path-breaking trade theories based on product differentiation, innovation, scale economies, monopolistic competition and similarity of income levels in an attempt to explain trade patterns between developed countries (see Leontief 1954; Johnson 1958; Posner 1961; and Linder 1961).[5]

Thus, evolutions in economic theories at the microeconomic, meso-economic and macroeconomic levels in the period surrounding the Second World War served to address a growing dissatisfaction with the empirical validity of mainstream orthodox theories based on neoclassical economics to explain the role of firms as an alternative locus of economic organisation apart from markets, to overcome the perceived lack of descriptive realism of the assumptions of the neoclassical economic model of the firm, to address the limitations of the theoretical model of competitive markets based on atomistic competition in the presence of oligopoly and monopolistic competition, and to explain the predominance of trade between developed countries.

Such revolution in economic thought exerted an important impact on the elaboration of a distinct body of theories dealing with international production, MNCs and foreign direct investment (FDI) starting with Hymer (1960). At the time of the emergence of the eclectic theory in 1976, various theories explaining international production, the MNC or FDI at the microeconomic, mesoeconomic and macroeconomic levels had already been developed. At the microeconomic level, the alternative theories of the MNC based on market power (Hymer 1960) and internalisation of markets (McManus 1972; Buckley and Casson 1976) had already been elaborated or were in the process of elaboration.[6] At the mesoeconomic level, the theories of industrial organisation that developed the work of Hymer (1960) advanced by Johnson (1970) and other scholars whose analysis emerged from the standpoint of the theory of trade and capital movements (Kindleberger 1969) or from the standpoint of industrial economics (Caves 1971, 1974) had similarly existed prior to the emergence of the eclectic theory in 1976, as have theories on oligopolistic rivalry between MNCs to include Hymer and Rowthorn (1970), Vernon (1971, 1974), Knickerbocker (1973) and Graham (1975). Such has also been the case with the product cycle model of Vernon (1966) and the theories dealing with financial factors in FDI (Aliber 1970, 1971; Ragazzi 1973; Rugman 1975) at the macroeconomic level.

The emergence of the eclectic theory in 1976 thus proceeded from the evolution in the development of economic theory in general, and in economic theories of international production, the MNC or FDI in particular, in combination with a number of empirical and largely country-specific studies on the factors influencing the location of FDI (Southard 1931; Southard *et al.* 1936; Barlow 1953; Dunning 1958) as well as Dunning's own study on the determinants of international production (Dunning 1973). His aim was not to explain the multinational firm, but 'the level and pattern of foreign value-added activities of firms, and/or of countries' (see Dunning 1991: 117). Thus, 'its unit of analysis is the totality of firms engaged in foreign value-adding activities' (Dunning 1988a: 39). The rapid development of various theories explaining either international production, or the MNC as the main institution for international production rather than international exchange, or FDI as the major means by which international production is financed at the time of the emergence of the eclectic theory in 1976 had already led in some cases to 'a fruitless confrontation between alternative theories setting out spuriously to encompass one another' (Cantwell 1991: 16). There had also been the recognition of the inadequacy of individual partial theories of international production, the MNC or FDI to explain all aspects of the phenomenon, the lack of a formal model relating international production financed by FDI to other crossborder modes of resource transfer, and the need to offer a holistic framework to identify and evaluate the significance of the factors influencing both the initial act of foreign production and the growth of such production that collectively led Dunning to favour an eclectic approach (Dunning 1979, 1988b).[7] These factors provided the rationale for constructing a general framework of analysis or a general paradigm of international production (Dunning 1988a) that fulfils the multiple objectives of allowing for the coexistence of alternative and complementary theories, and identifying the common basis or areas of disagreement between the different theories without being inextricably wedded to any one particular approach (Cantwell 1991).

The central thesis of the eclectic theory, or the eclectic paradigm since the mid-1980s, has always been that channels of international economic involvement or international economic transactions or the international competitiveness of a country's output of goods and services is determined by the possession of ownership-specific endowments of its enterprises, by the ability and desire of these enterprises to internalise these advantages or the markets to these advantages, and by comparative location endowments of home *vis-à-vis* foreign countries which are exogenous to firms (Dunning 1977).

The ownership-specific endowments of enterprises are the resource endowments or assets in the Fisherian sense (Johnson 1970) that are capable of generating a future income stream. Alternatively, ownership-specific endowments are the intermediate products to which a firm has privileged access, and are therefore the privileged property of the firm (Dunnin 1991).[8] These include tangible assets or Ricardian-type factor endowments such as natural resources, manpower and capital, or intangible assets such as knowledge, organisational and entrepreneurial skills and access to intermediate and final goods markets. The latter could also include the cultural, legal and institutional environment in which endowments are used, market structure and government legislation and policies (Dunning 1981a, 1988a). These ownership-specific endowments may be specific to the home or host country, the industries of enterprises, or to enterprises (Dunning 1981b, 1988a). Since the unique feature of the MNC is that it is both a multi-activity firm and engaged in the intra-firm transfer of intermediate products across national borders (Dunning 1988a), the very presence of MNCs is evidence of the transferability of at least some ownership advantages across space or national borders (Dunning 1979), or the capability of their use in the home country or elsewhere (Dunning 1988a), even though its origin may be linked partly or initially to location-specific endowments (Dunning 1981a). Such transferable assets may take the form of a legally protected right or a commercial monopoly, or may arise from size, diversity, or the technical characteristics of firms, the economies of joint production and/or marketing, and surplus entrepreneurial capacity (Kojima 1978).

Given their diverse nature, three kinds of ownership advantages have been distinguished since Dunning (1977). First, those ownership endowments which any firm may have over another *producing in the same location*. These advantages which stem from size, monopoly power, and better resource capability and usage derive from the concept of barriers to entry to new competition of Bain (1956).[9] Second, those which a branch plant of a national enterprise may have over a *de novo* enterprise (or over an existing enterprise breaking into a new product area) *producing in the same location*. Third, those which arise specifically from the multinationality of a company, and are an extension of the other two. Thus, while the first kind of ownership advantages explains how foreign firms have a competitive advantage over rival firms serving the same markets and represents the kind of ownership advantages that could in principle be sold arm's length to foreign firms (see Dunning 1991), the second kind describes the advantages of multi-plant firms and the advantages of common ownership (Caves 1980). Finally, the third kind of ownership advantages arises from operations in different economic environments across national borders, which enables the MNC to capitalise on different factor endowments and market situations, and to respond to changes in the different economic environments (Kogut 1985a). The second and third kinds of ownership advantages collectively provide ownership advantages of common governance (Dunning 1988a) that could only be realised through vertical or horizontal integration, or by the common ownership of related activities along a particular value-added chain, or across two or more value-added chains transcending national borders (Dunning 1991). In contrast to the first kind of ownership advantages, there is in general no external market for the second and third kinds of ownership advantages of a more collective kind.[10]

As the degree of multinationality of firms increases, ownership endowments become less specific to any particular location (home or host country) or to the industries of enterprises, but more specific to enterprises generating the assets (Dunning 1979).[11] Put another way, the increasing multinationality of firms leads to the greater importance of ownership-specific advantages of firms acquired as a result of FDI (the transaction ownership advantages) compared to those it possesses prior to the act of FDI (the asset ownership advantages) (Dunning 1991).[12] Related views in this context have been advanced by Rugman (1979), Buckley and Casson (1985) and Kogut (1985a, 1985b).

The evolution of the eclectic paradigm

As previously mentioned, both the emergence of the eclectic paradigm and its evolution as a paradigm represented a close intertwining of the relativist and absolutist approaches in the formulation of economic theory. The evolution of the eclectic theory into a framework or paradigm since the mid-1980s was a scholarly response to the need to continually adapt the eclectic theory as a tool to examine the economic questions raised by the increasingly widespread network of MNCs and the totality of their cross-border transactions in the world economy which has made these firms 'the new instrument of international economic integration' (Tolentino 1999).[13] But in addition, the evolution of the theory had been in response to the further rounds of debate associated with developments within the theory of international production, the MNC and FDI in much the same way that the growing debate in theoretical international economics generally sparked off the emergence of the eclectic theory in 1976. Some of the controversies surrounding the eclectic paradigm arise from the different choice of focus for the driving force underlying the growth and expansion of international production and the MNC, related partly to the different level of economic analysis used by various partial theories.

One of the theoretical debates revolves around the role of ownership advantages in the existence and growth of the MNC. Given that the eclectic paradigm is a holistic framework to identify and evaluate the significance of the factors influencing both the initial act of foreign production and the growth of such production as previously mentioned (Dunning 1988b), the theoretical perception of ownership-specific advantages of firms in the paradigm comprises both asset ownership advantages and transaction ownership advantages (Dunning 1988c).[14] Such distinctions in the types of ownership advantages and the differential ability and desire of enterprises to internalise their ownership-specific advantages as a modality to initiate as well as sustain international production in the eclectic paradigm are not of real interest to scholars advancing the internalisation theory of the MNC. Thus, while asset ownership advantages are necessary to explain at least the initial act of international production in the eclectic paradigm, their existence is not a necessary condition for international production in internalisation theory, despite an acknowledgement that this type of ownership advantages may help to explain international production (Buckley and Casson 1976; Casson 1987). The major contention of the internalisation theory advanced by Buckley and Casson (1976, 1985), Casson (1987) *et al.* has always been that the existence and growth of MNCs through the internalisation of intermediate product markets across national boundaries – a combination of internalisation and location factors – are both necessary and sufficient conditions for the existence and growth of the MNC.[15]

The controversy over the role of ownership advantages in the existence and growth of the MNC has centred on whether ownership advantages are exogenous or endogenous factors. Thus, on the one hand, asset ownership advantages are considered *endogenous to firms or MNCs* in the eclectic paradigm and the more dynamic theories of international production and the MNC, and *exogenous to the individual firm or MNC* in internalisation theory.[16] But perhaps a more appropriate manner in which to recast the debate is to focus the discussion instead on how asset ownership advantages are generated. Thus, in the eclectic paradigm and the more dynamic theories of international production and the MNC, asset ownership advantages can be regarded to proceed from the creation of endogenous structural imperfections by MNCs as a means to sustain the process of inter-firm competition in final product markets. These structural market imperfections 'arise from the proprietary ownership of specific assets by MNEs *vis-à-vis* those possessed by other enterprises' (Dunning 1988b: 2) or 'which arise from the actions of participants in or outside the market to distort the conditions of demand or supply' (Dunning 1995: 463) and 'which discriminate between firms (or owners of corporate assets) in their ability to gain and/or sustain control over property rights' (Dunning 1988a: 40). Thus in the view of the eclectic paradigm and the more dynamic theories of international production and the MNC, asset ownership advantages proceeding from the creation of endogenously generated structural market imperfections in final products are logically separate from internalisation (or the act of internalising the market for intermediate products across national borders). The latter is simply a modality by which asset ownership advantages are exploited intra-firm and transaction ownership advantages are achieved, and thus renders internalising firms at an advantage relative to non-internalising firms. Put another way, asset ownership advantages are ownership-specific endowments present prior to the act of internalisation.

By contrast, the long-run theory of the MNC of Buckley and Casson (1976) based on internalisation '. . . emphasises very general forms of imperfect competition stemming from the costs of organising markets, and it concentrates on imperfections in intermediate-product markets rather than in final-product markets' (Buckley and Casson 1976: 32–3). Thus, in their view and in the view of other scholars of internalisation theory, both asset and transaction ownership advantages arise as a direct consequence of internalising intermediate product markets or endogenising imperfections in intermediate product markets (see Rugman 1981) or overcoming exogenous transactional market imperfections in intermediate products. In this view, the multinationality, growth rate and profitability of firms since the Second World War are linked to the internalisation of knowledge or flows of knowledge owing to the need to bypass imperfect intermediate product markets for knowledge external or exogenous to the firm. Thus, ownership advantages in internalisation theory originate from, or are a result of, the internalisation and integration of the firm. As a result, not only are the distinctions between asset and transaction ownership advantages not of real interest to scholars of internalisation theory, but the entire concept of ownership advantages is rendered irrelevant or logically redundant (Buckley and Casson 1985; Casson 1987; Itaki 1991).

The debate between the two schools of thought concerning the role of ownership advantages unmasks the inherent differences of opinion with respect to the purpose behind the existence and growth of international production, the role of efficiency in the organisation of the firm, the relevance of competition between firms in final product markets, and the theoretical interpretations of the concept of ownership advantages

(Cantwell 1991). The existence and growth of international production in internalisation theory achieved through vertical and horizontal integration of the MNC across national borders is geared solely to reduce transaction costs or the costs of market transactions (Chandler 1977a, 1977b, 1980; Williamson 1971, 1975, 1985; and Teece 1980, 1981, 1987). As a result, the theory has assigned a more important role to efficiency of firms in terms of the organisation of the exchange of intermediate products, and a rather less important role to asset ownership advantages and inter-firm competition in final product markets (Cantwell 1991).[17] On the other hand, the other school of thought advanced by the eclectic paradigm and the more dynamic theories of international production and the MNC assert otherwise, and contend that apart from reducing *transaction* costs, firms may initiate and sustain international production as a means to lower *production costs per unit* (or unit costs), to improve the value-adding capabilities of the firm, in some cases to exploit a quasi-monopolistic position of a given firm relative to other firms in the same industry, or to capture the gains of restructuring production across national borders. It is the restructuring or rationalisation of production which creates the benefits of common governance rather than the transactions associated with it (Dunning 1988c). Perhaps more importantly, the existence and growth of international production on the basis of strengthening ownership advantages can enable an MNC to retain or increase its market share of final products, and thus to grow faster than its rival firms (Cantwell 1991).

Indeed, it was the theoretical perception of the market power theory of the MNC and theories of industrial organisation in relation to MNCs of asset ownership advantages as monopolistic advantages or anti-competitive devices that act as barriers to entry to other firms that led scholars of internalisation theory to claim that ownership advantages are not a necessary condition for international production (Tolentino 1993; Cantwell 1991). In fact, it does not help that the orthodox neoclassical economic theory of industrial organisation and the concept of barriers to entry to new competition of Bain (1956) continues to provide the basic framework behind asset ownership advantages in the emergence and evolution of the eclectic paradigm (see Dunning 1977, 1979, 1988a, 1988c). If the eclectic paradigm is to succeed as a general framework in arguing that asset ownership advantages are a necessary condition for international production, it must go beyond acknowledging that asset ownership advantages are both competitive advantages and monopolistic advantages (Dunning 1988b, 1991). The eclectic paradigm must allow for alternative conceptual interpretations within its theoretical structure by extending its conceptual interpretation of asset ownership advantages from a narrow emphasis on Bain-type monopolistic advantages which enable firms to erect barriers to entry to new competition and exercise monopoly power or market power in final product markets.[18] It must also accommodate the conceptual perception of ownership advantages as part of the rivalrous behaviour or competitive process consistent with the approach of Cantillon and the classical economists starting with Adam Smith, the Austrian economists such as Schumpeter and Hayek, as well as Penrose. Followers of this school of thought in the theory of international production include Vernon (1966, 1974) and the scholars behind the competitive international industries approach to international production that regard asset ownership advantages as competitive weapons which help sustain a process of competition between rival firms in an oligopoly (Cantwell 1991).[19] Unlike, *monopolistic* forms of ownership advantages which derive from the concept of barriers to entry to new competition of Bain (1954, 1956), *competitive* forms of ownership advantages derive from the concept of firms and entrepreneurs of unequal abilities of Schumpeter (1911, 1943) and Penrose (1959).[20] Competitive forms of ownership advantages enable a firm or entrepreneur to create a quasi-monopolistic position for itself, but dynamic competitive forces ensure that existing monopolies are eventually broken down by the constant struggle of other firms and entrepreneurs to enter markets that are seen to offer the prospect of profits; or alternatively in the development of further technological advances that result in new competition for producers that enjoy a degree of quasi-monopoly power (Tolentino 2000b).[21] Since such process of rivalrous competition is thwarted whenever non-market barriers block entry to potential competitors, its existence requires freedom of entry (Kirzner 1978). Competitive forms of ownership advantages thus derive from the differential ability of firms to discover profitable opportunities, and not from the presence of monopoly power, since the opportunities for profit accumulation are open to all firms.

In clearly distinguishing between the alternative interpretations of the concept of ownership advantages in the eclectic paradigm, the paradigm allows for the coexistence of the two different functions of the firm as envisaged by Penrose (1959). On the one hand, the firm or MNC is an entity that takes decisions over prices and output and earns higher profits from increasing its degree of market power. Alternatively, the firm is a device for innovation and knowledge creation and earns higher profits from creating 'new combinations' in production (Schumpeter 1911) or new areas of social or productive capability (Cantwell 2002). In this manner, the eclectic paradigm can synthesise classical, Schumpeterian and neoclassical analyses of the firm and the MNC in one cogent framework of economic analysis.

The theoretical debates since the emergence of the eclectic theory have also revolved around internalisation factors in the existence and growth of the MNC, related closely to the previous discussion on ownership advantages. To be precise, the theoretical discourse has been around the conceptual interpretation of internalisation as the 'internalisation of the ownership advantages or intermediate products' or the 'internalisation of the markets for ownership advantages or intermediate products'. The former has tended to be the theoretical concept of internalisation used by Dunning in the earlier versions of the eclectic theory (see Dunning 1977, 1979, 1981a).[22] Such theoretical perception is consistent with his earlier more narrow view of internalisation as a means to explain the intra-firm exploitation of ownership endowments present prior to the act of internalisation, and not as a means by which firms achieve hierarchical advantages by replacing markets post internalisation (Dunning 1991). In acknowledging the argument of scholars of internalisation theory whose theoretical perception of internalisation is the internalisation of the markets for intermediate products in response to transactional market imperfections rather than the internalisation of intermediate products *per se* in relation to structural imperfections in final product markets (see Casson 1986, for example) combined with the advancing process of globalisation of firms which render hierarchical ownership advantages relatively more important than asset ownership advantages, Dunning revised his theoretical interpretation of internalisation since the mid-1980s.[23] In particular, Dunning has since reformulated his theoretical interpretation of internalisation in the broader context of the firm *using* or *transferring* asset ownership advantages it possesses by itself, or the firm engaging in value-adding activities based upon these advantages it possesses, as well as a modality through which transaction ownership advantages are achieved (see Dunning 1988a, 1988b, 1988c, for example).[24] Thus, in this broader theoretical perception of the concept of internalisation advantages in the evolution of the eclectic paradigm since the mid-1980s, the eclectic paradigm hoped to reconcile its inherent differences with internalisation theory and to capture the full range of internalisation advantages that arise from the intra-firm exploitation of ownership endowments that the firm possesses prior to and after the act of internalisation.

Perhaps a more effective manner in which the eclectic paradigm as a general framework can demonstrate its ability to allow for the coexistence of alternative theoretical interpretations of internalisation or internalisation advantages, while at the same time addressing the controversy as to the proper theoretical interpretation is through the following. Where internalisation advantages derive from the greater ease with which an integrated firm is able to appropriate a full return on its ownership of distinctive assets (appropriability), then the relevant theoretical interpretation of internalisation advantages would be the internalisation of the ownership advantages or intermediate products. In this context, the conduct of the firm is the essential cause of *endogenous* structural market imperfections and failure in final products as described previously. The incentives for internalisation draw on the need of the firm to gain from the advantages of creating market imperfections or failure in an *internal* mechanism of surplus allocation.[25] In this respect, the firm or MNC *capitalises on* the creation of structural market imperfections. Put another way, owing to surplus or profits that could be achieved the firm chooses *not to externalise* structural market imperfections or failure that it creates itself. It explains why asset ownership advantages are exploited by one group of MNCs rather than another, or by MNCs rather than indigenous firms in the host country (Dunning 1988b).

On the other hand, where the internalisation advantages derive directly from the coordination of the use of complementary assets, subject to the costs of managing a more complex network or emanate from the need to *overcome* imperfections in intermediate product markets *exogenous* to the firm, then the relevant theoretical interpretation of internalisation advantages would be the internalisation of the markets for ownership advantages or intermediate products.[26] Internalisation in this context overcomes transactional, endemic or natural market imperfections 'where either, given the conditions of supply and demand, the market *qua* market is unable to organise transactions in an optimal way, or it is difficult to predict the behaviour of the participants' (Dunning 1995: 463). Market imperfections emerge from 'the failure of intermediate product markets to transact goods and services at a cost less than that which a hierarchy might incur by undertaking these transactions itself' (Dunning 1988a: 40). Thus, in this context it is an immutable factor – the intermediate product market *exogenous* to firms – that is the essential cause of market imperfections and failure. The incentives for internalisation draw on the need to *avoid or circumvent* the disadvantages of market imperfections or failure in *external* mechanisms of resource allocation – the price system and public fiat. It explains why the hierarchy of the firm rather than external mechanisms is the vehicle by which transactional ownership advantages are transferred across national borders (Dunning 1988b).

In clearly distinguishing between the alternative interpretations of the concept of internalisation in the eclectic paradigm, the firm or MNC can be regarded both as an *active agent* capitalising on the creation of endogenous structural imperfections in final product markets as well as a *passive product* or 'black box' responding to exogenous transactional imperfections in intermediate product markets. In this manner, the synthesis of the classical, Schumpeterian and neoclassical analyses of the firm and the MNC in one cogent framework of economic analysis is further reinforced. Although such conceptual distinctions of internalisation are rendered increasingly blurred with the advancing globalisation of firms (Teece 1990), the distinctions remain important in economic theory. Such refinement of the concept of internalisation in the eclectic paradigm could enable alternative theoretical perceptions of the concept to exist in a logically consistent manner, while also clearly distinguishing the circumstances by which a particular theoretical interpretation may be more relevant.

Conclusions

The emergence of the eclectic theory and its evolution as a paradigm represented a close intertwining of the relativist and absolutist approaches in the formulation of economic theory. This is because the history of the paradigm is a product of dynamic forces of interaction between forces external and internal to the economics discipline. The external forces were the historical economic circumstances brought about by the growth in importance of international production and the MNC in the world economy which led economic inquiry in the direction of explaining either international production, or the MNC as the main institution for international production rather than international exchange, or FDI as the major means by which international production is financed. The internal forces emerged from developments within economic theory in general, and within the theory of international production, MNC and FDI in particular, which led scholars to refine the theoretical structures of economic analysis. The eclectic theory emerged from a required shift in the emphasis of neoclassical economic theory on location factors towards newer theories incorporating factors associated with the ownership and organisation of economic activity in an attempt to address the changing patterns of international economic transactions after the Second World War. The theory was thus formulated based on a synthesis of the ownership, internalisation and location factors in explaining a broad range of international economic transactions. The relevance of the eclectic paradigm as a general framework in explaining changing patterns of international economic transactions derives from its role in overcoming the inadequacy of the various partial theories of international production, the MNC or FDI as a general explanation of the phenomenon, in relating international production financed by FDI to other crossborder modes of resource transfer, and in offering a holistic framework to identify and evaluate the significance of the factors influencing both the initial act of foreign production and the growth of such production. The aim was not to explain the multinational firm, but 'the level and pattern of foreign value-added activities of firms, and/or of countries' (see Dunning 1991: 117). Thus, 'its unit of analysis is the totality of firms engaged in foreign value adding activities' (Dunning 1988a: 39). In its objective to identify clusters of variables relevant to an explanation of all kinds of output of goods and services produced for foreign markets, the paradigm represents a general framework of analysis which – at least in the body of theory covering the field of international production, the MNC, or FDI – fulfils the multiple objectives of allowing for the coexistence of alternative and complementary theories, and identifying the common basis or areas of disagreement between different theories without being inextricably wedded to any one particular approach.

In seeking to enhance the broad appeal of the eclectic paradigm to various theoretical interpretations, the chapter suggested a refinement of its concepts of ownership advantages and internalisation. To support its view that asset ownership advantages are both competitive advantages and monopolistic advantages, the theoretical interpretation of asset ownership advantages in the paradigm needs to be broadened from a narrow emphasis on Bain-type monopolistic advantages which enable firms to erect barriers to entry to new competition and exercise monopoly power in final product markets. It must also accommodate the theoretical perception of ownership advantages as part of the rivalrous behaviour or competitive process between firms consistent with the approach of Cantillon and the classical economists starting with Adam Smith, the Austrian economists such as Schumpeter and Hayek, as well as Penrose. The eclectic paradigm must also effectively address the important distinction between 'internalisation of ownership advantages or intermediate products' and the 'internalisation of the markets for ownership advantages or intermediate products' within the context of *endogenous* structural market imperfections in final products and *exogenous* transactional market imperfections in intermediate products. In clearly distinguishing between the alternative interpretations of the concepts of ownership advantages and internalisation, the paradigm could more effectively synthesise neoclassical, classical and Schumpeterian theories of the firm and the MNC in one cogent framework of economic analysis.

Notes

1 The development of economic theory or the history of economic thought can be a product of forces *external* to the economics discipline (the relativist approach or the 'sociology of knowledge' approach) or forces *internal* to the economics discipline (the absolutist approach) (see Landreth 1976 and Ekelund and Hébert 1990). While the former approach examines the social, economic and historical milieu that shape economic ideas, the latter approach regards economic ideas as undergoing an evolution or progression of their own based on the efforts of scholars to refine the theoretical structures of economic analysis. The view of this author is that the emergence of the eclectic theory and its evolution as a paradigm is a product of the dynamic forces of interaction between forces *external* and *internal* to the economics discipline. The *external* forces were the historical economic circumstances brought about by the growth in importance of international production and the MNC in the world economy which led economic inquiry in the direction of explaining either international production, or the MNC as the main institution for international production rather than international exchange, or FDI as the major means by which international production is financed. The *internal* forces emerged from developments within economic theory in general, and within the theory of international production, MNC and FDI in particular, which led scholars to refine the theoretical structures of economic analysis.

2 From an opposite perspective, Dunning *et al.* (1986) analysed the conceptual antecedents of the theory of international production within the structure of Dunning's eclectic framework.

3 Although neoclassical or 'post-classical' economics did not become the mainstream orthodox theory until the 1890s, the foundations of neoclassical economics were established firmly in England and on the Continent by 1870 (Ekelund and Hébert 1990).

4 In providing a heterodox emphasis on transactions as the fundamental unit of economic investigation and the neglect of non-market transactions in particular

in neoclassical economics, Commons (1924, 1934) is a worthy antecedent of Coase (1937). See Dunning *et al.* (1986).

5 Although the remarkable presence of 'an organic interconnection of international trade, movement of productive factors, transport and market organisation' across national borders cited by Williams (1929) presupposed a recognition of a required modification to neoclassical trade theories based on factor immobility, there had been no attempt before the Second World War to develop other international trade theories driven by the presence of ownership advantages that are the privileged property of exporting firms (see also Dunning *et al.* 1986; Dunning 1977) or that addressed the organisation of economic activity (Dunning 1991). The more pathbreaking trade theories developed in the late 1950s and 1960s have been regarded to be more effective in explaining trade in goods between developed countries based on differences in ownership-specific endowments rather than location-specific factors based on differences in factor endowments between countries (Dunning 1977).

6 Although the MNC theory of internalisation of Buckley and Casson was not published until 1976, the same year that the eclectic theory was first presented at the Nobel Symposium, Dunning acknowledged the influence of these colleagues at the University of Reading in his analysis of internalisation advantages. However, it was the work of McManus (1972) that introduced Dunning to the concept of internalisation as applied to the MNC (see Dunning 1991).

7 Thus, the presence of ownership advantages and the absence of internalisation advantages and advantages of producing in a foreign location favour the route of contractual resource transfer in servicing foreign markets. Alternatively, the presence of both ownership and internalisation advantages and the absence of advantages of producing in a foreign location favour the route of trade in goods and services in servicing foreign markets. Finally, the presence of the three types of advantages – ownership advantages, internalisation advantages and advantages of producing in a foreign location – favour the route of international production in servicing foreign markets (Dunning 1988c, 1981b). Thus, no one advantage is both a necessary and sufficient condition to explain international production. However, in seeking to identify clusters of variables relevant to an explanation of all kinds of output of goods and services produced for foreign markets, the eclectic paradigm does not attempt to pinpoint or evaluate the specific parameters relevant to the explanation of particular types of MNC activities (Dunning 1988a). It is mainly for this reason that Dunning began to refer to the eclectic theory as an eclectic paradigm in the mid-1980s (see Dunning 1991).

8 Thus in the case of international production, FDI is the conduit for the transfer of these intermediate products across national borders.

9 In the evolution of the eclectic paradigm, this type of ownership advantages is referred to as intangible asset advantages that confer a property right to the firm possessing them (see, for example, Dunning 1988a, 1988c, 1995).

10 It was partly on the basis that the second and third kinds of ownership advantages have no external markets that scholars of internalisation theory came to conclude that the distinction between ownership and internalisation advantages may seem irrelevant or logically redundant (see Buckley and Casson 1985; Casson 1987).

11 Firm-specific ownership endowments are those assets which an enterprise may create for itself (e.g. certain types of knowledge, organisation and human skills) or purchase from other institutions, and as a result the enterprise acquires some proprietary right of use (Dunning 1981a).

12 Asset ownership advantages refer to the first kind of ownership advantages described by Dunning (1977), as previously mentioned. Transaction ownership

advantages or ownership advantages of a governance cost-minimising kind (Teece 1983) refer to the second and third kinds of ownership advantages described by Dunning (1977). Transaction ownership advantages arise from the joint ownership of complementary assets. Such distinction between asset and transaction ownership advantages first made in Dunning (1983) is indicative of his attempt to acknowledge the long-standing view of scholars in internalisation theory that ownership advantages may arise as a direct consequence of internalising intermediate product markets or endogenising imperfections in intermediate product markets (Rugman 1981). Notwithstanding this acknowledgement, Dunning continues to maintain that internalisation (or the act of internalising the market for intermediate products across national borders) is a modality by which asset ownership advantages are exploited intra-firm and transaction ownership advantages are achieved and thus renders internalising firms at an advantage relative to non-internalising firms, and that internalisation confers internalisation incentive advantages in addition to transaction ownership advantages.

13 A similar but more limited view is given by Julius (1990) who refers to a new level of international economic integration through FDI. But although MNCs are the agents behind international production financed by FDI, these firms also play a key role in international trade, international finance as well as in a variety of non-equity relationships such as turnkey contracts, management and service contracts, co-production agreements, subcontracting agreements, licensing and franchising agreements, arrangements concerning the transfer of technology and know-how, and other forms of collaborative arrangements with entities and enterprises in all parts of the world (Tolentino 1999).

14 Cantwell (1991) makes a similar argument by stating that advantages arising from the internalisation of intermediate product markets across national borders is complementary to, but not a substitute for, asset ownership advantages. In this view, he is in accord with Penrose (1959) in the consideration of internally generated growth associated with firm-specific ownership advantages as fundamental to both the existence and growth of the firm. Such views are fundamentally opposed to internalisation theory in which the impetus for the existence and growth is exogenous to the firm as it arises from the conditions of exchange in intermediate products in markets external to the firm. Such disagreements are indicative of differences in their perspective and time frame. Thus, if the perspective taken is explaining the position of some *countries* in the investment development path relative to other countries, other factors exogenous to firms but endogenous to countries must be accommodated. Alternatively, if the perspective taken is explaining the changing competitive position of *firms,* both endogenous factors relating to the creation of new ownership advantages by firms and exogenous factors associated with the changing characteristics of markets or hierarchies may be relevant (Dunning, 1991, 1988c). Some acknowledgement of the importance of ownership advantages in the dynamics of international production by scholars of internalisation theory is evident in Buckley and Casson (1985), Casson (1987) and Buckley (1988).

15 Such a view draws upon neoclassical economic theory that regards markets as the sole determinant of resource allocation, and thus the firm is regarded as a black box. In this role, the firm is a reactive or passive product of the changes in market structure and the transaction costs of markets (see also Coase 1937; Kindleberger 1969; Helpman and Krugman 1985).

16 A continuing uncertainty in the literature concerns whether or not all kinds of ownership advantages are assumed to be exogenous in internalisation theory, or only asset ownership advantages (Dunning 1991). The internalisation literature

appears to be ambiguous on this point, although Itaki (1991) seems to regard both asset and transaction ownership advantages as exogenous.

17 The existence of ownership advantages, or at least asset ownership advantages, is not viewed as a necessary condition for international production in internalisation theory owing to their overall emphasis on reducing transaction costs or the costs of market transactions in *intermediate products* in explaining international production. The definition of ownership advantages with reference to the attainment of faster growth, increased market share and higher profitability of given firms relative to other firms in *final product markets* (Cantwell 1991) helps to de-emphasise further the role of ownership advantages in relation to inter-firm competition in final product markets in internalisation theory. Besides, although market imperfections may exist in final products and intermediate products, firms cannot internalise final product markets in theoretical neoclassical economics since consumers are independent of producers or merchants (see Itaki 1991).

18 As mentioned previously, the concept of barriers to entry of Bain (1954, 1956) stemmed from the analysis of the implications of market power inherent in formal theoretical models of markets structures based on monopolistic competition and oligopoly.

19 This includes *inter alia* Graham (1975), Jenkins (1984, 1987), Pavitt (1988) and Cantwell (1989).

20 In Schumpeter's theory, the role of the dynamic innovative entrepreneur in introducing disturbances in the existing circular flow of income through 'new combinations' in production is the underlying factor in the process of economic development (Schumpeter 1911). With the rise of in-house corporate research and development in large firms in the twentieth century, the analysis of the role of the dynamic innovative entrepreneur was then extended to large firms in Schumpeter (1943). The 'new combinations' of production are manifested in the introduction of a new good or of a new quality of a good, the introduction of a new method of production, the opening of a new market, the conquest of a new source of supply of raw materials or intermediate products, and the formation of a new organisation of any industry. In such a process, the disequilibrium position brought about by the innovation of a single entrepreneur or technologically leading firm is imitated subsequently by a competitive fringe of firms which eventually catch up with the leading firm and, by increasing productivity growth and output, remove the leading firm's monopoly rent in order to return to an equilibrium position (Tolentino 1993). Thus, in Schumpeter's theory successful innovation requires an act of will, and depends on leadership (Ekelund and Hébert 1990). By incorporating the role of in-house research and development and endogenous innovation in large firms, the theory of the growth of the firm of Penrose (1959) applied and extended the original Schumpeterian theory (see Cantwell 2002).

21 Two major points need to be made about the monopoly position created in the market by the entrepreneur or large firm in the pursuit of opportunities for profit accumulation in original Schumpeterian theory: first, the monopoly position is merely an incidental by-product of the process of continuous innovation and is not the major source of profits for innovation; and second, the monopoly position created is quasi-monopolistic, i.e. the monopoly position is neither permanent nor absolute, owing to technological competition. Competition is not to be understood in the technical sense of a perfectly competitive market structure, but in the sense of rivalry that drives the market process down the road to coordination of individual plans of market participants. The competitive market process is required in the mobilisation of existing knowledge

and in the generation of awareness of new profitable opportunities to non-omniscient market participants. Original Schumpeterian theory is thus fundamentally opposed to the 'Schumpeterian hypothesis' in which the line of reasoning is reversed in a way that market power and the attainment of monopoly profits becomes the cause of innovation (see Cantwell 2002). Thus, in seeking to provide a theoretical framework to competitive asset advantages in the eclectic paradigm by the introduction of Schumpeterian theory, such framework is to be understood as the original theory as Schumpeter had developed it, and not the 'Schumpeterian hypothesis' which resulted from an attempt to recast the original insights of Schumpeter within the framework of orthodox theoretical neoclassical economics.

22 An exception is found in Dunning (1981b) where the second condition determining the propensity of a country's enterprises to engage in FDI arises from whether it pays enterprises to exploit their proprietary advantages themselves, that is, internalise their use or sell them, or the right to use them, to foreign firms.

23 To an extent, the polemic has arisen since the internalisation of ownership advantages in the early versions of the eclectic paradigm presupposed the applicability of the concept of internalisation in relation to the presence of structural market failure in final products. This is different from the major contention of scholars of internalisation theory that internalisation is intended specifically to denote a response to transactional market failure in intermediate products.

24 Perhaps a notable exception is found in Dunning (1991) where in a restatement of the central thesis of the eclectic paradigm, he refers to internalisation as 'the extent to which firms perceive it to be profitable to internalise the markets for these assets [referring to net competitive ownership advantages] and by so doing add value to their output' (p. 124).

25 Note that the internal mechanism of resource allocation is not the relevant parameter in this context. Rather, it is the internal mechanism of surplus allocation to secure the maximisation of future surplus or profits through appropriate capital accumulation. The latter is an approach associated with classical economic theory and with economic models linked to a 'classical revival' (see Cantwell 1984).

26 Teece (1990) discussed the significance of the distinction between appropriability and coordination. Similarly, Dunning and Rugman (1985) discussed the significance of the distinction between structural and transactional market imperfections and failure.

References

Alchian, A. and Demsetz, H. (1972) 'Production, information costs and economic organization', *American Economic Review,* 62: 777–95.

Aliber, R.Z. (1970) 'A theory of direct foreign investment', in C.P. Kindleberger (ed.), *The International Corporation: A Symposium,* Cambridge, MA: MIT Press.

Aliber, R.Z. (1971) 'The multinational enterprise in a multiple currency world', in J.H. Dunning (ed.), *The Multinational Enterprise,* London: George Allen & Unwin.

Bain, J.S. (1954) 'Economies of scale, concentration and condition of entry in twenty manufacturing industries', *American Economic Review,* 44(1): 15–39.

Bain, J.S. (1956) *Barriers to New Competition,* Cambridge, MA: Harvard University Press.

Bain, J.S. (1959) *Industrial Organisation,* New York: John Wiley.
Barlow, E.R. (1953) *Management of Foreign Manufacturing Subsidiaries,* Cambridge, MA: Harvard University Press.
Baumol, W. (1958) 'On the theory of oligopoly', *Economica,* 25(99): 187–98.
Baumol, W. (1967) *Business Behavior, Value and Growth,* New York: Harcourt, Brace and World.
Berle, A.A. and Means, G.C. (1932) *The Modern Corporation and Private Property,* New York: Macmillan.
Buckley, P.J. (1988) 'The limits of explanation: testing the internalization theory of the multinational enterprise', *Journal of International Business Studies,* 19(2): 181–93.
Buckley, P.J. and Casson, M.C. (1976) *The Future of the Multinational Enterprise,* London and Basingstoke: Macmillan.
Buckley, P.J. and Casson, M.C. (1985) *The Economic Theory of the Multinational Enterprise,* London: Macmillan.
Cantwell, J. (1984) 'The relevance of the classical economists to the theory of international production', University of Reading Discussion Papers in International Investment and Business Studies, No. 79.
Cantwell, J. (1989) *Technological Innovation and Multinational Corporations,* Oxford: Basil Blackwell.
Cantwell, J. (1991) 'A survey of theories of international production', in C.N. Pitelis and R. Sugden (eds), *The Nature of the Transnational Firm,* London: Routledge.
Cantwell, J. (2002) 'Innovation, profits and growth: Penrose and Schumpeter', in C.N. Pitelis (ed.), *The Growth of the Firm: The Legacy of Edith Penrose,* Oxford: Oxford University Press.
Casson, M. (1986) 'General theories of the multinational enterprise: their relevance to business history', in P. Hertner and G. Jones (eds), *Multinationals: Theory and History,* Aldershot: Gower.
Casson, M. (1987) *The Firm and the Market: Studies in Multinational Enterprise and the Scope of the Firm,* Oxford: Basil Blackwell.
Caves, R.E. (1971) 'International corporations: the industrial economics of foreign investment', *Economica,* 38: 1–27.
Caves, R.E. (1974) 'Causes of direct investment: foreign firms' shares in Canadian and United Kingdom manufacturing industries', *Review of Economics and Statistics,* 56: 272–93.
Caves, R.E. (1980) 'Investment and location policies of multinational companies', *Schweizerische Zeitschrift für Volkswirtschaft und Statistik,* 116: 321–7.
Chandler, A.D., Jr. (1977a) 'Institutional integration: an approach to comparative studies of the history of large scale business enterprise', in K. Nakagawa (ed.), *Strategy and Structure of Big Business,* Tokyo: University of Tokyo Press.
Chandler, A.D., Jr. (1977b) *The Visible Hand: The Managerial Revolution in American Business,* Cambridge, MA: Harvard University Press.
Chandler, A.D., Jr. (1980) 'The growth of the transnational industrial firm in the United States and the United Kingdom: a comparative analysis', *Economic History Review,* 33(3): 396–410.
Clapham, J. (1922) 'Of empty boxes', *Economic Journal,* Reprinted in American Economic Association (ed.), *Readings in Price Theory,* Chicago: R.D. Irwin.
Coase, R.H. (1937) 'The nature of the firm', *Economica,* 4(4): 386–405.
Cohen, K.J. and Cyert, R.M. (1965) *Theory of the Firm: Resource Allocation in a Market Economy,* Englewood Cliffs, NJ: Prentice Hall.

Commons, J.R. (1924) *Legal Foundations of Capitalism,* New York: Macmillan.

Commons, J.R. (1934) *Institutional Economics: Its Place in Political Economy*, New York: Macmillan.

Cyert, R.M. and March, J.G. (1963) *A Behavioral Theory of the Firm,* New York: Prentice Hall.

Dunning, J.H. (1958) *American Investment in British Manufacturing Industry*, London: George Allen & Unwin.

Dunning, J.H. (1973) 'The determinants of international production', *Oxford Economic Papers,* 25(3): 289–336.

Dunning, J.H. (1977) 'Trade, location of economic activity and the MNE: a search for an eclectic approach', in B. Ohlin, P.-O. Hesselborn and P.M. Wijkman (eds), *The International Allocation of Economic Activity,* London: Macmillan.

Dunning, J.H. (1979) 'Explaining changing patterns of international production: in defence of the eclectic theory', *Oxford Bulletin of Economics and Statistics,* 41(3): 269–95.

Dunning, J.H. (1981a) *International Production and the Multinational Enterprise,* London: George Allen & Unwin.

Dunning, J.H. (1981b) 'Explaining outward direct investment of developing countries: in support of the eclectic theory of international production', in K. Kumar and M.G. McLeod (eds), *Multinationals from Developing Countries,* Lexington, MA: Lexington Books.

Dunning, J.H. (1983) 'Market power of the firm and international transfer of technology', *International Journal of Industrial Organisation,* 1: 333–51.

Dunning, J.H. (1988a) 'The theory of international production', *International Trade Journal,* III(1): 21–63.

Dunning, J.H. (1988b) 'The eclectic paradigm of international production: a restatement and some possible extensions', *Journal of International Business Studies,* 19(1): 1–31.

Dunning, J.H. (1988c) *Explaining International Production,* London: Unwin Hyman.

Dunning, J.H. (1991) 'The eclectic paradigm of international production: a personal perspective', in C.N. Pitelis and R. Sugden (eds), *The Nature of the Transnational Firm,* London: Routledge.

Dunning, J.H. (1995) 'Reappraising the eclectic paradigm in an age of alliance capitalism', *Journal of International Business Studies,* 26(3): 461–91.

Dunning, J.H. and Rugman, A. (1985) 'The influence of Hymer's dissertation on the theory of foreign direct investment', *American Economic Review,* 75(2): 228–32.

Dunning, J.H., Cantwell, J.A. and Corley, T.A.B. (1986) 'The theory of international production: some historical antecedents', in P. Hertner and G. Jones (eds), *Multinationals: Theory and History,* Aldershot: Gower.

Ekelund, R.B., Jr. and Hébert, R.F. (1990) A *History of Economic Theory and Method,* New York: McGraw Hill.

Graham, E.M. (1975) 'Oligopolistic imitation and European direct investment', doctoral dissertation, Cambridge, MA: Harvard Graduate School of Business Administration.

Hébert, R.F. and Link, A.N. (1988) *The Entrepreneur: Mainstream Views and Radical Critiques*, New York: Praeger.

Helpman, E. and Krugman, P.R. (1985) *Market Structure and Foreign Trade,* Cambridge, MA: MIT Press.

Hilferding, R. (1910) *Das Finanzkapital: Eine Studie über die jüngute Entwicklung des Kapitalismus*, Vienna: Wiener Volksbuchhandlung. An English language translation is provided in T. Bottomore (ed.), (1981) *Finance Capital: A Study of the Latest Phase of Capitalist Development*, London: Routledge & Kegan Paul.

Hymer, S. (1960) 'The international operations of national firms: a study of direct foreign investment', doctoral dissertation, Cambridge, MA: Massachusetts Institute of Technology, published in Cambridge, MA: MIT Press, 1976.

Hymer, S. and Rowthorn, R. (1970) 'Multinational corporations and international oligopoly: the non-American challenge', in C.P. Kindleberger (ed.), *The International Corporation: A Symposium*, Cambridge, MA: MIT Press.

Itaki, M. (1991) 'A critical assessment of the eclectic theory of the multinational enterprise', *Journal of International Business Studies*, 22(3): 445–60.

Jenkins, R. (1984) *Transnational Corporations and Industrial Transformation in Latin America*, London: Methuen.

Jenkins, R. (1987) *Transnational Corporations and Uneven Development: The Internationalization of Capital and the Third World*, London: Methuen.

Johnson, H.G. (1958) *International Trade and Economic Growth*, London: George Allen & Unwin.

Johnson, H.G. (1970) 'The efficiency and welfare implications of the international corporation', in C.P. Kindleberger (ed.), *The International Corporation: A Symposium*, Cambridge, MA: MIT Press.

Julius, D. (1990) *Global Companies and Public Policy: The Growing Challenge of Foreign Direct Investment*, London: Pinter Publishers.

Kindleberger, C.P. (1969) *American Business Abroad: Six Lectures on Direct Investment*, New Haven, CT: Yale University Press.

Kirzner, I.M. (1978) 'The perils of regulation: a market process approach', occasional paper, Miami Law and Economics Center.

Knickerbocker, F.T. (1973) *Oligopolistic Reaction and the Multinational Enterprise*, Cambridge, MA: Harvard University Press.

Knight, F. (1921) *Risk, Uncertainty, and Profit*, Boston: Houghton Mifflin Company.

Kogut, B. (1985a) 'Designing global strategies: profiting from operational flexibility', *Sloan Management Review*, 27(1): 27–38.

Kogut, B. (1985b) 'Designing global strategies: comparative and competitive value-added chains', *Sloan Management Review*, 26(4): 15–28.

Kojima, K. (1978) *Direct Foreign Investment: A Japanese Model of Multinational Business Operations*, London: Croom Helm.

Landreth, H. (1976) *History of Economic Theory: Scope, Method and Content*, Boston: Houghton Mifflin Company.

Leontief, W.W. (1954) 'Domestic production and foreign trade: the American capital position re-examined', *Economia Internazionale*, 7(1): 3–32.

Levy, H. (1911) *Monopoly and Competition: A Study in English Industrial Organisation*, London: Macmillan. English Translation.

Liefmann, R. (1897) *Die Unternehmerverbände*, later enlarged upon in R. Liefmann (1932) *Cartels, Concerns and Trusts*, London: Methuen.

Liefmann, R. (1903) *Schutzzoll und Kartelle*, Jena: G. Fischer.

Liefmann, R. (1905) *Kartelle und Truste*, Stuttgart: E.H. Moritz.

Linder, S.B. (1961) *An Essay on Trade and Transformation*, New York: Wiley.

Marris, R. (1964) *The Economic Theory of Managerial Capitalism*, London: Macmillan.

McManus, J.C. (1972) 'The theory of the multinational firm', in G. Pacquet (ed.), *The Multinational Firm and the Nation State,* Toronto: Collier-Macmillan.

Ohlin, B., Hesselborn, P.-O. and Wijkman, P.M. (eds) (1977) *The International Allocation of Economic Activity,* London: Macmillan.

Pavitt, K. (1988) 'International patterns of technological accumulation', in N. Hood and J.E. Vahlne (eds), *Strategies in Global Competition,* London: Croom Helm.

Penrose, E. (1959) *The Theory of the Growth of the Firm,* Oxford: Basil Blackwell.

Posner, M.V. (1961) 'International trade and technical change', *Oxford Economic Papers,* 13(3): 323–41.

Ragazzi, G. (1973) 'Theories of the determinants of direct foreign investment', *IMF Staff Papers,* July, 471–98.

Rugman, A.M. (1975) 'Motives for foreign investment: the market imperfections and risk diversification hypothesis', *Journal of World Trade Law,* 9: 567–73.

Rugman, A.M. (1979) *International Diversification and the Multinational Enterprise,* Lexington, MA: Lexington Books.

Rugman, A.M. (1981) *Inside the Multinationals: The Economics of Internal Markets,* New York: Columbia University Press.

Schumpeter, J.A. (1911) *Theorie der wirtschaftlichen Entwicklung.* Translated as *The Theory of Economic Development* (1934), Cambridge, MA: Harvard University Press, 1934.

Schumpeter, J.A. (1943) *Capitalism, Socialism and Democracy,* London: George Allen & Unwin.

Simon, H.A. (1947) *Administrative Behavior,* New York: Macmillan.

Simon, H.A. (1955) 'A behavioral model of rational choice', *Quarterly Journal of Economics,* 69(1): 99–118.

Simon, H.A. (1959) 'Theories of decision-making in economics and behavioral science', *American Economic Review,* 49(3): 253–83.

Southard, F.A., Jr. (1931) *American Industry in Europe,* Boston: Houghton Mifflin Company.

Southard, F.A., Jr., Marshall, H. and Taylor, K.W. (1936) *Canadian American Industry,* New Haven: Yale University Press.

Sraffa, P. (1926) 'The laws of returns under competitive conditions', *Economic Journal.* Reprinted in American Economic Association (ed.) (1952) *Readings in Price Theory,* Chicago: R.D. Irwin, 1952.

Teece, D.J. (1980) 'Economies of scope and the scope of the enterprise', *Journal of Economic Behaviour and Organisation,* 1: 223–47.

Teece, D.J. (1981) 'The multinational enterprise: market failure and market power considerations', *Sloan Management Review,* 22(3): 3–18.

Teece, D.J. (1983) 'Technological and organisational factors in the theory of the multinational enterprise', in M.C. Casson (ed.), *The Growth of International Business,* London: George Allen & Unwin.

Teece, D.J. (1987) 'Profiting from technological innovation: implications for integration, collaboration, licensing and public policy', in D.J. Teece (ed.), *The Competitive Challenge,* Cambridge, MA: Ballinger.

Teece, D.J. (1990) 'Capturing value from technological innovation: integration, strategic partnering and licensing decisions', in F. Arcangeli, P.A. David and G. Dosi (eds), *Modern Patterns in Introducing and Adopting Innovations,* Oxford: Oxford University Press.

Tolentino, P.E. (1993) *Technological Innovation and Third World Multinationals*, London: Routledge.

Tolentino, P.E. (1999) 'Transnational rules for transnational corporations: What next?', in J. Michie and J. Grieve Smith (eds), *Global Instability: The Political Economy of World Economic Governance*, London: Routledge.

Tolentino, P.E. (2000a) 'The political economy of capitalism', mimeo.

Tolentino, P.E. (2000b) 'The conduct of industrial policy in an integrated Europe', *International Review of Applied Economics*, 14(3): 391–401.

Veblen, T. (1904) *The Theory of the Business Enterprise*, New York: Charles Scribner's Sons.

Veblen, T. (1919) *The Place of Science in Modern Civilization*, New York: B.W. Huebsh.

Vernon, R. (1966) 'International investment and international trade in the product cycle', *Quarterly Journal of Economics*, 80(2): 190–207.

Vernon, R. (1971) *Sovereignty at Bay*, Harmondsworth: Penguin.

Vernon, R. (1974) 'The location of economic activity', in J.H. Dunning (ed.), *Economic Analysis and the Multinational Enterprise*, London: George Allen & Unwin.

Williams, J.H. (1929) 'The theory of international trade reconsidered', *Economic Journal*, 39: 195–209.

Williamson, J. (1966) 'Profit, growth and sales maximization', *Economica*, 33(129): 1–6.

Williamson, O.E. (1963) 'Managerial discretion and business behavior', *American Economic Review*, 53(5): 1032–57.

Williamson, O.E. (1964) *The Economics of Discretionary Behavior: Managerial Objectives in a Theory of the Firm*, Englewood Cliffs, NJ: Prentice Hall.

Williamson, O.E. (1971) 'The vertical integration of production: market failure considerations', *American Economic Review*, 61: 112–23.

Williamson, O.E. (1975) *Markets and Hierarchies: Analysis and Antitrust Implications*, New York: Free Press.

Williamson, O.E. (1985) *The Economic Institutions of Capitalism*, New York: Free Press.

Williamson, O.E. (1986) *Economic Organization*, Brighton: Wheatsheaf Books.

8 Managerial beliefs, market contestability and dominant strategic orientation in the eclectic paradigm

Timothy M. Devinney, David F. Midgley and Sunil Venaik

Introduction

The eclectic paradigm as developed and expanded by Dunning (1977, 1979, 1988, 1995; Chapter 2) is an enduring and effective tool for understanding the factors leading to successful international expansion of the multinational corporation (MNC). Other chapters in this volume have gone into great detail about characteristics of Dunning's thinking and the reader should refer to them for a more complete overview. These chapters show both the power of the paradigm – due to its flexibility in integrating new theoretical and empirical developments – and its weaknesses – due to that same flexibility in many respects. The purpose of this chapter is to put the eclectic paradigm into a more formal structure and integrate it with recent work done by Devinney, Midgley and Veniak (2000) (hereinafter DMV) that looks at optimal strategic orientation. The reason this is an effective exercise is that the views of DMV provide a parsimonious way in which we can move Dunning's work into a more structured and more directly empirically verifiable framework that deals with many of the criticisms to which it is sometimes subject.

No paradigm, theory or framework is without criticism. Some criticisms are a matter of taste, others more substantive. The major criticisms of the eclectic paradigm as discussed in the literature are: (1) its failure to account for the role of managers, (2) its inability to handle the dynamic evolution of the MNC easily,[1] (3) an unclear specification of what can serve as measures of the major constructs in the paradigm and how those constructs are related, and (4) a limitation in dealing with the interaction between the policy environment and the firm (for a response on these, see Chapter 2).

Most of the attempts to deal with these issues have been constrained by accommodating them within the ownership, location, internalisation (OLI) structure – although modifications of that structure have been attempted, albeit at the margin. For example, entry mode choice has been looked upon as the managerial choice dimension in the structure although it is clear that in the background of the eclectic paradigm, ownership and internalisation advantages are also subject to a degree of managerial discretion. Exactly how much this is the case and to what degree such choices are changeable in short periods of time is open to argument. Dunning (1995) deals with it by creating a path-dependent structure where OLI today is related to OLI yesterday, past strategies and changes in exogenous factors, but the exact detail of how this might be accommodated in practice has yet to be resolved. Similarly, attempts to deal with the dynamic evolution of the MNC are somewhat awkward and the issue of the relationship between firm behaviour and policy choices by local governments is not wholly satisfactory. In the latter case, most work has focused on the investment development path, when, strictly speaking, the more relevant issues are those related to policy prescriptions that impact on particular firms (as shown in Chapter 4).

For us, these criticisms represent a more fundamental issue about the way in which the eclectic paradigm is formulated and how it might develop with time (we must keep in perspective the fact that theories are edifices both under construction and subject to destruction at the same time!). For example, according to the paradigm, ownership, location and internalisation advantages 'influence a firm's entry mode decision by affecting management's perception of asset power (ownership advantage), market attractiveness (location specific advantage) and costs of integration (internalisation advantage). Dunning's eclectic framework suggests that when OLI advantages are high, firms will prefer more integrated modes of entry (Brouthers *et al.* 1999: 832). What this fairly representative statement does not deal with, however, is the process by which these OLI advantages are to be integrated into a set of choices by the firm. In the case of the Brouthers *et al.* paper (which we use only as a convenient example), which builds predominantly on Agarwal and Ramaswami (1992), the assumption is that they are applied in a linear moderated way. In other words, each OLI factor affects entry mode but so, too, do interactions between the factors (although the ultimate three-way interaction is not considered). Also, since the number of OLI characteristics that they use is greater than three – size and experience, product differentiation, market potential, investment risk, contractual risk – we have to assume that they believe that the three primary pillars of the eclectic paradigm are not independent constructs.

This leads to a quandary and the dilemma is related to the endogeneity of the OLI advantages and who or what influences them. Some of this confusion is related to measurement – it is difficult to get measures of ownership, location and internalisation advantages hence any measure is a proxy and these surrogates do not always fit neatly into the theoretical constructs. However, a more important concern is that what we observe as representing the OLI advantages are themselves outcomes of the choices being made as well as choices in the past. We need to formulate a structure that allows us to separate these observed endogenous choices from the underlying exogenous influences.

Hence, although the eclectic paradigm provides great flexibility in understanding the antecedents and contributing factors to firms' choice of investment strategy the picture is not complete. We will argue that Dunning (1995; Chapter 2) was correct in recognising that the issue of exogeneity and endogeneity of components of the paradigm are critical to making it able to accommodate strategic decision-making. Equally, his view that one could exogenise components of the paradigm by separating past from contemporaneous influences is equally correct and we will make use of this thinking. However, where there has yet to be closure is how these components can be integrated into an approach that addresses not only what they are, but also how they interact to create empirically testable and managerially relevant structures that help us explain MNC strategy and structure. Only in this way will the power of the eclectic paradigm be realised and its main theoretical criticism – that it does not go beyond a convenient collection of theories – and empirical limitation – that it provides insufficient guidance as to what should be analysed – be muted.

Structuring the eclectic paradigm

According to DMV, the degree to which alternative MNC structures survive and thrive is determined by the interaction between a series of pressures and the ways in which managers react to these pressures strategically and operationally over time. The approach provides a rigorous way to develop a clearer understanding of the organisational phenomenon by separating the complex interaction between exogenous factors, firm constraints and managerial beliefs and reactions. What is useful about their approach is that they go beyond describing these factors, constraints and reactions and integrate them into an optimality-based framework, whereby distinct empirical implications can be derived and normative implications developed. In the main application of their approach, they create a more rigorous structure within which the integration-responsiveness framework of Bartlett and Ghoshal (1989) can be restated as an optimisation-based theory. However, the role of this chapter is not to just repeat that exercise but to see if similar thinking can provide a means by which the eclectic paradigm can answer some of its main criticisms.

The basis of DMV is five macro constructs that influence the choice of MNC strategic orientation:

1 The *overarching higher level pressures* associated with environmental, technological, economic and market conditions. These include the social, legal, business and governmental milieu within which commerce occurs.
2 The way these pressures serve to determine *the dimensional structure of the market*. Within the context of the eclectic paradigm the dimensional structure is represented by ownership, location and internalisation advantages.

3 The *set of strategic orientations possible* in any given macro environment. Based on the structure of the pressures in a market there will be a limit to what is possible, both physically and competitively. In the eclectic paradigm this would relate to the possible strategic investment options available to the firm, independent of whether or not the firm was cognisant that those options were available. This will be related directly to the policy choices made by governments.
4 The *influence of the firm's existing structure*. The path-dependent nature of the evolution of the firm will determine where, in any given time frame, it can operate effectively. Along with the pressures faced by the firm from the outside are the constraints the firm places on itself from its historic operational and strategic choices. This is encapsulated in what is called *the technological feasibility constraint*. In the eclectic paradigm this represents the realistic set of strategic investment options.
5 *Managerial beliefs* regarding what is best for the firm. Managers need to make choices about what is 'best' from what is 'possible' and this will be influenced by their perceptions of the nature of the pressures and what is the most advantageous for the firm. Within the eclectic paradigm this represents the managers' assessment of the OLI advantages available from any set of strategic investment options.

Hence we can think of examining the nature of the market and strategic structure of MNCs by asking a series of related questions. What does the environment look like? How might we represent the pressures from the environment in a parsimonious way? Given our parsimonious representation of the pressures being faced, what is possible (ignoring any one firm)? Examining the firm itself, there are two related questions. First, what is operationally and strategically feasible at any point in time? Second, what do the managers of the firm believe is the correct strategic orientation at any point in time? It is the interaction of all these questions, being answered by a host of related firms, that will ultimately determine the strategic orientation of the MNC and whether that orientation is survivable.

At one level DMV is a contingency-based approach where the 'optimal' orientation is determined by the fit between what is possible for a firm, what the environment allows and what the managers believe. However, what makes this approach unique is twofold: market contestability and the role of managerial beliefs. First, it is also an equilibrium approach based upon notions of market contestability. Although firms might have a specific orientation that is 'optimal' for them, it is not necessarily the case that their orientation will be sustainable in the market. Sustainability is determined based upon *market contestability*. Firms are competing continuously for both resources – which include financial, organisational and material components – and customers – who might be quite heterogeneous in terms of geographic location and tastes. A contestable structure is one in which no other structure can dominate it on these dimensions *jointly* while meeting a market profitability requirement (Baumol *et al.* 1982). More formally, what this means is that contestability is the 'equilibrium condition' imposed on the system outlined in the constructs above.

What this implies is that, in a truly global market, a frontier of different strategic positions can be sustained only if they are equally profitable in the long run.[2] Otherwise, one structure would come to dominate over time. Hence, firms faced with different histories, customers and organisational, financial and material resources will not necessarily be driven to one specific structure and no one structure may be found that dominates all others.

Second, contingency theory (Donaldson 1985) is based on the 'fit' between structure and environment. Good strategy requires, at a minimum 'fit' or 'alignment' with changing environmental conditions (Chandler 1962; Venkatraman and Prescott 1990). In simple terms, the proposition is that the fit between strategy and its context – whether it is environmental dynamism (Burns and Stalker 1961; Randolph and Dess 1984), organisational characteristics (Blau 1974), technological characteristics (Mintzberg 1979; Woodward 1965) or task attributes (Thompson 1967) – has significant and positive implications for performance. Contingency theory predicts, and to an extent empirical findings support, that business development will contribute more to company performance when aligned in harmony with environmental and organisational attributes (Henderson and Venkatraman 1992). Where DMV differs is that what matters is the fit between structure, environment and managerial beliefs about structure and environment. More formally they posit that the structure–environment fit only defines the set of feasible options for the firm at any point in time. What ultimately drives strategic orientation – that is, what the firm *actually* does – is the trade-off managers make between the various strategic dimensions. However, this is not to say that such choice is indeterminate. What forces managers to align their beliefs with what is feasible and how they change what is feasible given their beliefs is the continuing pressure to survive in markets that are being contested by competitors, and this requires efficiency and performance. In equilibrium such pressure will imply that the marginal beliefs of managers are aligned with marginal cost of making feasible changes within the current environment.[3,4]

Figure 8.1 outlines schematically how this thinking would be applied to the eclectic paradigm. What this figure covers is the nature of the composition and flow of the major points in our exposition of the eclectic paradigm. Within the figure the block arrows indicate direct linkages; the lined arrows indicate influences; and the dashed arrows indicate indirect feedback effects. An explanation of the components of the figure is given within the schematic. The empirical implications of all of the linkages – direct, influences, and indirect feedback – are identical; the distinctions we are making by using different terms are theoretical. We will explain the details of each of the components in the sections that follow. Here we give a cursory overview to keep each of the parts in perspective.

We can think of the environment of the MNC as being represented by the economic, political, geographic, social and cultural environment. This can be characterised within the eclectic paradigm and we will assume that it represents the relevant structure. Note that the point is that the structure is exogenous. We are not specifying how different strategic options sit in the environment, just how the environment is characterised. The characterisation of the environment directly affects the space of options that are available to all firms – *the set of all strategic options* – as well as the decisions of policy makers in regard to the composition of specific location advantages. The set of all strategic options can be further narrowed down based upon prior decisions of the firm – *the firm's existing asset structure* – to a more limited set of feasible options available to the firm – *technical feasibility*.[5] These will be based upon the both the environment and the historic decisions of the firm. Managers must ultimately make a choice of strategic orientation – where to locate and with what form – that encompass OLI advantages and do so by merging their beliefs about what is correct (*managerial beliefs*) with what is possible, technically, for the firm to achieve at any point in time. The environment, policy choices of governments, possible strategic orientations

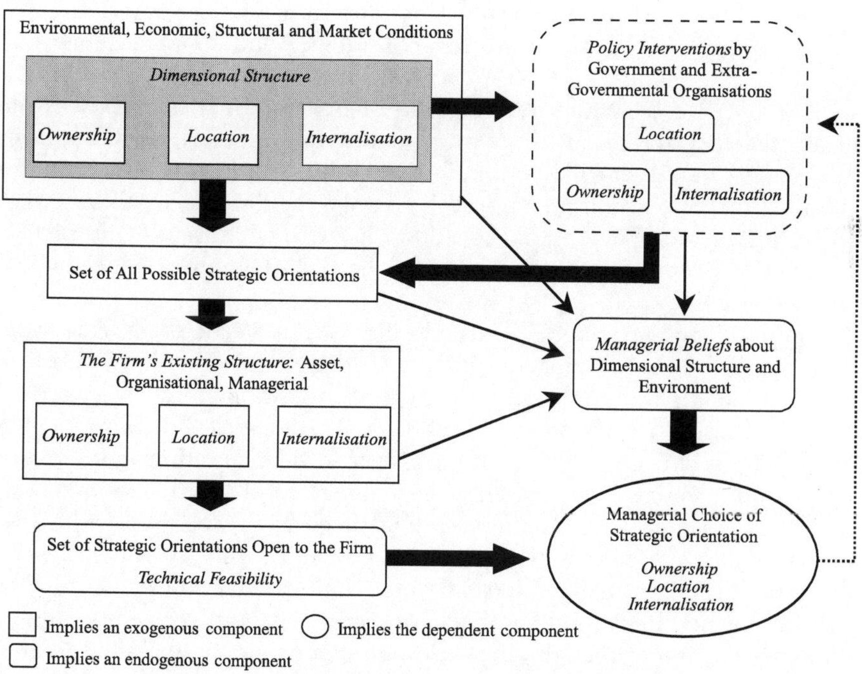

Figure 8.1 The structured approach to MNC strategic orientation.

and the firm's historic structure influence what managers believe. Note that in our model their beliefs are not influenced by what is feasible since that would confuse two contemporaneous factors in the model.

Strategic orientation – investment options and OLI orientation

Before we can proceed with expanding the eclectic paradigm we need to establish more clearly what the dependent variable is that we are interested in. Previous work on the eclectic paradigm has focused mainly on the mode and location of entry choice – sometimes separately, sometimes jointly. In some respects this is limiting, and we would suggest that the choice variable should be more broadly construed as *strategic orientation*. Strategic orientation will encompass two parts: (1) a portfolio of entry modes by locations conditional on the set of possible entry modes and locations, and (2) the components of OLI advantage that the organisation chooses or that follow directly from the entry portfolio choice. We can discuss each of these in turn.

We can characterise the entry mode by market choices of firm k at time t as simply the matrix $\mathbf{M}_{kt}^*$ with dimensionality markets (m) by entry modes (e). For simplicity, the elements of $\mathbf{M}_{kt}^*$ can be represented by a 0 or 1 to indicate whether a specific entry modes – e.g. greenfield, acquisition, joint venture, licensing, etc. – are chosen for a specific market. Because multiple mode entry is common, there is no reason to believe that one mode of entry alone is chosen, i.e. companies can operate in a market using licensing, exporting/importing, joint ventures, contracting and so on.[6] Hence, the decision to be made by a MNC is *not* 'Do I enter a market using a specific mode?' but 'Do I change my portfolio of international operations from what existed yesterday, $\mathbf{M}_{kt-1}^*$, to another orientation today, represented by $\mathbf{M}_{kt}^*$?' This may seem like a simple distinction but as we show how this thinking integrates with the approach of DMV the difference will be shown to be important. Such thinking allows us to deal with the interaction between different entry modes as well as the time-dependent nature of any firm's investment path.

The focus on entry mode choice also limits the eclectic paradigm by not always explicitly accounting for an intermediate choice made by the firm when it chooses to apply a specific ownership advantage or engage in an explicit decision regarding internalisation. The empirical assumption has been that specific entry mode choices carry with them either more or less ownership advantage or greater or lesser degrees of internalisation; however, this is not necessarily the case (and is clearly recognised within the literature at a conceptual level). For example, although a greenfield investment might imply greater internalisation than a joint venture or alliance but there is no indication that all alliance structures imply the same degree of internalisation or that all greenfield investments imply that all aspects of the investment are fully internalised (e.g. there could be a host of licensing and contractual arrangements that are clearly not internal). Hence, we can think of the firm not only making entry choices but also making joint entry–internalisation–ownership choices where these choices are all related endogenously in the sense that they are taken jointly. We will discuss the relationship between these shortly. What we can conclude now is that we can characterise the internalisation and ownership choices that firm k makes across markets at time t by two matrices, $\mathbf{O}_{kt}^*$ for the ownership choices and $\mathbf{I}_{kt}^*$ for the internalisation choices where the elements represent the degree to which specific ownership (OA) or internalisation advantage (IA) is taken up in a particular market or country.[7] Hence the dimensionality of $\mathbf{O}_{kt}^*$ is OA × m and $\mathbf{I}_{kt}^*$ is IA × m.

At this point we have left out localisation advantage since it is generally discussed as being outside the purview of the firm. Indeed this is a criticism sometimes levelled at the paradigm (see Chapter 2). We will cover this in discussion shortly, particularly the issue of co-evolution between environment and firm. However, from the standpoint of a firm's strategic orientation there is a decision the firm must make that has two components. In making a specific entry mode choice the firm has available to it, at least in theory, the localisation advantages of the markets in which it operates. This appears to be the general gist of most empirical OLI research, i.e. locate in country X get country X's advantages. But any firm may choose not to avail itself of the specific localisation advantages that are available. For example, Australia is sometimes considered to be the most Asian of Asian countries because it has a greater concentration of multiethnic residents than any other nation in the region. What this has led to is a rash of MNCs setting up call centres for the region in Australia. The advantages are clear: higher quality labour, low currency values, positive labour environment, wide-ranging linguistic skills, cheap rents (in outlying communities), and so on. However, many companies do not take advantage of all of these advantages and choose to operate call centres in other countries as well. This is mainly due to minimisation of risk of outages and the importance of specific countries in terms of volume of calls. But the main point is that firms do make location advantage choices and do so by deciding which advantages amongst those available they will apply to their business. For consistency we denote this $\mathbf{L}_{kt}^*$, which represents the degree to which specific location advantages in markets are absorbed into the firm's operations.[8] Like $\mathbf{O}_{kt}^*$ and $\mathbf{I}_{kt}^*$ the dimensionality of $\mathbf{L}_{kt}^*$ is LA (location advantages) × m.

The firm's overall *strategic orientation* at any point in time can now be encapsulated into

$$\mathbf{S}_{kt}^* = \{\mathbf{M}_{kt}^*, \mathbf{O}_{kt}^*, \mathbf{L}_{kt}^*, \mathbf{I}_{kt}^*\}.$$

The three of these choices are endogenous in the sense that they are: (1) contemporaneous in time, (2) subject to many of the same exogenous pressures, and (3) co-determined. It is the last point that is critical in the sense that specific ownership advantage choices cannot be made independent of entry mode choice, will not be made independent of location advantage choice and imply constraints on aspects of internalisation advantage.

Addressing MNC environment

As noted earlier, the eclectic paradigm provides a simple framework into which the entry mode choices are made by focusing on what it considers the three dimensions underlying the decision. What is important both conceptually and empirically is we can separate the OLI effects from those of other types of effects on the dependent variables of choice, change of strategic investment portfolio and the level of the OLI advantages utilised by the firm. This is Dunning's point when he says that the eclectic paradigm was never meant, 'to offer a full explanation of all kinds of international production' (see Chapter 2). The purpose of this chapter is not to question the degree to which the paradigm has been successful at remaining separate from other theories of entry mode choice; hence we will assume that these three dimensions are relevant. However, we will differ from Dunning in that we will argue that what determines these dimensions *should not* be context specific. It is true that the extent to which certain advantages are valuable to particular firms will vary and be subject to organisational, temporal and environmental and other sorts of contextual factors. Similarly, particular types of governments will exploit different location advantages in different institutional and market environments. However, this does not mean that the spectrum of what constitutes OLI advantage changes – just that certain agents will make choices as to relevance. This is subtle but important because it points to a confusion that appears in comparability of empirical findings and leads to the 'laundry list' of variables complaint. The paradigm needs a mechanism to reduce potential advantages into actual context specific advantages and this is what DMV's approach focuses on.

Another assumption we make, and one potentially subject to criticism, is that these dimensions are independent, *ex ante*. That is, when considering the dimensions, the factors that underlie them meet a criterion of discriminant validity. This does not imply that when managers make entry mode decisions they do not act as if combinations of the ownership, location and internalisation are important – indeed, this is why our specification of $\mathbf{S}_{kt}^{*}$ is central – but that the factors underlying these constructs are theoretically independent in measurement.[9] Although this assumption is not critical it simplifies much of our discussion and we will deal with the implication of easing this restriction at the end of the chapter.

The option space and feasibility constraint

Given the structure that we have imposed we can think of firms existing in an environment where a large set of possible investment modes, investment paths and location choices are theoretically available. All this does is define the space over which all firms competing can be thought of operating. This can expand and contract only to the extent that the environment changes to widen or constrain it. For example, the set of possible options can be restricted by ownership restrictions (such as exist in China or Vietnam), the opening of investment opportunities (should a country like North Korea decide to allow foreign investment), or the closing down of investment opportunities to specific groups (such as US restrictions on investments in Iran or Cuba). What this allows is for policy prescriptions to be entered into the model at the appropriate point.

It should also be clear that the path-dependent nature of a firm's strategic, organisational and physical structures makes it unlikely, practically, that this space is open to them. Hence, the importance of what is feasible for the firm at any point in time becomes important. Operationally, we can represent a specific feasible set of options by two components, the firm's location in the space at time t−1 – we can denote this as

$$\mathbf{S}_{kt-1}{}^{*} = \{\mathbf{M}_{kt-1}{}^{*}, \mathbf{O}_{kt-1}{}^{*}, \mathbf{L}_{kt-1}{}^{*}, \mathbf{I}_{kt-1}{}^{*}\}$$

and the cost of moving from that point to any other point in the space at time t – which we can represent as

$$\mathbf{C}_{kt} = \mathbf{C}(\mathbf{S}_{kt} \,|\, \mathbf{S}_{kt-1}{}^{*}).$$ [10]

$\mathbf{C}_{kt}$ encapsulates the notion of 'remedial efficiency'. For any specific strategic orientation to fall into the MNC's feasible strategic set it must fit with the constraints of the broader environment, the firm's existing structure and the transactional characteristics of exchange that are possible. In other words, the new orientation must be achievable and the MNC must be able to 'get' to its new strategic position from its existing point of operation. For most firms, the 'best' option is rarely available to them simply because they cannot organisationally move from where they are to where they might want to be. Carson *et al.* (1999) shows that remedial efficiency is made of three components: (1) joint profitability, (2) reallocation feasibility, and (3) switchover feasibility. Joint profitability refers to the fact that any new orientation is jointly profitable to all the players (e.g. subsidiaries of the multinational, local alliance partners, and so on). Reallocation feasibility implies that a rent allocation arrangement can be made that all relevant parties are not made worse off by the new orientation (e.g. if one party is made worse off side payments can be instituted to induce them to agree to the new orientation). Switchover feasibility refers to the costs of taking down the old orientation and building the new one (e.g. the cost of closing down a plant in one country and establishing operations in another as a means of moving capacity).

Location advantages and policy prescriptions

Location advantage falls into a more general category and can be thought about as the easing or increasing of constraints on the firm. In other words, if location advantages are a combination of environmental, and hence difficult to change, characteristics such as the age distribution of a country, and policy prescriptions, such as legislation on hiring and firing workers, we need to consider it in such a light. In the case of the latter there is a strong likelihood that the specific policy prescriptions would be taken by governments and others that are based upon their expectations of what the impact of those changes will be on the location choices of firms (hence the feedback effect shown in Figure 8.1). In the case of the former influences, there is little a government can do in the short term and we will consider these fully exogenous influences. Therefore, at one level any location at any point in time can be represented as having distinctive benefits along the dimensions of the location advantage attributes and we can think of governments, unions and relevant societal decision makers as determining what these are to some degree.

However, an added complication is that ownership and internalisation advantages can fall prey to the same sort of policy influences, although there is no reason to believe that they will be subject to exogenous location features. Hence, we can think of policy-makers as affecting all three characteristics of the option space available to firms but doing so in different ways with respect to different sources of advantage. In this respect, the set of options available will be defined by a policy space that is a restriction of the possible options open to firms.

$$\mathbf{P}_{kt}^{*}=\{\mathbf{pO}_{kt-1}^{*}, \mathbf{pL}_{kt-1}^{*}, \mathbf{pI}_{kt-1}^{*}\}$$

can be thought of as the policy-mediated option space within which distinctive benefits to different firms and certain structures that the policy-makers consider to be important will exist.

The issue of the endogeneity of the policy environment is an interesting one and although we are not in a position to engage in a full-blown discussion in this chapter it is worth covering how it would fit into our thinking. Perhaps the best explanation is found in Ozawa's work (for an example, see Chapter 4). Although we would agree with their general statement that '[MNCs and governments] are *the chief drivers* of endogenous growth', the question from our perspective is the form that this would take. It is unlikely that we can say that the choices of managers today are determinants of the policy choices of governments today. Although one can certainly find examples where a specific investment choice by a company is conditional on a 'deal' with a specific government relating to things like tax concessions, grants and so on, this is not necessarily representative of endogenous choice *per se*, especially that we would need to account for in a large-scale empirical model. What is more likely true is that governments attempt to anticipate firm reactions and it is this that is most relevant for general policy orientation. However, this does not lead to endogenous choice since expectations can be characterised on the basis of exogenous factors. In addition, although we can think of MNCs and governments being joint drivers of growth we cannot necessarily see them being motivated by the same factors, hence the underlying models cannot be thought as being co-determined in any way. Therefore, from our perspective policy choices available to firms in time t, represented by $\mathbf{P}_{kt}{}^{*}$, are made based on anticipation of effects in time t but determined in time t−1, hence the components $\{\mathbf{pO}_{kt-1}{}^{*}, \mathbf{pL}_{kt-1}{}^{*}, \mathbf{pI}_{kt-1}{}^{*}\}$.

Managerial beliefs

Although the eclectic paradigm ultimately deals with strategic decisions, its 'rationalist' orientation tends to make it deterministic in its approach, hence the criticism that there is a lack of latitude for managerial discretion in the decisions it is modelling (Johanson and Vahlne 1990). This is an important criticism for both practical and theoretical reasons.

Managerial beliefs and actions occupy a prominent position in strategic thinking as they provide a means through which organisations respond and maintain alignment with shifting market, technological and socio-political environments (see, e.g., Rajagopalan and Spreitzer 1996). Numerous findings (e.g. Barr *et al.* 1992; Lant *et al.* 1992; Smith *et al.* 1991; Webb and Dawson 1991) show that managerial interpretations of organisational conditions directly influence the need for strategic change. Indeed, the basic statements found in support of the eclectic paradigm indicate that just such logic is assumed to exist; otherwise, managers would not be able to make optimal decisions regarding the right investment alternatives. However, the role of managers is downplayed and this is most evident when comparing the eclectic paradigm with the Upsalla internationalisation model (Johanson and Vahlne 1977, 1990).

Ultimately what are held to be location, ownership and internalisation advantages are the responsibility of the purview of managers. This goes beyond just saying that managers decide on the levels of the OLI advantages that their firms choose and states that managers have an explicit trade-off that they make when deciding what is conditionally optimal for their organisation at any point in time. How managers decide this is not known since no one has attempted to discern what the marginal rate of substitution is between specific types of advantages. However, theoretically it is not unreasonable to believe that managers have preferences for specific types of advantages given the conditions their firms have faced over time.

Hence we can think of characterising managers' preferences for specific structures as being represented as a function of the alternatives available in the option space.

$$\mathbf{B}_{kt}^{*}=\{\mathbf{bO}_{kt}^{*}, \mathbf{bL}_{kt}^{*}, \mathbf{bI}_{kt}^{*}\}=\mathbf{G}[\mathbf{S}_{kt-1}^{*}\,|\,\mathbf{P}_{kt}]$$

where $\mathbf{P}_{kt}$ represents the option space, and $\{\mathbf{bO}_{kt}^{*}, \mathbf{bL}_{kt}^{*}, \mathbf{bI}_{kt}^{*}\}$ represents the managers' beliefs about the specific advantages. Note that these beliefs are not market-specific. They represent the managers' view of the basic value of one type of advantage relative to another, not an assessment (that may be in error) about the level of each of the advantages in each market or available through each entry mode. Hence, although we are assuming that managers do not make errors we do allow them to have biases. These biases are characterised by the relationship between the preferences and the prior choices of the management (represented by $\mathbf{S}_{kt-1}^{*}$). Table 8.1 summarises the key components of our approach and how they are represented.

Strategic orientation and dominant structures

The final question we need to address is how to utilise these components in a way that allows us to better understand the components of the eclectic paradigm and the relationship between its structure and managerial choice. The structure we have presented can be thought of in two parts. The first part entails the characterisation of the landscape. This is done by selective reduction of the 'universe' of options through policy orientation. We have purposely left out the decision model behind this since it is less relevant to our goals. The landscape is further reduced through the mediating effects of managerial orientation toward the relevant decision variables and the ability of the organisation to institute specific options. The final piece is how this leads to the choice of a specific strategic outcome for the firm that is sustainable. This is where the notion of dominant structures comes in.

Dominant structures arise in two contexts – one is related to firm optimisation, the other to an equilibrium of competing firms. For both we need to introduce a profit function that allows us to view options in terms of their value. Let us define this as $\Pi(\bullet)$. The first question is whether any new alternative available to any one firm is better than any other alternative. Any structure $\mathbf{S}_{kt}^{*}$ will be said to dominate another structure $\mathbf{S}_{kt}'$ for firm k at time t when the following conditions are met:

(O1) Managers believe that the shift to $\mathbf{S}_{kt}^{*}$ is superior to a shift to $\mathbf{S}_{kt}'$. This implies that

$$\mathbf{G}[\mathbf{S}_{kt}^{*}\,|\,\mathbf{P}_{kt}]-\mathbf{G}[\mathbf{S}_{kt-1}^{*}\,|\,\mathbf{P}_{kt-1}]>\mathbf{G}[\mathbf{S}_{kt}'\,|\,\mathbf{P}_{kt}]-\mathbf{G}[\mathbf{S}_{kt-1}^{*}\,|\,\mathbf{P}_{kt-1}]$$

or that the marginal value of the individual components in terms of OLI advantages of $\mathbf{S}_{kt}^{*}$ is superior in the eyes of managers when compared to the existing position of the firm $\mathbf{S}_{kt-1}^{*}$.

Table 8.1 An overview of the key components of the approach

Construct	*Representation*	
Global MNC orientation	Portfolio of mode of operation by market – $\mathbf{M}_{kt}$* Ownership portfolio by market – $\mathbf{O}_{kt}$* Internalisation portfolio – $\mathbf{I}_{kt}$* Localisation portfolio – $\mathbf{L}_{kt}$*	All choices are endogenous. The dimensionality of **M*** is mode × market. The elements of **M*** are {0,1} indicating the use of a specific mode. Multiple modes are possible. The dimensionality of **O***, **L***, and **I*** are OA, OL and OI × market (m) respectively. The elements of **O*** and **I*** are cardinal orderings over [0,1] by specific advantage.
Set of all possible strategic orientations	$\mathbf{P}_{kt}^*=\{\mathbf{pO}_{kt-1}^*, \mathbf{pL}_{kt-1}^*, \mathbf{pI}_{kt-1}^*\}$	Determined by a combination of the exogenous environment moderated by the policy choices of governments, supranational trading organisations, and social decision-makers. Because policy prescription choices are not contemporaneous with the other decisions in the model, all effects are exogenous, but affected by anticipation of effects of decisions. $\mathbf{P}_{kt}$* is the environment as seen by the firm at time t but is dependent on decisions made by others in t−1.
Firm's existing structure: asset, organisational and managerial	A firm's current location – $\mathbf{S}_{kt-1}^*=\{\mathbf{M}_{kt-1}^*, \mathbf{O}_{kt-1}^*, \mathbf{L}_{kt-1}^*, \mathbf{I}_{kt-1}^*\}$	All effects are exogenous.
Managerial beliefs about environmental and dimensional structure	$\mathbf{B}_{kt}^*=\{\mathbf{bO}_{kt}^*, \mathbf{bL}_{kt}^*, \mathbf{bI}_{kt}^*\}=\mathbf{G}[\mathbf{S}_{kt-1}^* \mid \mathbf{P}_{kt}]$	$\{\mathbf{bO}_{kt}^*, \mathbf{bL}_{kt}^*, \mathbf{bI}_{kt}^*\}$ represents the managers' beliefs about the specific advantages. Managerial beliefs are exogenous in that they are, like policy prescriptions, formed independent of the choice at hand.
Set of strategic orientations open to the firm – feasibility constraint	The cost of moving from the firm's current location to any other point in the space – $\mathbf{C}_{kt}=\mathbf{C}(\mathbf{S}_{kt} \mid \mathbf{S}_{kt}^*)$.	All effects are exogenous.

(O2) Organizationally the shift to $\mathbf{S}_{kt}^*$ is superior to the shift $\mathbf{S}_{kt}'$ and that this is better than the status quo. This implies that

$$\Pi(\mathbf{S}_{kt}^*)-\mathbf{C}(\mathbf{S}_{kt}^*|\mathbf{S}_{kt-1}^*)>\Pi(\mathbf{S}_{kt}')-\mathbf{C}(\mathbf{S}_{kt}'|\mathbf{S}_{kt-1}^*)>\Pi(\mathbf{S}_{kt-1}^*|\mathbf{P}_{kt}).$$

This is a relatively simple interpretation that follows clearly from the structure we have discussed. The more important question is the issue of market contestability and dominance, and the two are related. *A strategic orientation will be said to meet the requirements of market contestability when there is no dominant structure that outperforms it within the set of options available to all firms in the market at a specific point in time.*[11] More formally $\mathbf{S}_{kt}^*$ is a dominant strategic orientation when:

(D1) Any other firm (denoted j) that chooses that orientation cannot outperform firm k when it chooses $\mathbf{S}_{kt}^*$. This implies that

$\Pi(\mathbf{S}_{kt}^*)-\mathbf{C}(\mathbf{S}_{kt}^*|\mathbf{S}_{kt-1}^*)>\Pi(\mathbf{S}_{kt}^j)-\mathbf{C}(\mathbf{S}_{kt}^j|\mathbf{S}_{jt-1}^*)$, for all j, where we denote firm j's mimicking of firm k's strategic orientation as $\mathbf{S}_{kt}^j=\{\mathbf{M}_{kt}^*, \mathbf{O}_{jt}^j, \mathbf{L}_{kt}^j, \mathbf{I}_{jt}^j\}$. $\mathbf{M}_{kt}^*$ is firm k's investment mode choice and $\{\mathbf{O}_{jt}^j, \mathbf{L}_{kt}^j, \mathbf{I}_{jt}^j\}$ is the set of OLI advantages associated by that choice for firm j.[12]

(D2) Any other firm (denoted j) that chooses a sub-component of that orientation cannot outperform firm k when it chooses $\mathbf{S}_{kt}^*$. If we denote firm j's mimicking any sub-component of firm k's strategic orientation as $\mathbf{sS}_{kt}^j=\{\mathbf{sM}_{kt}^*, \mathbf{sO}_{jt}^j, \mathbf{sL}_{kt}^j, \mathbf{sI}_{jt}^j\}$ this implies that $\Pi(\mathbf{S}_{kt}^*)-\mathbf{C}(\mathbf{S}_{kt}^*|\mathbf{S}_{kt-1}^*)>\Pi(\mathbf{sS}_{kt}^j)-\mathbf{C}(\mathbf{sS}_{kt}^j|\mathbf{S}_{jt-1}^*)$, for all j and all s.[13]

Note that this has some interesting implications. First, $\mathbf{S}_{kt}^*$ must be dominant within the firm in the sense that conditions (O1) and (O2) are met. This almost goes without saying but it does require that: (1) it is profitable, (2) it meets the requirements of remedial efficiency, and (3) managers believe it is the right orientation. Second, $\mathbf{S}_{kt}^*$ must dominate all MNC (D1) and local (D2) competition. For condition (D1) to fail implies that if someone else does exactly what I do they beat me, even when taking into account the cost they have of switching to a new structure. We can think of this as the MNC dominance condition. However, we must account for local competition as well and this is the point of (D2). If, for any subset of my market/entry mode choices (e.g. in any one market or in any one group of markets) there is a structure that outperforms my orientation, then it is dominated. This is the local dominance condition since it implies that either local competitors or MNCs with less grand investment portfolios can pick my strategy off piecemeal.

A third, but slightly different implication is related to whether firm j's dominance is applied or not. Although $\mathbf{S}_{kt}^*$ might be contestable in the sense that there is another structure that dominates it in the market (conditions (D1) and (D2) are failed), this does not imply that $\mathbf{S}_{kt}^*$ would not be chosen (by firm k) or, more importantly, that the competing focal firm (j in our parlance) would choose to apply its dominance by operating with $\mathbf{S}_{jt}^*=\mathbf{S}_{kt}^j$. In other words, firm j may find that the structure that dominates firm k is not the most profitable or the most managerially desirable orientation for itself when firm j applies conditions (O1) and (O2) to its own choices. However, the fact that $\mathbf{S}_{jt}^*=\mathbf{S}_{kt}^j$ exists as a viable option is both a competitive threat capable of use by firm j and an indication of market vulnerability facing firm i.

Finally, note that these four conditions are all inclusive in the sense that there is no need to ask the question 'how would firm k fare if entry were taken into a different set of markets or into those markets in which firm k operated but via a different mode?' Since the choice of which firm is k is arbitrary all firms are making optimal choices using (O1) and (O2). So for every firm the question of optimality and dominance is being asked. If there were another configuration that configuration would be optimal for some firm (which we could call k) and all other firms and positions (which we would call j) would be checked for dominance. Hence, there is no need for any additional conditions to ensure stability and period-by-period local equilibria.

The eclectic paradigm reframed

We began this chapter by looking at the criticisms of the eclectic paradigm and have attempted to reframe the paradigm in a way that allows for a structured reaction to these criticisms. In other words, our main point has been to show that these criticisms can be dealt with by keeping the essential character of the eclectic paradigm but by viewing it from a slightly different angle with marginally different tools. We were further motivated by our own view of the paradigm's limitations. If we have our own criticism of the eclectic paradigm it is not in what it is in the sense of Dunning's original conception, but what it has attempted to become when applied and expanded over the last 25 years. This has led to three problems. First, a mixture of levels of analysis, in particular macro country-level FDI flows versus micro firm-level FDI choices. Second, concerns about unending endogenous feedback where network structures, government policy and competitive reaction depend on company choices and company choices on competitive reaction, network structures, and government policy. Third, attempts to expand a rather narrow component of firm decision-making into areas of firm investment for which it may not be the most effective theoretical structure. Hence, by addressing these issues we hope to have not just restated the eclectic paradigm in a different language but added to its richness. To conclude, we turn to some of the implications of our approach.

The eclectic paradigm is a 'rationalist' approach to investment choice. However, within a generally rationalist framework it has never been formalised to an extent that allows us to ask the basic question of how one structures what is optimised and how that is sustained. The traditional approach is to discuss OLI advantages and seek a relationship between those and investment choice (type, country or both) empirically. What we have argued is that the application of OLI advantage and investment choice are endogenous outcomes that are decided jointly and hence should be structured not as OLI advantage leads to investment choice ($\{\mathbf{O}_{kt}^*, \mathbf{L}_{kt}^*, \mathbf{I}_{kt}^*\} => \mathbf{M}_{kt}^*$) but that the firm's decision variable is the joint OLI, investment option ($\{\mathbf{O}_{kt}^*, \mathbf{L}_{kt}^*, \mathbf{I}_{kt}^*\} <=> \mathbf{M}_{kt}^*$). This subtle but important implication indicates that most empirical eclectic paradigm research has been focusing on the final endogenous relationship in the chain exemplified by Figure 8.1, and not with the entire process as it should be modelled.

A second related issue is that of what determines which strategies and firms survive in the market. Most work somewhat vaguely discusses competition between MNCs and between MNCs and local firms but how this competition plays out is never clearly specified. In our case, the notion of dominant strategic orientation allows for a more complete understanding of sustainability by comparing adjacent structures. This gets around the concerns of Madhok and Phene (Chapter 5) as to who it is the MNC competes with. In our case, everyone competes with everyone else (whether this is known or not) since that determines the frontier over which specific structures are considered to be sustainable. The fact that firms may consciously (i.e. for rational reasons) or unconsciously (i.e. based on managerial bias or lack of information) choose not to institute a specific orientation does not imply that that decision is costless in an opportunity cost sense since it presents an option to someone than can be executed. In addition, there is no concern in our formulation as to whether competition is over profits or for resources. As long as that competition creates a dominant strategic orientation it is *per se* good.

Perhaps the most general and, indeed, all-encompassing criticism of the eclectic paradigm is that it lacks dynamic character and fails to deal with the role of managers in deciding strategy. This is best embodied in the work of Johanson and Vahlne (see Johanson and Vahlne 1977) and their comparison between the two paradigms (Johanson and Vahlne 1990). As our approach shows, this viewpoint loses its validity when one structures the eclectic paradigm slightly differently. As we have demonstrated, the eclectic paradigm can deal with circumstances where managers operating in different firms use different assessments of what is correct. This can be based on how they form conjectures as well as knowledge biases. All that is required is that such viewpoints be put up to scrutiny in a contestable environment. In Johanson and Vahlne's work there is no such rigour leading to a cycle of endogeneity that is difficult to disentangle.

A fourth point about our structure is that it allows for a specific government (or quasi-governmental) policy model to be added without huge concerns. Hence, one can think of extending this approach to question how specific governments might engender FDI or other development into their region. And this can be based on any structure one wishes – political science, economics, sociology, etc. According to our framework, government policy influence occurs at two levels. First, it can influence the policy space. Second, it can work to alter the beliefs of managers. In the former case, governments can move to open up the feasible space available to firms but this needs to be done with a viewpoint as to which firms are relevant and how the changes envisioned allow for adjustment in dominant strategic orientations. If the expansion of the policy space does not allow for relaxation of the 'technical feasibility' constraint to the dominant firms – which is determined by these firms' own internal constraints – then there is likely to be little effect. Similarly, it is possible that the technical feasibility constraint can be relaxed for some firms but that it has no effect on which orientations are dominant. In this case, there will also be no impact. Finally, with respect to the latter issue – managerial beliefs – a necessary requirement is that policy changes should change the viewpoint of firms' management as to which advantages (and which components of which advantages) are most relevant. Again, it is not sufficient to just loosen the technical feasibility constraint: managerial beliefs must be affected as well. The reverse is also true.

Finally, our approach has quite specific empirical implications. First, it implies that the relevant dependent variables include both entry mode choice and OLI advantages as an endogenously linked package. Research to date has only really examined this component of the equation but our discussion indicates that this is only one important part of the puzzle. Second, it implies that the mode of entry is best represented as a portfolio representing entry mode(s) by markets rather than singular and independent entry mode choices. This follows from the fact that firms receive a competitive advantage not from a singular choice to enter a market or not but in how that next entry fits with the complex mix of its prior entry choices. Third, the nature of market entry analysis is best represented as a frontier not as a linear regression, implying that data envelope analysis or stochastic frontier analysis is a more relevant approach than linear regression. This follows from the fact that there can be quite large variance in strategic orientations but similar overall effects on performance. Linear regression assumes that the average tendency matters. Frontier analysis assumes that it is the maximum of the dependent variable that matters. Fourth, it is difficult to determine the source of the heterogeneity in strategic orientation without independent assessments of the firm's current operations and the costs of changing those operations (which determine its feasibility constraint) as well as the beliefs of managers. This is an onerous task given that no research to date has dealt with how managers trade off specific strategic advantages let alone the dimensions of the eclectic paradigm. Finally, different policy environments need to be characterised and their influence on the beliefs of managers and the landscape facing firms need to be built into the model. All of these issues will require a reassessment as to how we examine market entry choices empirically.

Conclusion

The eclectic paradigm has been an enduring approach to our understanding of the market entry strategies of firms. It has allowed us to look at these choices in a simple yet powerful manner, which is consistent with a large body of both neoclassical and institutional economic thinking. What we have attempted to do in this chapter is to use a different perspective by which we might restructure the eclectic paradigm to take into account the criticisms levelled at it. Our approach does not require any radical re-evaluation of the paradigm and hence can stand with the large body of supporting research that already exists. Yet our take on the eclectic paradigm is somewhat different in that we emphasised the importance of managerial beliefs, remedial efficiency and contestability as the ultimate drivers of which market entry choices are ultimately made and which strategic orientations have a chance of surviving. In this sense, we have indeed expanded on the paradigm, if only in a small way.

Notes

1 Note that Dunning (Chapter 2) merges our criticisms 1 and 2. This is due to the fact that we view the issue of strategy in two ways, the discretion of the managers and the role they play and the evolution of the firm with time. Each will be discussed shortly.
2 Market contestability is generally considered in a competitive context where the operative profit constraint is that average economic profits are zero; that is, once taking into account a normal rate of return for risk. In our case, we will leave this open and simply argue that firms attempt to maximise profits and have the minimum requirement of zero profits.
3 One area of simplification is that by considering all possible options, we are assuming that managers have complete information about what is possible. This is clearly not the case but a simplification both for tractability and reduction in the number of factors that must be considered in the model. Otherwise, an additional component of the model would have to deal with how managers know about what the possible options are in the market.
4 Theoretically, this can be stated as saying that the marginal beliefs of managers between any set of dimensions (e.g. between more or less ownership advantage and less or more internalisation) are equated with the organisations' physical marginal rate of substitution that is possible between these dimensions (e.g. the degree to which it could substitute ownership advantage and internalisation advantage).
5 We use the term 'technical' to mean 'structural and organisational'. Hence, it is a broader term than just what is technical from a operational point of view.
6 This form is used for simplicity but can be complicated. For example, one can think of adding a third dimension to indicate different operational levels of the value chain, such as would be the case when a greenfield investment is used for production, importing for components, and a joint venture for distribution and

contracting for retailing and service. The main point is that **M** is a characterisation of the entry mode choices.

7 Again we will make some expositional simplifications. Rather than deal with the two distinct types of ownership advantages, we will keep them together in a single construct. There is no loss of generality by doing this. Second, in both the case of **O** and **I** we can identify the level of ownership or internalisation advantage along a spectrum [0, 1]; from no use (0) to full use (1). Both **O** and **I** are represented by a matrix of advantages (OA and IA) by markets (m).

8 Without specifying what actual location advantages are we can consider the elements of **L**, identifying the extent to which a specific advantage is absorbed; which is represented along a spectrum [0, 1] indicating no use (0) to full use (1).

9 This is quite a complex issue that appears, for the most part, to be ignored by most empirical literature in international business and strategy. For example, let's assume that all managers make decisions only in environments where OLI advantages are all high or they are all low. Also, assume that OLI advantages are latent, meaning that they can only be revealed through actions – i.e. decisions. Even though, theoretically, we would expect each dimension to be independent we have insufficient variance in the decisions to get a range in the independent constructs that allows for a full examination of what would happen when for example ownership advantage is low and location advantage high. Hence, our analysis would indicate that rather than three dimensions with independent OLI constructs we are seeing one dimension with the three constructs being heavily correlated. For a discussion of this, see Venaik *et al*. (2003).

10 We are assuming, without loss of generality, that this cost function is the same for all firms.

11 We have used a more restrictive notion than is necessary by imposing the 'at time t' constraint.

12 Note that the OLI advantages of firm j and firm k will differ when different investment portfolios are made. We should consider $\{\mathbf{O}_{jt}^{j}, \mathbf{L}_{kt}^{j}, \mathbf{I}_{jt}^{j}\}$ as the optimal OLI advantages associated with $\mathbf{M}_{kt}$* when it is chosen by firm j.

13 **s** can be considered a transposed vector of 0s and 1s that reduces **S*** to a single market or narrower group of markets.

References

Agarwal, S. and Ramaswami, S.N. (1992) 'Choice of foreign market entry mode: impact of ownership, location and internalization factors', *Journal of International Business Studies*, 23: 1–27.

Barr, P.S., Stimpert, J.L. and Huff, A.S. (1992) 'Cognitive change, strategic action and organizational renewal', *Strategic Management Journal*, 13: 15–36.

Bartlett, C., and Ghoshal, S. (1989) *Managing Across Borders: The Transnational Solution,* Boston: Harvard Business School Press.

Baumol, W.J., Panzar, J.C. and Willig, R.D. (1982) *Contestable Markets and the Theory of Industry Structure*, New York: Harcourt Brace Jovanovich.

Blau, P.M. (1974) *On the Nature of Organizations*, New York: Wiley.

Brouthers, L.E., Brouthers, K.D. and Werner, S. (1999) 'Is Dunning's eclectic framework descriptive or normative?' *Journal of International Business Studies*, 30: 831–44.

Burns, T. and Stalker, G.M. (1961) *The Management of Innovation,* London: Tavistock Publications.

Carson, S., Devinney, T., Dowling, G. and John, G. (1999) 'Understanding institutional designs within marketing value systems', *Journal of Marketing*, 63: 115–30.

Chandler, A.D. (1962) *Strategy and Structure: Chapters in the History of the Industrial Enterprise*, Boston: MIT Press.

Devinney, T.M., Midgley, D.F. and Venaik, S. (2000) 'The optimal performance of the global firm: formalizing and extending the integration-responsiveness framework', *Organization Science*, 11: 674–95.

Donaldson, L.D. (1985) *In Defence of Organization Theory: A Reply to the Critics*, New York: Cambridge University Press.

Dunning, J.H. (1977) 'Trade, location of economic activity and the MNE: a search for an eclectic approach', in B. Ohlin, P.O. Hesselborn, and P.M. Wijkman (eds), *The International Allocation of Economic Activity*, London: Macmillan.

Dunning, J.H. (1979) 'Explaining changing patterns of international production: in defence of the eclectic theory', *Oxford Bulletin of Economics and Statistics*, 3: 461–91.

Dunning, J.H. (1988) *Explaining International Production*, London: Unwin Hyman.

Dunning, J.H. (1995) 'Reappraising the eclectic paradigm in an age of alliance capitalism', *Journal of International Business Studies*, 3: 461–91.

Henderson, J.C. and Venkatraman, N. (1992) 'Strategic alignment: a model for organizational transformation through information technology', in T.A. Kochan and M. Useem (eds), *Transforming Organizations*, New York: Oxford University Press.

Johanson, J. and Vahlne, J.-E. (1977) 'The internationalisation process of the firm – a model of knowledge development and increasing foreign market commitment', *Journal of International Business Studies*, 8: 12–24.

Johanson, J. and Vahlne, J.-E. (1990) 'The mechanism of internationalisation', *Management International Review*, 7: 12–24.

Lant, T.K., Milliken, F.J. and Batra, B. (1992) 'The role of managerial learning and interpretation in strategic persistence and reorientation', *Strategic Management Journal*, 13: 585–608.

Mintzberg, H. (1979) *The Structuring of Organizations: A Synthesis of the Research*, Englewood Cliffs, NJ: Prentice-Hall.

Rajagopalan, N. and Spreitzer, G.M. (1996) 'Toward a theory of strategic change: a multi-lens perspective and integrative framework', *Academy of Management Review*, 22: 48–79.

Randolph, W.A. and Dess, G.G. (1984) 'The congruence perspective of organization design: a conceptual model and multivariate research approach', *Academy of Management Review*, 9: 114–27.

Smith, C., Child, J. and Rowlinson, M. (1991) *Reshaping Work: The Cadbury Experience*, Cambridge: Cambridge University Press.

Thompson, J.D. (1967) *Organizations in Action: Social Science Bases of Administrative Theory*, New York: McGraw-Hill.

Venaik, S., Midgley, D.F. and Devinney, T.M. (2003) 'A new perspective on the integration-responsiveness pressures confronting multinational firms', *Management International Review*, in press.

Venkatraman, N. and Prescott, J.E. (1990) 'Environment-strategy co-alignment: an empirical test of its performance implications', *Strategic Management Journal*, 11: 1–24.

Webb, J. and Dawson, P. (1991) 'Measure for measure: strategic change in an electronic instruments corporation', *Journal of Management Studies*, 28: 191–206.

Woodward, J. (1965) *Industrial Organization: Theory and Practice*, New York: Oxford University Press.

9 The micro-mechanics of foreign operations' performance

An analysis based on the OLI framework

Gabriel R.G. Benito and Sverre Tomassen

Introduction

During the last decades in international business research, an impressive stock of knowledge has been accumulated regarding many aspects of MNCs' structures and behaviours, including issues such as why firms go abroad, why they choose particular locations for certain activities, how interdependencies between activities in various locations are organised, and which methods of foreign operations they select for conducting those activities. All in all, that research has been crucial in providing a better understanding of the international firm.

Business research deals to a large extent with performance issues: what is performance, how and why does it vary, and what are the main drivers of performance? Such issues are given substantial underlying emphasis in practically every major theoretical tradition in the field of international business, but our knowledge of them remains sketchy and unsystematic. Theoretical frameworks in international business typically assume that by making given *ex ante* choices regarding, say, the location, scope and organisation of foreign activities, firms increase the probability of obtaining certain *ex post* performance levels (see, for example, Anderson and Gatignon 1986). In other words, performance is supposedly linked to the actual choices made by firms. However, when explicitly discussed, the connections between actions and performance are rarely fully worked out. Sometimes performance issues never really enter into the core of a study, be it theoretical or empirical; they remain implicit or as part of the general backdrop.

Perhaps because international business researchers have tended to focus on decisions regarding structural aspects of firms' activities abroad, such as where activities are located and how they are organised, the issues related to outcomes and subsequent adaptation and change have received less attention. The implicit assumptions are that the initial choices provide the basic set-up within which satisfactory solutions to problems that may arise within the relevant time horizon can be worked out (or even be prevented in the first place), or that the competitive and institutional environments are so static that adaptation never really becomes a major concern. This helps explain the preoccupation in many earlier studies with the structural aspects of international business: there is a widespread belief that structural choices largely solve the adaptation problems associated with, and consequently the performance of, foreign operations (Benito *et al.* 1999). Hence, while there is an abundance of studies of choices between different entry modes, different foreign operations methods, different locations, etc., there is a corresponding lack of studies looking into actual performance achieved *across* and *within* the structural choices made by firms. Our assertion is that performance can be expected to differ substantially even within a given generic category of operations, for example activities undertaken in wholly owned subsidiaries (in contrast to, say, joint ventures or contractual arrangements of various kinds).

This chapter deals with determinants of performance and the variation between firms regarding the performance of their operations abroad. There are several reasons for variation in performance, ranging from the fairly obvious, for instance that some initial decisions may perhaps not have been the most appropriate, and that environmental contingencies may have changed in substantial ways, to ones that may be less easy to predict, identify and unravel because they are the result of human behaviour in complex dynamic situations. Human behaviour, whether proactive or reactive, explorative or adaptive, involves a significant element of choice which in turn necessarily leads to variation in outcomes and hence performance. Albeit arguably difficult to predict with great accuracy, such behaviour is shaped by the institutional, economic and social contexts within which it takes place. In international business, contexts are usefully described as consisting of ownership, location and internalisation issues as set out in the OLI framework (Dunning 1981, 1988, 2000).

The OLI framework not only draws on but also provides a very constructive synthesis of various central contributions to a theory of the multinational enterprise. In many respects, it comprises the resource-based approach (O factor), the product life cycle model (L factor), economies of scale and scope (L and I factors), and internalisation and transaction cost approaches (I factor). Performance issues are central in all these approaches, albeit in different ways. Taken together they provide the basic building blocks for understanding how international business settings work. The aim of this chapter is to extend the OLI framework by explicitly bringing in the issue of performance. Specifically, we sketch out a framework of the micro-mechanics of performance in international business, which includes elements from the resource-based approach, production and location economics, and transaction cost theory.

The performance of international business activities

Linking performance of international business activities to the OLI framework can be problematic in many respects, the first and foremost challenge being the ambiguity of the 'performance' term. Many, and often divergent views, can be found in both theoretical and empirical works on the topic. Just looking at empirical research in international business, one can easily observe this ambiguity through the use of a variety of performance measures ranging from financial and accounting terms such as parent company stock prices, ROI, ROA and ROS (Chatterjee *et al.* 1992; Goethals and Ooghe 1997; Gómez-Mejia and Palich 1997; Shrader 2001), via subjective measurements of various critical aspects regarding international business activities (such as market share, marketing, market access and distribution) (Chowdhury 1992; Geringer and Herbert 1991; Woodcock *et al.* 1994), to using proxies, such as longevity (Barkema *et al.* 1997), divestment (Benito 1997), and survival (Hennart *et al.* 1998; Li 1995).

Partly because of the ambiguity of the concept, but also because both short-term and long-term strategic dynamics need to be understood and assessed within the MNC, some researchers propose using multiple measurements of performance (Ramaswamy 1992). The assessment of performance of international operations must sometimes be taken beyond a simple revenue minus cost type of calculation since the rationale for certain operations can be to provide, for example, a bridgehead for other international operations, to gain market access, or to challenge competitors on their own home market. However, our intentions with this chapter do not require an all-encompassing, thorough and detailed account of performance. In fact, an overly complex concept of performance would make it virtually impossible to get much headway in discussing the performance ramifications of the OLI framework. An aggregate, yet simple, view of performance can serve our purposes well at this stage. Taking the OLI framework as our starting point, we posit that at a basic level performance (P) can simply be stated as a function of OLI factors:

$$P=f\,[O, L, I] \qquad (1)$$

where O, L, and I can be conceived as vectors consisting of a range of ownership, location and internalisation advantage elements, respectively. Looking at individual firms i, equation (1) can generally be interpreted as how the specific set of advantages being used by a given firm in the set of operations it conducts in various locations leads to a certain level of performance. Of course, simply taking a snapshot of a particular OLI-configuration disregards the fact that OLI advantages seldom occur instantaneously; they take time to develop. At any given point in time t, an existing OLI-configuration could be thought of as being the stock of OLI advantages, which again reflects the outcomes of a myriad of actions and events that may have occurred a long while ago. Performance differences at the firm level can consequently to some extent be attributed to differences between firms with regards to their OLI-configuration, i.e. to varying stocks of O, L, and I advantages.

While L advantages are predominantly exogenously given, stocks of O and I advantages are constantly under threat because they can be copied or surpassed by competitors. A principal reason for differences in performance between competing firms is therefore likely to be their ability to utilise, or mobilise, their OLI configurations at any point t. Such abilities, or capabilities, are dependent on prior OLI advantages within the firm and there are obviously major differences in how the firms are capable to exploit these advantages (Amit and Schoemaker 1993; Madhok 1997). Even if firms within the same industry have more or less equal access to technological and human resources, enjoy largely the same location advantages, and have organised their international activities in basically the same manner, substantial differences in performance are nevertheless observed. Firms' unique abilities, which we here denote as α, to take advantage of their OLI stock should hence also be taken as a crucial part of their OLI configurations. For given firms, we hence have the following general performance function:

$$P_i = g\ [O_i,\ L_i,\ I_i, \alpha_i] \tag{2}$$

The degree to which the three OLI components are effectively utilised probably hinges on different abilities depending on the types of OLI advantage: for example, mobilising ownership factors may require a different set of organisational routines and processes than those required to deal with location and/or internalisation issues. Hence, α_i should be regarded as a set consisting of the various capabilities of a firm that are especially suited for utilising its O, L, and I advantages, $\{\alpha_i^O,\ \alpha_i^L,\ \alpha_i^I\}$. Function (2) can thus be re-written as,

$$P_i = g\ [O_i\ \alpha_i^O,\ L_i\ \alpha_i^L,\ I_i\ \alpha_i^I\ ] \tag{3}$$

Looking specifically at economic performance, π, at a basic level it can be defined as a compound of revenues and costs, where the cost part of the equation can be further divided into production costs and transaction costs. Hence, for a given foreign operation j the performance relationship can be written as:

$$\pi_j = \text{Revenue } (R_j) - \text{Costs [Production costs } (PC_j) + \text{Transaction costs } (TC_j)] \tag{4}$$

The theoretical perspectives that provide the building blocks of the OLI framework give important contributions, though with different points of departure, on the various elements of this equation. For example, the resource-based perspective claims that successful firms create rents due to their ownership and/or access to heterogeneous and unique resources, which can be used to develop and implement different successful and sustained strategies (Barney 1986; Rumelt 1984; Wernerfelt 1984).

Traditional economic reasoning concerning production and location issues focus on the revenues and costs of different locations and production technologies. The transaction cost approach mainly accentuates the costs of organising economic activities in various governance structures (Buckley and Casson 1976; Hennart 1982; Williamson 1975, 1985).

In the following, we first thoroughly review the three theoretical building blocks for the OLI framework in order to obtain a more detailed, albeit not all-embracing, depiction of the factors driving the performance of foreign operations. Based on the OLI framework, the components of (4) can be decomposed into the followings sets $R_j=\{R_j^O, R_j^L, R_j^I\}$, $PC_j=\{PC_j^O, PC_j^L, PC_j^I\}$, and $TC_j=\{TC_j^O, TC_j^L, TC_j^I\}$. Thereafter, we move to discussing why performance may differ across international operations and firms.

Ownership advantages and performance

According to the resource-based view (Barney 1986; Penrose 1959; Peteraf 1993; Rumelt 1984; Wernerfelt 1984), resources are the fundamental determinants of a firm's performance. The concept of resources is broad, as it has been proposed to comprise all those assets, capabilities, information and knowledge, organisational processes, competencies, firm characteristics, reputation, etc. that are controlled or owned by the firm and that improve its efficiency and effectiveness (Barney 1991; Daft 1983). Hence, the ability to generate revenues depends on the nature of firms' resources. When resources are difficult to imitate and trade and there are few substitutes, firms are in a better position to secure their revenues. In particular, resources of a tacit nature (such as technology and know-how), and developed within the firm over a long time, are of special importance because they are so difficult to transfer, re-deploy, and imitate (Dierickx and Cool 1989).

According to Peteraf (1993), it is possible under certain circumstances for a firm to create persistent above-normal rents, and these rents can be of both Ricardian type (Ricardo 1817) and of monopoly type (Bain 1956). Ricardian rents are created when superior productive assets are limited in supply, which can lead to firms having lower average costs than their competitors, and/or firms that are better able to meet customers' needs. Such resources include ownership of valuable land, patents and copyrights. Monopoly rents result from restrictions of output, which lead to higher prices. Hence, monopoly profits are certainly created out of market power rather than being obtained as a result of the firm's possession of unique resources.

Profits can also be made due to *ex ante* uncertainty and information asymmetry. If all firms had equal access to information, competition for superior resources would have reduced anticipated profits to zero. Conversely, strategic factor markets will be imperfectly competitive when different firms have different expectations about the future value of a strategic resource. However, the economic performance of firms does not depend simply on whether their strategies create imperfectly competitive product markets, but also on the cost of implementing those strategies (Barney 1986). Imperfections in a strategic factor market can result both in higher returns from pursuing particular strategies and in lower costs of implementing those strategies.

In cross-border activities, central factors that disrupt the information symmetry between firms and markets are factors such as differences in language, culture, and political systems (Welch and Luostarinen 1988). In a world where transfer of such knowledge is done at zero costs, more knowledge will always be better than less. However, in a world where knowledge accumulation has a cost, it is far from obvious that this remains true. Frequently, there will be a trade-off between the costs of acquiring new knowledge and the benefit a firm will have by holding that specific knowledge. According to Kogut and Zander (1993), the multinational firm emerges not out of the failure of markets for the trading of knowledge, but out of its superior efficiency as an organisational instrument through which the transfer of knowledge can take place across borders. The dynamic processes of exploiting existing knowledge and exploring new knowledge are therefore a necessity in the accumulation and development of capabilities for firms venturing abroad.

The development of the resource-based perspective can in many ways be regarded as a response to the somewhat narrow and simplistic views of the firm provided by conventional economic theory (Penrose 1959), including early views on firms' international activities. Attempts at explaining international production primarily as a function of macroeconomic factors, thereby indirectly assuming that the determinants of firms' performance more or less were localised outside the firm, produce caricature-like descriptions of the modern firm. Parallel (in time) to the seminal works of Edith Penrose, Hymer (1960 [1976]) sought to explain firms' internationalisation as a function of their market power. Firms increased their domestic market power by mergers and acquisitions as well as by expansion of capacity. When few competitors are left in the local market, the profits earned by a high degree of monopoly power get invested abroad in order to develop the firm's position in foreign markets (Cantwell 2000). It is worth noting that the higher profits earned in the later stages of growth are not necessary generated by higher efficiency (in many cases, the firm can be less efficient than its few remaining competitors), but by higher prices in the final product market. As pointed early on by Bain (1956), building effective entry barriers becomes, of course, a prerequisite for the sustainability of supra-normal profits.

Industrial organisation scholars have emphasised that internationalisation is not only a consequence of market power and monopoly rents earned in a local market, but also a consequence of firms' wish to strengthen their bargaining power towards trade unions and various local governments

(Cowling and Sugden 1987). First, outsourcing of activities to several and smaller subcontractors reduces the power of formerly large trade unions within the company. Second, the ability to shift between different production locations increases the bargaining power *vis-à-vis* both local government and trade unions regarding wages and conditions of work (Cantwell 2000). Hence, both the options to shift and to outsource may have effects on performance by their reduction of production costs.

The market power argument has been criticised for taking a one-sided and static view on firm behaviour, especially in the case of internationally competitive industries. Venturing into foreign markets will almost invariably expose firms to a higher degree of competition. Given that at least some competitors exist somewhere, efficiency then becomes a prerequisite for survival for a firm. Ownership advantages that strengthen the efficiency of a firm, through, for example, patents, can lower unit costs and thereby increase the profit margin. Such advantages are hence obviously of major importance for future growth and survival (Cantwell 2000). With few or no ownership advantages, its competitors will most likely conquer a firm in the long run. Monopoly rents created in one market – usually the home market – through the use of market power, can seldom be sustained when the firm internationalises its market activities.[1]

As mentioned earlier, performance-driving factors based upon ownership advantages have first and foremost been the domain of the resource-based perspective, but they are of course also reflected in industrial organisation's early preoccupation with issues concerning market power and monopolistic advantages (Hymer 1970). Important insights can also be found elsewhere, especially in transaction cost theory, which, points out that certain kinds of ownership advantages should lead to higher revenues as well as lower costs. For example, using the terminology of transaction cost theory, resources such as knowledge and reputation would be termed as specific assets, which in most, albeit perhaps not all, cases are roughly equivalent to ownership advantages.[2]

It must be emphasised that according to transaction cost theory the linkage between asset specificity and internalisation is a symbiotic one. When the degree of asset specificity is low (for example when standard technology is used), the firm experiences a production cost penalty if it chooses to carry out business activities inside the firm instead of procuring them from external suppliers. An outside supplier can serve a larger number and a wider variety of customers using the same type of technology, and thereby achieve scale, scope and learning economies more easily. Conversely, when assets are highly idiosyncratic, there are no longer any scope and/or scale incentives to externalise the transactions, and production can take place within the firm without a production cost punishment (Riordan and Williamson 1985). In addition, transaction costs will be reduced due to better control with opportunism (Williamson 1975, 1985). The rent potential created by a high degree of asset specificity can only be realised through internalisation. Hence, to give an unambiguous answer on the real sources for rents in such cases, is like answering 'which came first, the chicken or the egg?'

Some transactions cost (or internalisation) theorists argue that the notion of ownership advantages is more or less superfluous since such advantages originate from the internalisation and integration of assets (Buckley 1988; Buckley and Casson 1976; Itaki 1991). They claim that ownership advantages are economic assets whose value equal future expected supra-normal profits, and which result from the firm's organisational ability to internalise and integrate – the addition of ownership advantages is therefore 'double counting' (Buckley 1988: 182). A counter-argument is that it is not the ownership advantage that is internalised *per se*, but the market for such advantages: multinational companies internalise imperfect markets, either in order to acquire knowledge and other assets or to exploit them (Hennart 2001). Cantwell (2000: 39) points out furthermore that: 'For firms to grow relative to their competitors in final product markets, ownership advantages are necessary. The generation of ownership advantages, achieved mainly through innovations, is necessary for competitive success and indeed survival.' In other words, in order to survive it is not sufficient to internalise markets, say a market for some intermediate input. A firm without ownership advantages will simply not survive in a competitive environment. According to transaction cost reasoning, knowledge and reputation are proprietary assets that should be regarded as clear examples of ownership advantages. Dependent on whether or not it is fruitful to carry out transfers internally, proprietary assets may create rents through licensing and similar types of contracts (protected by patent systems) and/or through vertically integrated systems. The use of restricted licensing agreements in various countries gives the patent owner some discretion in exploiting separated markets, and that can in turn potentially result in monopoly rents. Due to differences in the elasticity of demand across markets prices will vary, and maximisation of royalties is hence dependent on how effectively the patent owner can prevent the licensees from invading each other's markets. In some cases, transportation costs, tariffs, and governmental regulation can prevent such invasions, but in most cases, barriers to trade are low and the only solution will be vertical integration and less competition (Hennart 2000). The main performance elements of ownership advantages are summarised in Table 9.1.

Location advantages and performance

Location advantages have always been at the core of economic approaches to internationalisation; gains from trade between nations arise from differences across various locations with regards to cost and demand characteristics, which in turn reflect inter-country variation in terms of natural resource endowments and a wide range of socially created assets. Internationalisation at the firm level can be explained in similar ways. For

Table 9.1 An inventory of the performance (R^O, PC^O, TC^O) implications of *O* advantages

Resource-based perspective	*Location and production economics*	*Transaction cost theory*
Revenue: • Superior products that are limited in output create monopoly rents through increased prices (Peteraf 1993) • Information asymmetry creates above normal returns (Barney 1986) • Tacit resources (non-imitable and non-substitutable) create rents through higher prices (Barney 1991; Dierickx and Cool 1989; Peteraf 1993) • Ability to develop and replicate knowledge that is difficult to imitate (Kogut and Zander 1993, 1992)	Revenue: • Proprietary assets linked to location advantages such as sophisticated demand (market seeking firms) (Vernon 1966) • Monopoly rents are created from market power (Hymer 1960, 1976)	Revenue: • Proprietary assets combined with internalisation create potential revenues due to monopoly and restriction of competition (Buckley 1985; Hennart 2000)
Production costs: • Superior technology in limited supply creates Ricardian rents through lower production costs (Peteraf 1993) • The ability to transfer knowledge (across borders) (Kogut and Zander 1993) • Information asymmetry leads to lower production costs (Barney 1986)	Production costs: • Proprietary assets linked to location advantages, such as labour costs and supply capabilities (Vernon 1966) • Organisation of work between different locations may reduce the costs of wages and conditions of work through stronger bargaining power (Cowling and Sugden 1987; Cantwell 2000) • Advantages such as patents can lower unit costs (Cantwell 2000) • Scale economies lowering production costs (Vernon 1971)	Production costs: • When asset specificity is high there is no production cost penalty of internalising the transaction (Riordan and Williamson 1985)
		Transaction costs: • Specific assets combined with internalisation (Williamson 1975)

example, some firms move production abroad due to increased competition (and thereby also lower profit) in their home market. The size and growth of foreign markets are then important pull factors (Buckley and Casson 1981).[3] Some firms seek immobile assets, such as labour, land and infrastructure, that particular countries can offer, being attracted by the quality, availability and/or price of the resources that they depend upon.[4]

A question that intrigued the early contributors to the theory of FDI was how a MNC could compete in a foreign market against local based companies (Hymer 1960 [1976]; Caves 1971; Horst 1972). They proposed that the possession of ownership advantages is a necessary condition for neutralising their (initial) competitive disadvantage. Firms' ownership advantages, which are partly generated by investments in knowledge and R&D and internalised through the use of subsidiaries and partly a result of large size *per se* (i.e. scale), usually reflect the market structure and resource availability in their home country, but it is when they get combined with local resources (e.g. access to inputs, lower costs, access to distribution channels) that superior competitive advantage emerges. Hence it is the combination of those two types of advantages that make it possible for the MNCs to create rents by internationalising. This was also noticed by Kindleberger (1969) who regarded the MNC as a product of monopolistic competition driving firms beyond the borders of their countries of origin.

Agmon and Lessard (1977), de Meza and van der Ploeg (1987), Rangan (1998), and Rugman (1979) have pointed out the potential benefits of operating in several locations. First, multi-location firms diversify away location-specific risks, minimising their risk exposure and smoothing out fluctuation in their revenue streams. To the extent that investors recognise and prefer such diversified companies compared to single-country based companies, this will be reflected in higher share prices and in a lower cost of capital. Second, by having alternative production sites firms gain some flexibility, which in turn gives them the opportunity to shift production according to exchange rate movements, resource availability and/or changes in input prices.

Based on a well-known typology of FDI motives, Rugman and Verbeke (2001a) work out some important location factors contributing to firm performance (or competitiveness). For resource-seeking FDI, it is of course particularly important to seek out those locations that have natural resources at the lowest real cost, although additional factors including effective institutions, proper legal frameworks and high-quality transportation infrastructure are also important. For market-seeking FDI, host country market characteristics, trade barriers, investment climate, cost factors, etc. are important (Dunning 1973). The efficiency-seeking MNCs are searching for location advantages that are complementary to their own specific advantages such as appropriate infrastructure, appropriate levels of technology development and supporting institutions. Also, plant-level scale economies are easier to achieve if MNCs have a network of units where the various units specialise on the basis of the best possible matches between the resources available in the MNC's internal network and the specific advantages of the different locations (Rugman 1990). Finally, FDI motivated by strategic asset-seeking would lead to a search for areas where research and development activities are highly developed (Dunning 1996). While companies can, in principle, access global markets for a large range of tangible assets, the intangible assets that are critical to such activities as R&D, design and core manufacturing are typically embedded in local clusters (see, for example, Porter and Sölvell 1998). Localising in such areas may also provide spill over effects to the MNC through linkages with local innovation systems (Rugman and Verbeke 2001b).

The last point is also touched upon from a competence-based approach (Cantwell 1989). Cantwell argues that when firms reach a sufficient level of technological strength, they get more eager to locate their production in those areas where their major competitors originate. Such moves offer firms access to alternative sources of complementary innovations. Also, in locating production to innovative areas in the industry, Silicon Valley being the typical example, firms may gain access to resources that give them opportunities in directions that they would have difficulty in developing in their original locations (Cantwell 2000). Such opportunities may create above-normal rents through unique product innovations, but also through more efficient production technologies.

A competitive advantage may also create rents if firms enjoy a privileged market position due to, for example, network externalities (Cool *et al.* 2002). When consumers' benefit from using a product increases with the number of other consumers using compatible products, which are created from the same technology base such as in parts of the software industry, the strategic interest for other firms to compete in the industry gets radically reduced. Firms with already large customer bases then have an advantage that is due to industry structure more than to superior technology *per se* (Katz and Shapiro 1992). Because such positions often imply a lack of competition, they can of course be used to create monopoly rents.

According to Hennart (2000), becoming a MNC, i.e. extending firm activities across borders, is dependent both on governance and location considerations. The location decision (i.e. choosing the best location) is based on those factors just mentioned regarding location and production economies; relative labour cost comparisons, transportation costs, resource availability, tariffs and non-tariffs barriers to trade, political risk, and so on (i.e. those factors that may reduce production costs). The governance decision, on the other hand, is largely driven by the potential transaction costs that occur by doing business in the local market. Well functioning legal institutions that effectively enforce various instruments established to protect property rights issues related to trademarks and patents, financial transactions, etc. as well as competitive markets, are location factors that help reduce transaction costs (Hennart 2000). Table 9.2 summarises the main performance elements of location advantages.

Internalisation advantages and performance

Transaction cost economics deals primarily with the economising consequences of aligning different types of transactions to genuinely different

Table 9.2 An inventory of the performance (R^L, PC^L, TC^L) implications of *L* advantages

Resource-based perspective	*Location and production economics*	*Transaction cost theory*
Revenue: • Access to alternative sources of complementary innovations (Cantwell 1989) • Capture of revenues due to network externalities (Cool *et al.* 2002)	Revenue: • Lower revenues due to location disadvantages (Hymer 1960, 1976) • Higher revenues due to innovations ('spill-over' from innovative areas) (Cantwell 2000; Rugman and Verbeke 2001b) • Lower tax, higher subsidies, etc. (see for example Rugman and Verbeke 2001a) • Higher revenue due to network externalities (Katz and Shapiro 1992) Production costs: • Higher costs due to location disadvantages (Hymer 1960, 1976) • Lower production costs due to technological innovations (Rugman and Verbeke 2001b) • Lower input prices (Dunning 1973) • International diversification lowers risk, increases flexibility, and leads to lower cost of capital and production costs (de Meza and van der Ploeg 1987; Rangan 1998; Rugman 1979) • Higher scale (Rugman 1990) • Synergies with local firms (Dunning 1996)	Production costs: • Lower input prices (labour, transportation, tariffs, political risk) (Hennart 2000) Transaction costs: • Lower transaction costs in locations with well functioning institutions (Hennart 2000)

governance structures, in particular the discrete structural alternatives of markets, hybrids and hierarchies. The transaction cost literature seldom puts explicit focus on the performance of the companies, but implicitly performance goes as a thread through the whole logic of the theory. According to the framework, organisational forms differ with respect to their ability to solve adaptation problems of external as well as internal kinds, in the use of incentives and control mechanisms, and with respect to transaction costs (Hennart 1982; Williamson 1985). Since a basic contention of the theory is that the most efficient solutions are the ones that minimise transaction costs in the long run, there is obviously a need for a detailed description of such costs. Earlier literature has sketched out both *ex ante* and *ex post* types of transaction costs. *Ex ante* types of transaction costs are those related to 'drafting, negotiating, and safeguarding agreements' (Williamson 1985: 20), whereas *ex post* transaction costs are principally the following:

1 Mal-adaptation costs basically arise from communication and coordination failures between contracting parties which in turn make them unable to react rapidly to changing conditions (Dahlstrom and Nygaard 1999). Adaptation problems are the order of the day when the environment is uncertain. Appropriate responses to environmental changes require prompt and correct information, but typically much of the information received from, say, a foreign unit is incomplete, or too voluminous, or too poorly formulated to provide a proper basis for decision-making regarding adequate courses of action. Mal-adaptation costs are simply the opportunity costs of ineffective and inappropriate responses.
2 Bargaining costs is a general term for expenses related to negotiations between different parties, including costs incurred as a result of the need to renegotiate due to unclear contract formulations or make changes to the contract. According to Milgrom and Roberts (1992), such costs include time spent on bargaining, resources used during bargaining, and losses that occur as a result of failure in reaching efficient agreements.
3 Monitoring costs occur when resources are used to secure the fulfilment of contractual commitments (Dahlstrom and Nygaard 1999).
4 Bonding costs occur due to the necessity of completing secure commitments. Bonding includes a variety of activities that are believed to contribute positively to increased commitments in a relationship: for example, developing personal ties between parties, developing common company cultures, building incentive systems, time spent together to solve third-party problems, and development of career possibilities within the MNC (Heide and John 1988).
5 Rindfleisch and Heide (1997) argue that the opportunity costs of productivity losses stemming from effort adjustments are also a type of

ex post transaction cost. For instance, prior to entering a foreign market a MNC faces asymmetric information regarding the true characteristics of various actors that may have a role to play in the MNC's decision to enter a market and how to enter it, for example information about potential agents and distributors, or about joint-venture partners. The adverse selection problem is probably most acute whenever MNCs contemplate acquisition as the mode of entry. Opportunity costs in the form of productivity losses are often observed in such situations: for instance, lack of skills and motivation among employees in the target firm require that time is spent on training them and that effort is put into motivating them, recruitment of new personnel may be needed, and it is likely to take time before the employees of the acquired unit are equally productive as other employees in the MNC.

Even though it is the minimisation of transaction costs that is at the centre of the stage, it must be recognised that additional elements need to be taken into account in order to get the proper picture of performance, i.e. revenue as well as production cost implications need also to be analysed. First, there is the possibility that MNCs create welfare losses by maximising profits through restriction of output of goods and services. Vertical integration can work as an instrument for creating barriers to entry and thereby monopoly profit at the expense of the costumers in the final product market (Buckley 1985). Second, vertical integration can make scale economies possible. The argument is that the cost of internalised operations will be easier to recover if large transactions are of a repetitive character. Hence, higher levels of transaction frequency provide an incentive for firms to employ hierarchical governance structures (Rindfleisch and Heide 1997; Williamson 1985).[5] Third, potential cross-border scope economies can be exploited more easily and cheaply (with regard to transaction costs) within the framework of an organisation, i.e. an MNC (Galbraith and Kay 1986).

According to the resource-based view, organisations have some key features that contribute to their performance. First, organisations are able to pursue so-called dynamic efficiency, which means that firms can create new options based on their superior technology and expand the scope of activities beyond those activities that are efficiently coordinated by the market (Moran and Ghoshal 1996). The ability to hinder market forces temporarily opens up possibilities to pursue innovative activities. Many of the activities that are associated with innovations occur within the firm, and since innovations often are faced with a weakly functioning price mechanism, missing markets and high degrees of uncertainty and ambiguity, markets are not well suited to take care of these activities even though they may create rents for the firm at later stages. Second, a feeling of shared purpose makes organisations able to create an atmosphere that shapes the values and goals of their members. That, in turn, leads to the development of trust and commitment, which is of major importance in reducing opportunism and transaction costs. Further, as already mentioned above, it is an efficient way of transferring knowledge across borders (Kogut and Zander 1993).

As already touched upon above, industrial organisation scholars argue that the use of O advantages through subsidiaries combined with a utilisation of local resources lead both to increased revenues and to reduced costs (production costs as well as transaction costs). The first is due to internalisation of competition; the latter are due to less haggling (Caves 1971; Hymer 1960 [1976]; Kindleberger 1969), the use of internal transfer prices (Hymer 1970), and the reduced market power of suppliers, including trade unions (Cantwell 2000). The implications of internalisation advantages for performance are summarised in Table 9.3.

Why does performance vary?

Stocks of OLI advantages are generated and accumulated over a period of time. However, OLI advantages (especially of the O and I kinds) are also

Table 9.3 An inventory of the performance (R^I, PC^I, TC^I) implications of *I* advantages

Resource-based perspective	*Location and production economics*	*Transaction cost theory*
Revenue: • Dynamic efficiency (Moran and Ghoshal 1996)	Revenue: • Internalisation of competition (Hymer 1960, 1976; Caves 1971; Kindleberger 1969)	Revenue: • Restriction of output and thereby higher prices (Buckley 1985)
Production costs: • Efficient way of transferring knowledge (Kogut and Zander 1993)	Production costs: • Weakened effectiveness of labour unions (Cantwell 2000) • Internal transfer prices (Hymer 1970)	Production costs: • Scope economies across borders (Galbraith and Kay 1986) • Scale economies (Rindfleisch and Heide 1997; Williamson 1985)
Transaction costs: • Trust developed through relationship building likely to be even stronger for internal relations (Moran and Ghoshal 1996)	Transaction costs: • Reduction of haggling due to bilateral monopolies (Hymer 1960, 1976)	Transaction costs: • Reduction of transaction costs (Dahlstrom and Nygaard 1999; Heide and John 1988; Hennart 2000; Rindfleisch and Heide 1997; Williamson 1985)

under constant threat of being copied and/or improved upon by competitors. Hence, for these resources to be the basis for sustained competitive advantages, four necessary conditions must be present (Peteraf 1993). First, there must be some resource heterogeneity across firms regarding their resource bundles and the capabilities that are the basis for their productive activities. This is a basic condition because if many firms in the same industry have access to the same type of valuable resources, then there is no basis for competitive advantages. Second, resources must be difficult to trade, for example because they are highly idiosyncratic, or are jointly owned and/or employed. Such imperfect mobility of resources locks resources on to specific firms. Third, there must also be *ex ante* restrictions to competition, which means that there must be a positive difference between the *ex post* net present value of the resources and the *ex ante* costs of acquiring the resources. Resources are considered valuable when the firm can exploit them, such as when new strategies are developed or implemented in such a way that the firm can be more efficient and effective. Fourth, *ex post* restrictions to competition (i.e. sustained imitation or substitution problems for rivalry firms) must also be present. A firm's ability to compete in its chosen product market is consequently attributable to qualities of the resources a firm employs to support its competitive position.

It follows from this set of conditions that variation in the performance of international operations hinges on the ownership of resources that have disparate productivity. Hence, the only way to get hold of these resources is to outwit the resource market by developing expectations about the future values of potential resources that are more precise than the competitors are able to do (Barney 1986). Since this would have to happen before the actual acquisition of a resource, it also implies that not acquiring a resource can also put the firm in a better position than its competitors (Makadok 2001).

The stock of resources (or OLI advantages) is consequently an important prerequisite for variation in performance among international ventures. However, a firm may create superior performance not only by its superior stock of OLI advantages, but also out of its distinctive competence in making use of these resources (see Chapter 8). These capabilities are closely intertwined with the prior stock of OLI advantages since they have been developed over time, in close and complex interaction with the stock of OLI advantages already accumulated, but they are also unique compared to the traditional stock of resources.

Why make this distinction? The purpose of these capabilities (i.e. α_i^O, α_i^L, α_i^I) is to improve the productivity of the stock of OLI advantages possessed by the firm. Since resources such as O and I advantages are constantly under pressure from the product and organisational innovations of competitors, it is crucial to create capabilities that are difficult to imitate. Otherwise, competition will increase through the launch of similar products and substitutes, total output will increase and monopoly rents will diminish since the demand curve has become more elastic (Barney 1991; Peteraf

1993). As a result, in order to effectively protect their rents, firms must successfully develop capabilities that are non-substitutable as well as non-imitable. In addition, since organisational capabilities often involve a closely interrelated mix of routines, tacit knowledge and organisational memory that is difficult to transfer through arm's-length trade from one organisation to another, non-tradable capabilities may create so-called appropriable quasi-rents (Klein *et al.* 1978).[6] The firm cannot capture rents by selling the capability, but the capability can be a source of sustained competitive advantage.

As pointed out earlier, the dynamic processes of exploiting existing knowledge and exploring new knowledge are crucial for the accumulation and development of capabilities for firms venturing abroad. Hence, performance differences between firms partly reflect the problems of transferring, redeploying and imitating these resources (see Chapter 5) – which is a good example of what we call α^O capabilities. A firm that is able to transfer knowledge at a lower cost than its competitors, and a firm that creates and replicates resources more quickly than the market is able to imitate them, faces, everything else being equal, lower production costs and obtains higher revenues through information asymmetry, lags and imitative-hampering frictions (Kogut and Zander 1992, 1993). Why are some firms better than others at solving this? Former experience with similar types of knowledge transfers appears to be one important factor (Welch and Luostarinen 1988). The ability to manage knowledge transfer processes is typically learned by doing, as shown, for example, in a study of Norwegian companies' entry into India (Tomassen *et al.* 1998). Some of the firms in that study had to transfer technology from Norway to be able to manufacture and compete in the local Indian market. Former experience from similar developing markets that had also involved technology transfer through turnkey contracts linked to payback agreements and licensing contracts turned out to be of utmost importance in order to succeed when entering the Indian market. However, this is not only an example of α^O type of capability, but also an example of capabilities linked to I advantages (α^I). The ability to replicate certain types of organisational structures may create operational efficiencies and reduce set-up costs. Fast food chains like McDonalds and Burger King are good examples. Products such as fries and hamburgers are not particularly sophisticated, but these companies have developed tremendous experience in rolling out their total concept through franchising contracts and wholly owned subsidiaries, creating rents through most of the value chain.

Because α^O and α^I capabilities are closely connected to various activities that take place inside the companies, they probably constitute the most likely avenues through which the distinctive nature of a firm develops and strengthens (Coombs and Metcalfe 2000). That, in turn, is of course a prerequisite for differences between firms with regard to their performance.

Since L advantages are in most cases exogenously given, the α^L parameter may at first glance seem to be of little importance. L advantages such as easy access to key markets and locations in a country, possibilities for further expansion, good access to distribution channels, labour and raw materials are nevertheless often critical aspects of international operations. How well companies succeed in obtaining a superior position with regard to L factors depends on their former experience in the country in question, visible presence in that market and/or close connection with key firms and people that know the market well. Although market presence may not in itself contribute much to performance differences since such action is quite easy to copy, the ways firms accumulate and develop knowledge about a local market, and how strong and qualified the business network is, definitely can. Pepsi's activities in India are a case in point with regard to the importance of α^L capabilities. Pepsi entered the Indian market in 1988 through a 36.89 per cent equity joint venture. It was obviously not an ideal situation for Pepsi, but hands-on experience with the market was perceived as being a prerequisite for further expansion in the market (to gain, *inter alia*, first-mover advantages such as brand recognition and loyalty, and access to major distribution channels). In the following years, Pepsi in India was successful in achieving significant brand loyalty, and its market share in the country gave Pepsi a major advantage during the 1990s *vis-à-vis* Coca-Cola, which had withdrawn from the Indian market in 1977 due to governmental restrictions towards wholly owned operations. In 1993, when wholly owned operations again were allowed, Coca-Cola decided to re-enter the Indian market (Singh 1997). However, because Pepsi's marketing activities had over the years created a superior reputation to those of local brands and Coca-Cola, market shares have not come easily for Coca-Cola. As pointed out by Rumelt (1987), reputation is a type of capability that may create rents. In addition, goodwill from government institutions was also developed due to the fact that Pepsi, by establishing a joint venture as its initial set-up, agreed upon the rules set by the Indian government. Coca-Cola did not.

The study by Tomassen *et al.* (1998) gives a good example of the importance of networking. One of the cases that were studied involved a Indo-Norwegian joint venture, and the interviewee, one of the partners in a small consumer goods company, made it clear that the critical factor for the Norwegian company's successful business in India was that he had developed a close relationship with the Indian manager in the local partner company over a long period of time. Without that personal tie, it would have been close to impossible to fulfil the company's aims in the market.

To conclude this brief account of α capabilities associated with OLI advantages, it seems that they hinge on: (1) firms' ability to accumulate knowledge from learning (by doing) and hence their experience levels, (2) the usage and diffusion of that knowledge throughout the organisation, and (3) the networks that link firms to both people and other organisations. Nevertheless, and this cannot be over-emphasised, there is a profound dependence between the resources accumulated (i.e. OLI advantages) and the capabilities developed. Firms' capabilities can only create superior performance given that the other OLI resources already are in place (Makadok 2001). Hence, firms have to acquire OLI advantages as well as develop their α_i^O, α_i^L, and α_i^I capabilities.

Performance components, decision-making and company performance

In the chapter so far we have largely treated the various components (i.e. revenue and various types of costs) of performance separately. We did so in order to highlight the specific aspects of the different components but, while focusing on one component at a time certainly makes the presentation easier to follow, in reality the different components are of course interconnected. Not taking into account how they relate to and are affected by each other implies a simplification that comes at the expense of giving a less adequate description of the real issues that decision-makers must try to resolve on a recurring basis, including finding reasonable solutions to the potential dilemmas and trade-offs they frequently face when trying to balance the quest for larger revenues while keeping cost of various kinds at bay. For example, as Baumol (1959) and others demonstrated many years ago, firms that maximise sales revenues operate at a point that is beyond the optimal scale for a firm. Also, Williamson (1985) asserts that trade-offs between production costs and transaction costs may exist. Because outside suppliers usually enjoy production cost advantages (e.g. due to larger scale, learning curve economies, etc.) it often makes sense to let them take care of a variety of activities including supplies of various components and other intermediate inputs. Market and contractual solutions can be entrusted upon to carry out the required transactions, especially as long as specific investments are not needed in order to perform the activities in question. While this is true for a wide range of relatively standardised activities and/or inputs, occasionally the degree of asset specificity, while non-trivial, may fall short of providing a sufficiently clear transaction cost incentive to perform a given activity in-house.

Although it is outside the scope of this chapter to attempt even a superficial coverage of the issues involved, a few points should be mentioned. First, companies obviously ought to keep a focus on the actual performance of their operations abroad. They must keep in mind that revenues and costs are just elements in the performance equation, not ends in them themselves, and should avoid falling into the traps of both revenue myopia and cost myopia.

Second, and a somewhat less trivial point, is that many companies have several foreign operations, and the question then arises about how to manage the performance of the resultant portfolio of activities, i.e. the total performance (TP_i) of the company. If each and every operation were totally independent, company performance would simply be the sum of the performance of all its operations, i.e. $TP_i = \Sigma\ \pi_j$. For many, if not most, companies that is not the case, and the task of figuring out how the various operations are linked to, depend on and influence each other with regard to revenues and costs becomes a daunting one.

A third issue is whether the various performance components differ in terms of their importance as perceived by decision-makers. Are they given neutral treatment in terms of emphasis and attention or, in other words, do managers generally consider the various types of revenues and costs on equal terms? Because some components are more visible, easier to measure, or simply provide more positive connotations, managers may be led to emphasise them – revenues perhaps being the prime example. Some times that situation is reinforced by incentive schemes that focus on achievements in the market place, such as the number of units sold or total sales. Whereas the revenue side is probably the favourite among managers, transaction costs are the most likely candidate for losing out in the competition for managerial attention. Such costs are hard to measure, and do not show up directly in company accounts.

It cannot be taken for granted that companies are fully capable of analysing performance issues in a rational manner. If let alone, i.e. if they operate in an environment that gives no corrective feedback, companies are likely to make mistakes repeatedly, and of course, if they really are alone in a market (i.e. there are prohibitive barriers to entry) mistakes will not be punished, which in turn would perpetuate the status quo. What this suggests is that external forces are needed to discipline the behaviour of firms, and that attention must be given to the selection mechanisms whereby firms that perform poorly are outcompeted and replaced by those that are high-performers.

Economists generally believe that market competition is possibly the most effective mechanism for selecting survivors on the basis of achieved performance. The role of the market as a selection mechanism has been especially emphasised by transaction cost scholars who argue that even if transaction costs may be hard to observe and measure, the efficiency (i.e. cost minimisation) implications of choice and design of governance structures are such that competitive processes will drive out those companies that are inefficient (Anderson 1988; Williamson 1985). Hence, even though transaction costs may generally receive little explicit attention in the sense that they are consciously taken into account in strategic decision-making, the solutions and choices that are likely to win over time are supposedly nevertheless those that, albeit implicitly, best reduce transaction costs. The problem with this line of reasoning is that while transaction costs are expected to be at their highest in market failure situations, it is above all in those same situations that competitive market processes will be most inactive as a selection mechanism.

Conclusion

The performance of firms' international activities is an issue that has attracted interest ever since the very first studies of the internationalisation of firms, but the issue remains evasive. There is no agreement among scholars about what factors drive performance, how performance considerations enter decision-making processes, or even how the concept of performance should be interpreted and measured in the first place. In this chapter we have made an attempt to advance the current state of knowledge by looking at performance through the lenses of the OLI framework. That framework, which is the undisputed leading theoretical framework in the field of international business studies, provides a fruitful synthesis of several theoretical strands in economics and business studies.

Based on a simple decomposition of the performance concept, this chapter has, primarily, given an overview of the main performance implications entrenched in the OLI framework and discussed the reasons for why performance may vary across firms. In addition, the connections between the various performance elements and the ways performance can affect and be affected by decision-making in firms has briefly been touched upon. Even though the treatment of these issues has been cursory and must be regarded as preliminary, hopefully we have provided a basis for future work in the area.

Our analysis demonstrates the usefulness of using the OLI framework as an analytical lens through which the performance of MNCs and their operations can be examined. While the research agenda arising from our analysis is unavoidably a broad one, the following questions appear especially pertinent. First, in order to investigate whether some companies consistently outperform others, and if so to better understand why, we need to separate the performance drivers that are company-specific from those that are not. The distinction made here between general OLI factors and their corresponding company-specific α parameters should be given further scrutiny. Second, the concept of performance could be further analysed. The aim should not be to arrive at some undisputed all-embracing measure of performance. Such a goal would be, we believe, both illusive and unfeasible: performance is both a lot of things and different things to different actors. There is nevertheless a need to sort out the various components of performance (as we have sought to do in this chapter), to discuss alternative interpretations of performance, and to compare different measures of performance against each other. Third, transaction costs are a mainstay of international business theory and they are evidently central to the real-life organisation and performance of foreign operations. Yet, the measurement of such costs remains rudimentary and taking them properly into account continues to be easier said than done. A major advance in the analysis of performance would call for a clearer grasp of their nature.

Notes

1 The line of reasoning presented here echoes the qualifications to simple market power explanations provided by the product life cycle approach (Vernon 1966, 1971). In later stages of the product life cycle, firms preserve their competitive position through scale economies as opposed to superior products and/or technologies, which were the basis for the above normal rents created in early stages of the product life cycle.

2 Williamson (1989) describes six types of specific assets: (1) site specificity, (2) physical asset specificity, (3) human asset specificity, (4) dedicated assets, (5) brand name capital, and (6) temporal specificity.

3 According to Vernon (1966) process and product innovations are most likely to occur at the home base since that is where demand signals are most easily interpreted and where resources are most readily available from the outset. The initial production of innovations also typically takes place at home, near the first customers, and largely irrespective of cost. Producers initially face a demand that is insensitive to price, but as the market first grows and then matures, competition increases and products become more standardised. From the initial focus on the home market, exports start gaining momentum – frequently after being initiated by unsolicited orders from abroad – but because the output sold on foreign markets is usually quite small for yet some time, export markets are often serviced from the production base at home given that existing production capacity is sufficiently large. At that point, establishing production facilities abroad would only make sense if marginal production costs at home plus transport costs exceed the average production cost in the foreign country. Later on, as total demand increases and as competition increases globally, the emphasis on lowering production costs may lead to moving production to less developed countries that have a labour-cost advantage in producing standardised products. Thus, as a product reaches the mature stages of its life cycle, it tends to be imported by countries enjoying higher standards of living and higher labour costs, i.e. the very same markets that once provided the context for the initial appearance of the product.

4 Because factors of production can often be acquired at a lower cost in the developing parts of the world, one would expect capital to flow mainly from developed economies to less developed economies. However, this has not been the case since the Second World War. The bulk of FDI has occurred between developed countries (Narula and Dunning 2000). Hymer (1960, 1976) was among the first to recognise that firm- and industry-level explanations were needed in order to capture the essence of FDI: (1) firms that transfer and exploit proprietary resources abroad face disadvantages due to language, cultural barriers and lack of knowledge about local business practices and systems, and (2) to meet such disadvantage firms must have some advantages *vis-à-vis* local firms. Such advantages – for example economies of scale, superior technological knowledge, product differentiation, access to capital, etc. – allow MNCs to invest abroad. However, because FDI also provides opportunities to close markets and increase the market power of a firm, it can work as a way of perpetuating the monopolistic role of the MNC in foreign markets.

5 According to Williamson (1979), the characteristics of the investment must also be taken into account. If the investments in question are of a non-specific character, the best solution remains to carry out transactions in the market place.

6 Such resources, which are defined by Peteraf (1993: 194) as the 'excess of an asset's value over its value to the second-highest valuing potential user or bidder for the resource', are bound to the firm and cannot be easily used by others (hence the term 'non-tradable').

References

Agmon, T. and Lessard, D.R. (1977) 'Investor recognition of corporate international diversification', *Journal of Finance*, 32(4): 1049–55.

Amit, R. and Schoemaker, P.J.H. (1993) 'Strategic assets and organizational rent', *Strategic Management Journal*, 14(1): 33–46.

Anderson, E. (1988) 'Strategic implications of Darwinian economics for selling efficiency and choice of integrated or independent sales forces', *Management Science*, 34(5): 599–618.

Anderson, E. and Gatignon, H. (1986) 'Modes of foreign entry: a transaction cost analysis and propositions', *Journal of International Business Studies*, 17(3): 1–26.

Bain, J.S. (1956) *Barriers to New Competition: Their Character and Consequences in Manufacturing Industries*, Cambridge, MA: Harvard University Press.

Barkema, H.G., Shenkar, O., Vermeulen, F. and Bell, J.H. (1997) 'Working abroad, working with others: how firms learn to operate international joint ventures', *Academy of Management Journal*, 40(2): 426–42.

Barney, J.B. (1986) 'Strategic factor markets: expectations, luck, and business strategy', *Management Science*, 32(10): 1231–41.

Barney, J.B. (1991) 'Firm resources and sustained competitive advantage', *Journal of Management*, 17(1): 99–120.

Baumol, W.J. (1959) *Business Behavior, Value and Growth*, New York: Macmillan.

Benito, G.R.G. (1997) 'Divestment of foreign production operations', *Applied Economics*, 29(10): 1365–77.

Benito, G.R.G., Pedersen, T. and Petersen, B. (1999) 'Foreign operation methods and switching costs: conceptual issues and possible effects', *Scandinavian Journal of Management*, 15(2): 213–29.

Buckley, P.J. (1985) 'The economic analysis of the multinational enterprise: Reading versus Japan?', *Hitotsubashi Journal of Economics*, 19(2): 117–24.

Buckley, P.J. (1988) 'The limits of explanation: testing the internalization theory of the multinational enterprise', *Journal of International Business Studies*, 19(2): 181–93.

Buckley, P.J. and Casson, M.C. (1976) *The Future of the Multinational Enterprise*, London: Macmillan.

Buckley, P.J. and Casson, M.C. (1981) 'The optimal timing of a foreign direct investment', *Economic Journal*, 91(361): 75–87.

Cantwell, J. (1989) *Technological Innovation and Multinational Corporations*, Oxford: Blackwell.

Cantwell, J. (2000) 'A survey of theories of international production', in C.N. Pitelis and R. Sugden (eds), *The Nature of the Transnational Firm, 2nd edition*, London: Routledge.

Caves, R.E. (1971) 'International corporations: the industrial economics of direct foreign investments', *Economica*, 38(149): 1–27.

Chatterjee, S., Lubatkin, M.H., Schweiger, D.M. and Weber, Y. (1992) 'Cultural differences and shareholder value in related mergers: linking equity and human capital', *Strategic Management Journal*, 13(5): 319–34.

Chowdhury, J. (1992) 'Performance of international joint ventures and wholly owned subsidiaries', *Management International Review*, 32(2): 115–33.

Cool, K., Costa, L.A. and Dierickx, I. (2002) 'Constructing competitive advantage', in A. Pettigrew, H. Thomas, and R. Whittington (eds), *Handbook of Strategy and Management*, London: Sage Publications.

Coombs, R. and Metcalf, J.S. (2000) 'Organizing for innovation: co-ordinating distributed innovation capabilities', in N.J. Foss and V. Mahnke (eds), *Competence, Governance, and Entrepreneurship: Advances in Economic Strategy Research*, Oxford: Oxford University Press.

Cowling, K. and Sugden, R. (1987) *Transnational Monopoly Capitalism*, Brighton, Sussex: Wheatsheaf Books.

Daft, R.L. (1983) *Organization Theory and Design*, New York: West Publishing.

Dahlstrom, R. and Nygaard, A. (1999) 'An empirical investigation of ex post transaction costs in franchized distribution channels', *Journal of Marketing Research*, 36(2): 160–70.

De Meza, D. and van der Ploeg, F. (1987) 'Production flexibility as motive for multinationality', *Journal of Industrial Economics*, 35(3): 343–51.

Dierickx, I. and Cool, K. (1989) 'Asset stock accumulation and sustainability of competitive advantage', *Management Science*, 35(12): 1504–11.

Dunning, J.H. (1973) 'The determinants of international production', *Oxford Economic Papers*, 25(3): 289–336.

Dunning, J.H. (1981) *International Production and the Multinational Enterprise*, London: George Allen & Unwin.

Dunning, J.H. (1988) *Explaining International Production*, London: Unwin Hyman.

Dunning, J.H. (1996) 'The geographical sources of competitiveness of firms: The results of a new survey', *Transnational Corporations*, 5(3): 1–30.

Dunning, J.H. (2000) 'The eclectic paradigm as an envelope for economic and business theories of MNE activity', *International Business Review*, 9(2): 163–190.

Galbraith, C.S. and Kay, N.M. (1986) 'Towards a theory of multinational enterprise', *Journal of Economic Behavior and Organization*, 7(1): 3–19.

Geringer, J.M. and Herbert, L. (1991) 'Measuring performance of international joint ventures', *Journal of International Business Studies*, 22(2): 249–63.

Goethals, J. and Ooghe, H. (1997) 'The performance of foreign and national takeovers in Belgium', *European Business Review*, 97(1): 24–37.

Gómez-Mejia, L.R. and Palich, L.E. (1997) 'Cultural diversity and the performance of multinational firms', *Journal of International Business Studies*, 28(2): 309–35.

Heide, J.B. and John, G. (1988) 'The role of dependence balancing in safeguarding transaction-specific assets in conventional channels', *Journal of Marketing*, 52(1): 20–35.

Hennart, J.-F. (1982) *A Theory of Multinational Enterprise*, Ann Arbor: University of Michigan Press.

Hennart, J.-F. (2000) 'Transaction cost theory and the multinational enterprise', in C.N. Pitelis and R. Sugden (eds), *The Nature of the Transnational Firm, 2nd edition*, London: Routledge.

Hennart, J.-F. (2001) 'Theories of the multinational enterprise', in A.M. Rugman and T.L. Brewer (eds), *The Oxford Handbook of International Business*, New York: Oxford University Press.

Hennart, J.-F., Kim, D.-J. and Zeng, M. (1998) 'The impact of joint venture status on the longevity of Japanese stakes in U.S. manufacturing affiliates', *Organization Science*, 9(3): 382–95.

Horst, T. (1972) 'Firm and industry determinants of the decision to invest abroad: an empirical study', *Review of Economics and Statistics*, 54(3): 258–66.

Hymer, S.H. (1960 [1976]) *The International Operations of National Firms: A Study of Direct Foreign Investment*, Cambridge, MA: MIT Press (originally presented as the author's thesis, Massachusetts Institute of Technology, 1960).

Hymer, S.H. (1970) 'The efficiency (contradictions) of multinational corporations', *American Economic Review*, 60(2): 441–8.

Itaki, M. (1991) 'A critical assessment of the eclectic theory of the multinational enterprise', *Journal of International Business Studies*, 22(3): 445–60.

Katz, M.L. and Shapiro, C. (1992) 'Product introduction with network externalities', *Journal of Industrial Economics*, 40(1): 55–83.

Kindleberger, C.P. (1969) *American Business Abroad: Six Lectures on Direct Investment*, New Haven: Yale University Press.

Klein, B., Crawford, R.G. and Alchian, A.A. (1978) 'Vertical integration, appropriable rents, and the competitive contracting process', *Journal of Law and Economics*, 21(2): 297–326.

Kogut, B. and Zander, U. (1992) 'Knowledge of the firm: combinative capabilities, and the replication of technology', *Organization Science*, 3(3): 383–97.

Kogut, B. and Zander, U. (1993) 'Knowledge of the firm and the evolutionary theory of the multinational corporation', *Journal of International Business Studies*, 24(4): 625–45.

Li, J. (1995) 'Foreign entry and survival: effects of strategic choices on performance in international markets', *Strategic Management Journal*, 16(5): 333–51.

Madhok, A. (1997) 'Cost, value and foreign market entry mode: the transaction and the firm', *Strategic Management Journal*, 18(1): 39–61.

Makadok, R. (2001) 'Toward a synthesis of the resource-based and dynamic-capability views of rent creation', *Strategic Management Journal*, 22(5): 387–401.

Milgrom, P. and Roberts, J. (1992) *Economics, Organization and Management*, Englewood Cliffs, NJ: Prentice-Hall International.

Moran, P. and Ghoshal, S. (1996) 'Theories of economic organization: the case for realism and balance', *Academy of Management Review*, 21(1): 58–72.

Narula, R. and Dunning, J.H. (2000) 'Industrial development, globalisation and multinational enterprises: new realities for developing countries', *Oxford Development Studies*, 28(2): 141–67.

Penrose, E.T. (1959) *The Theory of the Growth of the Firm*, London: Basil Blackwell.

Peteraf, M.A. (1993) 'The cornerstones of competitive advantage: a resource-based view', *Strategic Management Journal*, 14(3): 179–91.

Porter, M.E. and Sölvell, Ö. (1998) 'The role of geography in the process of innovation and the sustainable competitive advantage of firms', in A.D. Chandler, P. Hagström and Ö. Sölvell (eds), *The Dynamic Firm: The Role of Technology, Strategy, Organization, and Regions*, Oxford: Oxford University Press.

Ramaswamy, K. (1992) 'Multinationality and performance: a synthesis and redirection', in S.B. Prasad and R.B. Peterson (eds), *Advances in International Comparative Management*, London: JAI Press.

Rangan, S. (1998) 'Do multinationals operate flexibly? Theory and evidence', *Journal of International Business Studies*, 29(2): 217–37.

Ricardo, D. (1817) *On the Principles of Political Economy and Taxation,* London: J. Murray.

Rindfleisch, A. and Heide, J.B. (1997) 'Transaction cost analysis: past, present, and future applications', *Journal of Marketing*, 61(4): 30–54.

Riordan, M.H. and Williamson, O.E. (1985) 'Asset specificity and economic organization', *International Journal of Industrial Organization*, 3(4): 365–78.

Rugman, A.M. (1979) *International Diversification and the Multinational Enterprise*, Lexington, MA: Lexington Books.

Rugman, A.M. (1990) *Multinationals and Canada–United States Free Trade*, Columbia, SC: University of South Carolina Press.

Rugman, A.M. and Verbeke, A. (2001a) 'Location, competitiveness, and the multinational enterprise', in A.M. Rugman and A. Verbeke (eds), *The Oxford Handbook of International Business*, Oxford: Oxford University Press.

Rugman, A.M. and Verbeke, A. (2001b) 'Subsidiary-specific advantages in multinational enterprises', *Strategic Management Journal*, 22(3): 237–50.

Rumelt, R.P. (1984) 'Towards a strategic theory of the firm', in R.B. Lamb (ed.), *Competitive Strategic Management*, Englewood Cliffs, NJ: Prentice-Hall.

Rumelt, R.P. (1987) 'Theory, strategy, and entrepreneurship', in D.J. Teece (ed.), *The Competitive Challenge: Strategies for Industrial Innovation and Renewal*, Cambridge, MA: Ballinger.

Shrader, R.C. (2001) 'Collaboration and performance in foreign markets: the case of young high-technology manufacturing firms', *Academy of Management Journal*, 44(1): 45–60.

Singh, K. (1997) *Broken Commitments: The Case of Pepsi in India*, New Delhi: Public Interest Research Centre.

Tomassen, S., Welch, L.S. and Benito, G.R.G. (1998) 'Norwegian companies in India: operation mode choice', *Asian Journal of Business & Information Systems*, 3(1): 1–19.

Vernon, R. (1966) 'International investment and international trade in the product cycle', *Quarterly Journal of Economics*, 80(2): 190–207.

Vernon, R. (1971) *Sovereignty at Bay: The Multinational Spread of U.S. Enterprises*, London: Longman.

Welch, L.S. and Luostarinen, R. (1988) 'Internationalisation: evolution of a concept' *Journal of General Management*, 14(2): 36–64.

Wernerfelt, B. (1984) 'A resource-based view of the firm', *Strategic Management Journal*, 5(2): 171–80.

Williamson, O.E. (1975) *Markets and Hierarchies: Analysis and Antitrust Implications*, New York: Free Press.

Williamson, O.E. (1979) 'Transaction-cost economics: the governance of contractual relations', *Journal of Law and Economics*, 22(2): 233–61.

Williamson, O.E. (1985) *The Economic Institutions of Capitalism*, New York: Free Press.

Williamson, O.E. (1989) 'Transaction cost economics', in R. Schmalensee and R. Willig (eds), *Handbook of Industrial Organization*, Amsterdam: North-Holland.

Woodcock, C.P., Beamish, P.W. and Makino, S. (1994) 'Ownership-based entry mode strategies and international performance', *Journal of International Business Studies*, 25(2): 253–73.

10 The eclectic paradigm and the recognition of finance-specific factors*

Lars Oxelheim, Trond Randøy and Arthur Stonehill

Introduction

The major proposition of this chapter is that a firm's financial strength affects its ability to engage in foreign direct investment (FDI). Proactive financial strategies aimed at enhancing a firm's financial strength are leading indicators of FDI. Such strategies range from pursuing globally recognised accounting and disclosure, listing and selling the firm's equity on prestigious foreign equity exchanges, to the implementation of cross-border debt/equity swaps. We argue that by having a superior proactive financial strategy a firm is able to minimise its cost of capital and maximise its availability of capital relative to its competitors, both domestic and worldwide. By lowering the discount factor for any investment (both domestic and global) the firm's likelihood of engaging in FDI would be enhanced. We suggest that finance-specific factors are not merely a by-product of a firm's competitive strength or weakness, but constitute a distinct set that deserves attention when investment patterns are to be interpreted.

Finance-specific strategies are important to all firms but are particularly important to multinational companies (MNCs) resident in small, illiquid industrial or emerging capital markets. In the existing literature, these firms are treated in the same manner as MNCs resident in the most financially developed markets such as the United States, the United Kingdom and Japan. However, MNCs from the illiquid capital markets must first pursue a proactive financial strategy to gain a competitive cost and availability of capital. This is needed to provide a level playing field on which to compete with MNCs that already enjoy a competitive cost and availability of capital because of their residence in the most developed and liquid capital markets.

Our objective is to provide a theoretical bridge between the OLI paradigm on FDI (Dunning 1977, 1988, 1995, 1998, 2000) on the one hand, and international cost of capital research on the other hand (e.g. Stulz 1996; Rajan and Zingales 1998). The OLI paradigm has predominantly focused on how FDI relates to the asset side of a non-financial firm's balance sheet. One

* This chapter is based on Oxelheim, Randøy and Stonehill (2001).

exception is Dunning's (1993: 150) discussion of a 'financial asset advantage' that concerns 'firms' superior knowledge of, and access to, foreign sources of capital'. He points out how this financial asset advantage usually is a by-product of the size, efficiency, and knowledge of multinational firms. Dunning provides a point of departure. However, he does not identify specific proactive strategies that firms can undertake in order to create such an advantage, apart from becoming more multinational. In a recent overview article of international business research, and the evolution of the eclectic paradigm in particular, Dunning (2000) does not provide the specifics that need to go into the 'financial asset advantage'.

In order to illustrate the linkage between finance-specific factors and FDI within an OLI framework, let us assume that a firm introduces a financial 'blueprint' of how to lower its cost of capital. We also assume that it takes time for the competitors to imitate the chosen strategy. When the firm is a bank then the case fits well into the traditional OLI framework (Gray and Gray 1981). This has also shown to be the case for other financial institutions or portfolio investors (Dunning and Dilyard 1999). If it is a non-financial firm, however, then the case deserves special recognition. How and to what extent can the financial 'blueprint' be seen as an ownership advantage for non-financial firms? To qualify as such an advantage it should still be unique to the firm and not easily copied. But what should be put in these requirements? As in the case of a patent in the goods market, where the period during which the firm enjoys a shelter is fixed by the legislators, the answer is discretional. Hence, as long as the financial 'blueprint' remains unique to the firm and not copied by others – be it one year or 17 years – it provides an ownership advantage that accrues to the firm in terms of a cost of capital lower than that of its domestic and foreign peers.

We emphasise the distinction between the creation of an ownership advantage and the elimination of an ownership disadvantage. For instance, an Eastern European firm making its way out of a thin and regulated market by an innovative financial strategy may have eliminated a disadvantage versus its peers in developed countries. But, more importantly in this context, it may also have created an ownership advantage versus its peers in Eastern Europe to be exploited by FDI during a period, however, limited.

We use existing literature and case-based evidence to identify finance-specific strategies that influence the FDI decision.[1] These financial strategies are grouped into two categories: (1) proactive strategies under the MNC's control; (2) reactive strategies in response to financial market imperfections. Financial reactive strategies that respond to financial market imperfections have already been partly incorporated into the existing FDI and OLI literature. Our contribution is to show that proactive financial strategies can create O, L, and I advantages. When these strategies are incorporated into the OLI framework, new insights are created from which we generate hypotheses for further testing.

The chapter is organised as follows. The following section summarises the OLI paradigm with special emphasis on its finance-specific content. The next section shows how recognition of proactive and reactive financial strategies adds to the relevance of the OLI framework. In this section, we generate eight testable hypotheses related to the influence of proactive financial strategies on the FDI decision. The final section presents a summary and conclusions.

FDI, OLI and finance-specific advantages

Early research on FDI identified the role played by research and development. Large, research-intensive firms, typically resident in the most developed capital markets, were observed to dominate FDI (Vernon 1966; Gruber *et al.* 1967; Hirsch 1967). The decision to undertake FDI was a stage in their growth strategy (Buckley and Casson 1976). These firms were able to create differentiated products that could be competitive abroad (Vernon 1966; Caves 1971; Hymer 1976). The ability of a firm to utilise its competitive advantage through FDI was said to depend on discovering product, locational or financial market imperfections that encourage FDI. Dunning (1958), Vernon (1966), Caves (1971), Hymer (1976), Buckley and Casson (1976), Dunning (1977 and 1988), Rugman (1980) and Hennart (1989) pioneered the research to find a comprehensive framework for explaining FDI. This line of research is what we define as the OLI paradigm.

The OLI paradigm attempts to answer three questions about FDI: (1) Based on present and potential *ownership* advantages, should a particular firm be involved in foreign markets? (2) Based on *location* advantages, where should the firm invest abroad? And (3), how should the firm serve foreign markets? Should it be through *internalisation* (FDI or sales subsidiaries) or through arm's-length arrangements (such as licensing or export through intermediates)? The OLI paradigm offers a framework for answering these questions. Dunning (2000: 163) points out that '. . . foreign production undertaken by MNCs is determined by the interaction of three sets of interdependent variables – which, themselves, comprise the components of three sub-paradigms'. Hence, when we mention the importance of finance-specific factors in the rest of the chapter we assume that they appear as one leg, maybe two, to be supported by the other variables of the triad.

Ownership-specific advantages and finance-specific factors

The 'O' in the OLI paradigm relates to ownership-specific (firm-specific) advantages. In deciding whether to undertake FDI, a firm must have developed firm-specific characteristics that enable it to be competitive in the home market. These characteristics must be transferable abroad and strong enough to compensate for the extra costs and barriers that confront those who try to do business abroad. Firm-specific characteristics typically possesed by successful MNCs are the proprietary knowledge or know-how incorporated in: (1) economies of scale and scope; (2) managerial and marketing expertise; (3) advanced technology stemming from a heavy emphasis on research; and (4) differentiated products.

The generic term 'financial strength' may be included under the heading of 'economies of scale and scope' with the implicit assumption that large, research-intensive MNCs are located in liquid, unsegmented markets and have unlimited access to capital, as suggested by Dunning (1993). This is of course not true for MNCs resident elsewhere or for those MNCs that do not have financial capabilities such as access to foreign equity markets, well-developed international banking relationships, globally recognised accounting and disclosure, etc.

Location-specific advantages and finance-specific factors

The 'L' in the OLI paradigm stands for location-specific advantages that skew FDI to a particular market. The theory of internationalisation and its corollary, network theory, attempt to answer the question of *where* to invest. Aharoni (1966) initiated the behavioural explanation of FDI, especially the initial decision of where to locate FDI. The behavioural approach has been extended and improved by a formalised theory of the process of internationalisation that explains not only the initial FDI decision but also the subsequent reinvestment decisions (Randøy and Dibrell 2002). Network theory explains how the MNC and its subsidiaries interact and compete for power (Johanson and Wiedersheim-Paul 1975; Johanson and Vahlne 1977; Forsgren 1989; Kogut and Zander 1993; Chen and Chen 1998; and Pedersen and Petersen 1998).

Early on, Aliber (1970) suggested that some FDI is motivated by imperfections in the foreign exchange markets. The OLI framework recognises such financial market imperfections. For instance, Dunning (1993) states explicitly that the propensity of firms to own foreign income-generating assets may be influenced by financial and exchange rate variables. Our interpretation is that this influence is a reactive strategy aimed at benefiting from these market imperfections (misalignments). However, the FDI decision can also be influenced by a forward-looking offspring of this strategy, namely, active management of foreign exchange and other financial risks (operating exposure management). In this case, the FDI provides an option in terms of production flexibility in handling exchange rate fluctuations (Kogut and Kulatilaka 1994). Here we have a borderline case related to the proactive financial strategies that we argue should be explicitly recognised in the OLI framework.

In addition to studies of the role of market imperfections, many studies have been published on the relationship between FDI and individual financial variables related to a specific location. Among the most prominent of these factors are the ones related to exchange rate and political risks. However, such factors are mostly used either as control variables or as part of 'risk diversification' as an explanation for outward FDI. Although they are relevant to the cost of capital they are generally not treated as part of the overall concept of financial strength as a key factor for FDI.

Internalisation advantages and finance-specific factors

The current theory of 'internalisation' holds that it is critical for a firm to constantly upgrade proprietary information and control the human capital that discovers it (Buckley and Casson 1976, 1998; Dunning 1977, 1981, 1988, 1993, 1995, 1997, 1998; Rugman 1980, 1981; Krugman and Venables 1994; Caves 1996; and Gray 1996). The 'I' (internalisation) factor in the OLI paradigm explains why a firm would choose to serve a foreign market through FDI rather than pursue alternative modes without ownership control of foreign activity. This is done through extensive research to develop expertise in technology, coordinated with expertise in management and marketing. Finance-specific information is ignored, as is the relevance of finance-specific expertise.

Without total ownership of its foreign subsidiaries the MNC would face higher transactional monitoring costs (or transaction cost) of its relationships with its subsidiaries. Ownership control through FDI is thus a response to market imperfections in the market for intermediates, such as knowledge, management and corporate control. One option for the MNC is to sell its expertise to foreign firms. However, the intermediate markets for such transfers are imperfect and would undervalue the potential value of the transfer. Therefore, an MNC would find it more profitable to exploit its ownership-specific advantages through FDI. In this manner, a larger value-added potential from the output of the firm's research could be retained in the MNC. The OLI paradigm does not explicitly address how finance-specific agency costs might affect FDI.

Finance-specific strategies that create FDI advantages

In this section, we will identify the finance-specific factors that deserve an explicit consideration within the OLI framework because they create advantages relevant for the FDI decision. We believe the importance of finance-specific advantages is much greater now then 20 years ago. The wave of restructuring that is currently sweeping over global business is to a large extent built upon cross-border mergers and acquisitions (M&A). Increased regionalisation in Europe, Asia and on the North and South American continents in terms of the EU, ASEAN, NAFTA and

MERCOSUR has increased competition leading to anticipation of squeezed profit margins from commercial activities. This has triggered a restructuring involving massive flows of FDI (UNCTAD 2000). We foresee that the stiffer competition on the commercial side should spill over to the financial side in terms of an increased relative importance of financial factors for the overall competitiveness of the firm. The result will further enforce the current emphasis on managing the cost of capital as a key to value creation. Performance measures such as EVA (Economic Value Added), MVA (Market Value Added) and SVA (Shareholder Value Added) focus the attention of managers on value creation for both MNCs and domestic firms. Consequently, the spotlight in creating value for the shareholders of the firm is on the combination of FDI and finance-specific factors.

What is then to be seen as an example of a financial ownership advantage? For instance, assume a US firm engages in a proactive financial strategy with the aim of lowering its cost of capital. It considers entering Venezuela by a direct investment. Its financial strategy is to prepare and implement a debt/equity swap. In this way, it can create a shelter to a low cost of capital for a longer or shorter period. The shelter is achieved by the firm being the first to execute the debt/equity swap (as was possible in the late 1980s and beginning of the 1990s), or by being the only one among its competitors that is allowed to proceed (sometimes as a result of successful lobbying) and win by handing in the lowest bid. The implementation of the financial strategy starts out with the firm buying a claim on the secondary market for claims on indebted countries. It pays, in our example, a market price of 40 per cent of the claim's face value in US dollars. Next, it presents its investment plan to the host country government and awaits an approval to proceed to a swap-auction. As a winner of that auction (with the lowest bid) it receives, in our example, the counter-value of 65 per cent of the US dollar face value in local currency. The cost of a US$ 65 million FDI has by financial creativity been reduced to US$ 40 million. The result does not violate an assumption of market efficiency. Moreover, the gap should not be seen as reflecting a market imperfection and thus a basis for a location advantage. Whereas the competitors may have found the higher cost prohibitive, our firm finds out that the favourable cost resulting from its financial creativity more than compensates for all other costs related to an FDI-entry into the Venezuelan market.

The finance-specific factors are here recognised within a framework of proactive and reactive strategies. Table 10.1 relates these strategies to the OLI paradigm. As mentioned earlier, the reactive strategies are partly covered by the current OLI framework. What is new here is the role of four proactive financial strategies as drivers of FDI. Below we will discuss these strategies one by one and generate hypotheses about how they motivate FDI within an OLI framework.

Table 10.1 Finance-specific factors and the OLI paradigm

Financial strategies	*Ownership advantages*	*Location advantages*	*Internalisation advantages*
Proactive strategies			
1. Gaining and maintaining a global cost and availability of capital:			
a. Competitive sourcing of capital globally	X	X	
b. Strategic preparatory cross-listing	X		
c. Providing accounting and disclosure transparency	X		
d. Maintaining competitive commercial and financial banking relationships	X		
e. Maintaining a competitive credit rating	X		
2. Negotiating financial subsidies and/or reduced taxation to increase free cash flow	X	X	
3. Reducing financial agency cost through FDI			X
4. Reducing operating and transaction exposure through FDI	X		
Reactive strategies			
1. Exploiting undervalued or overvalued exchange rates		X	
2. Exploiting undervalued or overvalued stock prices		X	
3. Reacting to capital control that prevent the free movement of funds		X	
4. Minimising taxation		X	X

Note: The crosses in the boxes indicate where we argue that there is a connection between FDI and finance-specific strategies.

Gaining and maintaining a global cost and availability of capital

With reference to Table 10.1, one category of proactive financial strategies by MNCs is designed to enhance financial capabilities in order to gain and maintain a competitive cost and availability of capital. For example, a company in a former Soviet republic lowers its cost of capital by breaking out from its segmented, illiquid home capital market by listing on a less prestigious Western European equity market. By this strategy, it does not achieve a global cost of capital but the strategy helps the firm to be ahead of its major peers and to be the first to enter a foreign market by FDI. For a Western European-based company, competitive can mean the achievement of a global cost of capital by some means that are unique among its peers.

This may also motivate them to enter a foreign market by FDI. Moreover, for a US firm already enjoying a global cost of capital there is still room for an advantage achieved by the use of some financial innovation or blueprint. This was exemplified in terms of an innovative use of debt/equity swaps earlier in this chapter. We list in Table 10.1 a number of important interrelated sub-elements of this strategy. All of them should be seen as instrumental to the creation of an ownership advantage.

Competitive sourcing of capital globally

The ability of a firm to minimise its cost of capital and maximise its availability should be seen as an ownership advantage. A firm that has chosen a proactive financial strategy to achieve this objective has a competitive advantage in future bidding to acquire international assets. This also provides the firm with a partial protection from being acquired by another firm that also has a competitive cost and availability of capital.

Research-intensive firms, such as those in pharmaceutical, biotechnical, telecommunications and information technology industries, have a particular problem in raising adequate debt financing. They lack sufficient collateral because of holding a high proportion of intangible assets (patents and discoveries) that have potential future value but little present liquidation value. If such firms are resident in small, illiquid industrial or emerging markets, their ability to achieve competitive financial strength is dependent on following a proactive strategy to gain and maintain access to global equity markets. For example, during the 1980s a number of Nordic MNCs in research-intensive industries utilised proactive strategies to gain a global cost and availability of capital, usually as a prelude to acquisitions or greenfield investments abroad (Oxelheim *et al.* 1998). These MNCs included such global leaders as Ericsson (telecommunications, Sweden), Hafslund Nycomed (pharmaceutical, Norway), Nokia (telecommunications, Finland), and Novo Industri (pharmaceutical and biotechnology, Denmark).

Hafslund Nycomed, a pharmaceutical firm from Norway, provides an illustrative case example. Now the firm named Nycomed Amersham has evolved into an Anglo-Norwegian company headquartered in Buckinghamshire. As part of an earlier internationalisation effort, the firm cross-listed its shares on the New York Stock Exchange (NYSE) in 1992. This was done to prepare the way for a US$ 74.7 million US equity issue. In 1994, Hafslund Nycomed was able to acquire Sterling Winthrops, its former US distributor and a significant research company in its field. The US acquisition was financed by a targeted US bond issue. The financing for the acquisition would have been unattainable without the prior US equity offer and NYSE listing. At a first glance, the development could be interpreted as if a finance-specific *ownership disadvantage* had been eliminated. However, it should rather be seen that a supportive financial ownership advantage had been created in relation to its major competitors from

Germany and Italy. Without this proactive financial strategy, Hafslund Nycomed would not have been able to undertake its foreign expansion.

Access to competitive sourcing of capital globally provides MNCs with resources beyond mere financial strength. An international equity issue can help growing firms to overcome their lack of global commercial visibility when located in a small or underdeveloped home market. A typical equity issue abroad helps the firm not only to become more visible but also more knowledgeable about particular foreign markets (Saudagaran 1988; Modén and Oxelheim 1997). Hence, firms that have recently made an international competitively priced equity issue have invested in an ownership advantage that is signalling future FDI.

H1: A firm is more likely to engage in FDI when it has access to competitively priced equity.

Strategic preparatory cross-listing

The proactive strategy of raising capital can be pursued in many ways. If the ultimate goal is to raise equity on a prestigious equity market, such as in London or New York, an MNC can follow a cautious track as a learning process. This involves cross-listing and raising equity in less prestigious markets. The alternative is to implement the strategy in one step.

The literature shows a positive effect on the market value of foreign firms listing on US stock exchanges (Sundaram and Logue 1996; Foerster and Karolyi 1999). However, several studies have shown an even more favourable effect on market value for firms following a dual strategy of simultaneously cross-listing and raising new equity on a foreign stock market (Modén and Oxelheim 1997; Miller 1999).[3]

Daimler-Benz (now DaimlerChrysler) provides case evidence. In 1993, Daimler-Benz became the first German company to list on the NYSE. On the day of the listing announcement its stock price increased by as much as 30 per cent. The Daimler-Benz cross-listing was done to prepare the way for a multibillion dollar euro-equity issue to help finance FDI, namely their new automobile manufacturing plant in Alabama, USA. The NYSE cross-listing and US equity issue also paved the way for the subsequent merger with Chrysler in 1998. The Daimler-Benz case clearly suggests that a proactive financial strategy was instrumental in making it economically advantageous to make a FDI in the US.

The high frequency of Israeli companies listed on NASDAQ (number three in this respect) as a prelude to strategic acquisitions is another case evidence of the connection between an international listing and FDI. Many of these firms, predominantly high-tech companies, have actually set a trend by making their initial public offering international.

Whether or not the strategy of cross-listing has as its ultimate aim to float new equity abroad, it should be seen as an investment in a financial ownership advantage. The advantage materialises once the company has managed to attract the attention and support of international portfolio investors. Since the cross-listing contributes to a better global recognition of the company the advantage will eventually accrue to the firm in the present value calculations of a potential foreign target company or greenfield investment. Hence, it will provide the company an opportunity to offer, in a sustainable way, a higher price than competitors when bidding for acquisitions. However, cross-listing is often limited by a firm's small size and the heavy cost and listing requirements of the most liquid stock exchanges, particularly the New York Stock Exchange. Since cross-listing is a prelude to an equity issue as a means for financing FDI, we suggest that:

H2: A firm is more likely to engage in FDI when it is cross-listed in a prestigious capital market.

Gaining access to capital at a competitive rate is very much a matter of global recognition. Interrelated to the two previously mentioned sub-elements of a proactive strategy to raise such capital are three sub-elements clearly aimed at bridging cross-border information gaps. These sub-elements are: (1) providing information to investors using globally recognised accounting and disclosure standards; (2) maintaining strong bank relationships at home and abroad; and (3) maintaining a strong credit rating.

Providing accounting and disclosure transparency

A firm attempting to gain a global cost and availability of capital needs to attract international investors to purchase and hold its securities. One of the keys to this strategy is to provide transparent accounting and disclosure of information. This implies that a firm's financial statements must meet international accounting standards, either US Generally Accepted Accounting Practices (US GAAP) or International Accounting Standards (IAS). The firm must also commit to an on-going investor relations programme to keep investors informed about the firm and allow investors personal access to the firm's key executives (Useem 1998). These activities can be expensive in terms of time and money but are an investment in an ownership advantage.

Evidence of the necessity to invest in transparency is provided by a recent comprehensive survey of German accountancy (Glaum 2000). It finds that more than 80 per cent of German firms (end of 1997, early 1998) expect to prepare financial statements in accordance with international accounting standards (either US GAAP or IAS) within year 2003. One of the major concerns of German managers is the ability to attract international investors. A case in point was the restating of Daimler-Benz's financial statement in connection with its 1993 cross-listing on the NYSE. Its restated earnings for the first half of 1993 dropped from a profit to a

$592 million loss. Nevertheless, investors responded favourably (30 per cent increase in its stock price). Increasingly, German companies have recognised that the ability to provide transparent, timely and reliable accounting information is necessary in order to achieve a global cost and availability of capital for the eventual undertaking of FDI.

H3: A firm is more likely to engage in FDI when it is following globally recognised accounting and disclosure standards.

Maintaining strong commercial and investment banking relationships

Gaining recognition by international investors is not accomplished in a vacuum. The key for a firm to attract international investors is to maintain a close relationship with the most important international banks and investment firms (Useem 1998). These global players underwrite and syndicate almost all debt and equity issues that are sold abroad. They also control and advise the leading trust funds, pension funds, mutual funds and other institutional sources of funds. The successful sales of a firm's securities both abroad and domestically depends on the marketing efforts and perceived quality of its financial sponsors.

Novo Industri (now Novo Nordisk) pioneered the efforts by a number of Nordic MNCs to escape dependence on segmented and illiquid home capital markets (Stonehill and Dullum 1982). Prominent international financial institutions orchestrated the Nordic securities issues. In Novo's case, Goldman Sachs (US) and Morgan Grenfell (UK) were the lead sponsors. Morgan Stanley (US) was the lead sponsor for a number of other Nordic equity issues abroad (Oxelheim *et al.* 1998). Having a close international banking relationship creates an ownership advantage in terms of lower agency costs stemming from reduced information asymmetries and access to a broad investor clientele (see, for example, Sharpe 1990; Boot and Thankor 2000).

H4: A firm is more likely to undertake FDI after it has retained as advisors one of the prominent international banking institutions.

Maintaining a competitive credit rating

In order to gain a competitive cost and availability of capital, a firm must gain access not only to foreign equity markets but also to international debt markets. These include the euro-syndicated loans, Eurobond, euro medium-term loan and euro commercial paper markets. Such a strategy creates a viable alternative to total dependence on the firm's commercial bank. Escaping total dependence is especially important for firms resident in bank-dominated economies such as Germany, Japan and many emerging market countries.

Gaining access to international debt markets relies on a firm's credit rating. Most, but not all, firms that hope to tap these markets ask for a credit rating from Moody's, Standard & Poor, Fitch or other credit rating services. The higher the credit rating, the lower the interest rate. Without an 'investment grade' rating (usually A or better) it is in most cases not possible to tap the international debt markets for significant amounts of capital. Proactive strategies aimed at creating an ownership advantage in terms of a strong credit rating also help to establish a firm's credibility in equity markets and with government officials in potential markets for future FDI. Anecdotal evidence in terms of a surging demand for credit rating services points to an increased use of this kind of proactive strategy.

H5: A firm is more likely to engage in FDI when it enjoys a strong investment grade credit rating.

Negotiating financial subsidies and/or reduced taxation to increase free cash flow

In addition to raising equity and debt in global capital markets, the MNC also relies on internally-generated free cash flow. With reference to Table 10.1 again, our second proactive financial strategy deals with activities aimed at negotiating financial subsidies and/or reduced taxation to increase free cash flow (Oxelheim 1993). In case of positive outcomes, these benefits will accrue in terms of ownership and locational advantages. A positive outcome materialises in a company-designed package of benefits. The low cost of capital following from the subsidy and/or reduced taxation should be seen as an ownership advantage. Being a company-designed benefit makes it disputable whether one could classify the negotiated outcome as a locational advantage as well. However, in our view, the mere propensity of the host government to negotiate signals that the benefit also may fulfil the criteria for being a locational advantage, being available to everyone on an *ex ante* basis.

Case evidence is provided by the aforementioned Daimler-Benz. It was able to negotiate favourable financial and operating concessions as a pre-requisite for locating its new automobile manufacturing plant in Alabama. Other anecdotal cases are the firm-specific tax packages negotiated by BMW in South Carolina, Disney in France and Hyundai in Oregon.

H6: A firm is more likely to engage in FDI when the firm is able to negotiate reduced taxation and/or to attract subsidies for financing it.

Reducing financial agency cost through FDI

Our third proactive financial strategy is based on *financial* agency costs and how they can be transformed into OLI advantages. MNCs can reduce financial agency costs through FDI, while exploiting an ownership advantage. For example, if an MNC enjoys a low global cost and high availability of capital, why should it let other firms with higher cost of capital provide the funding for part of its foreign presence (for example through joint venture partners or other local intermediates)? The imperfect market for financial intermediates encourages the MNC with a globally competitive cost of capital to utilise this advantage within the firm. In other words, a FDI that capitalises on the financial capabilities and resources of an MNC leads to lower and more dependable financial transaction costs, as well as lower monitoring costs than would be the case with potential licensees, joint venture partners or independent distributors.

The financial agency costs of undertaking FDI can be further reduced in cases where the host country also provides a supportive investment industry. When the aforementioned Hafslund Nycomed made an equity issue and cross-listing in the US market in 1992, the company was able to attract favourable attention from the leading pharmaceutical investment industry in New York, more so than would have been the case in other locations. We emphasise the fact that the combination of international equity issues, international listings and FDI can create lower monitoring costs for MNCs. Furthermore, a recent study by Oxelheim and Randøy (2003, forthcoming) suggests that Scandinavian firms can reduce agency costs (and thus boost firm value) by including board members from more 'demanding' corporate governance systems (the Anglo-American). Thus, financial agency costs should be added to the list of transaction costs that are featured in the internalisation theory of FDI. We hypothesise:

H7: A firm is more likely to engage in FDI when the firm is able to reduce financial monitoring costs through such an investment.

Reducing operating and transaction exposure through FDI

MNCs are subject to operating (foreign exchange, interest rate and price) exposure, as well as foreign exchange transaction exposure. The proactive strategy to reduce this exposure is a borderline case between the reactive strategies discussed below. The proactive strategies we urge should be included in the OLI framework as drivers of FDI. As was previously noted, this strategy is an offspring of the general recognition of market imperfections within the OLI framework. A successful risk management programme is assumed to create value and lower the cost of capital. It is seen as an ownership advantage and as such creating an incentive for MNCs to undertake FDI.

H8: A firm is more likely to engage in FDI when the firm has implemented a successful programme to reduce financial and operating exposure.

Table 10.1 also lists strategies that react to perceived market imperfections. Most of these reactive strategies have already been researched, although not always within an OLI framework. The following strategies deserve attention.

Foreign exchange rate market imperfections

The OLI framework has been criticised for not giving sufficient attention to exchange rate variables (Itaki 1991). We believe that misaligned exchange rates affect the *timing* of FDI, rather than being the main motivation for a particular FDI. Overall, there have been mixed empirical results on the question of whether there is a universal link between exchange movements and FDI. For example, Swenson (1994) found a positive correlation between dollar depreciation and inward FDI. On the other hand, Stevens (1998) failed to identify such an effect. Part of the reason why the results are inconclusive could be that most research on this issue has been conducted at the macro level of countries, rather than at the firm level of actual FDI decisions. Research on MNCs suggests that exchange rate volatility, rather than exchange rate misalignments, affects FDI decisions (see, for example, Ekström 1998).

Misaligned stock markets

Case evidence also suggests the possibility of a timing impact on FDI from an undervalued or overvalued national stock market. Some of the largest recent financial 'bubbles' have been the overvalued Japanese stock market in the 1980s (Ito and Iwaisako 1995), and the undervalued post-crisis emerging markets of Southeast Asia in 1998 (Johnson *et al.* 2000). Both scenarios have provided unique investment opportunities for MNCs. During the 1980s, the Japanese MNCs increased their FDI, especially acquisitions, in the United States and Europe. In 1998 and 1999, MNCs throughout the world undertook considerable FDI (acquisitions) in depressed asset markets in Asia. This 'bottom-fishing' strategy has shown how industrial investors have been able to make timely FDIs which are not easily accessible to portfolio investors. One such example is when Tektronix, Inc. bought its key supplier for $10 million in cash for a 160,000-square-foot facility and some industrial equipment in Penang, Malaysia, from CAM Advanced Technologies. The purchase price was way below the pre-crisis level, and according to Daniel Kunstler of J.P. Morgan Securities in San Francisco: 'Let's face it, asset values in Asia have come way down. That can't be all bad' (Oregonian 1998).

During the last decade, FDI through acquisitions in the United States has been motivated partly by a rising US stock market, but also by good growth prospects and a low US rate of inflation. These acquisitions occurred despite a potentially overvalued US dollar, although the strong dollar effect was partly alleviated by a heavy use of new stock issues by foreign MNCs to finance these acquisitions. As anecdotal evidence of the link between stock market misalignments and FDI, we refer to the new pattern of foreign equity issues that has emerged during the later part of the 1990s. Directed and euro-equity cash issues have to a large extent been replaced by directed non-cash issues used as payment for FDI acquisitions. It should be noted, however, that these new stock issues were very dependent on the foreign MNCs' themselves achieving and maintaining a low cost and high availability of capital.

Restrictions on the movements of funds

A finance-specific locational factor that is often neglected in the OLI framework is the need for an MNC to have unrestricted ability to transfer funds internationally. This right is necessary in order to attract international investors to buy and hold the MNCs' securities, which is the key to enjoying a global cost and availability of capital. If a country were to impose capital controls on the transfer of funds or restrict ownership of its own MNCs, it would risk losing the competitiveness of its own economy. MNCs from outside the country can react to the financial result of political risk with a variety of strategies, but MNCs resident in that country have fewer strategic options available. However, before controls are imposed resident MNCs sometimes engage in FDI in 'safe' countries, i.e. they become political risk safety seekers. One example was the transfer of some MNC headquarters from Hong Kong to 'safe' countries prior to the Chinese takeover. Hong Kong and Shanghai Banking Corporation (now HSBC) adopted this strategy.

FDI to minimise taxation

In an effort to minimise taxes, an MNC might undertake FDI in a tax haven country, or at least in a country with a relatively low tax rate. Foreign subsidiaries located there create the possibility of earning lower-taxed income or at least deferred tax income. Such income can be enhanced through transfer pricing strategies, although taxation authorities usually carefully scrutinise transfer prices. Minimising taxation should be considered not only a locational but also an internalisation advantage. The ability to earn income in the right locations is much easier to accomplish with foreign subsidiaries that are fully owned than if the MNC must deal with outside partners.

Summary and conclusions

The theme of this chapter is that the OLI paradigm should be enriched by an explicit recognition of the FDI-drivers that emerge from four finance-specific proactive strategies. We argue that firms that successfully follow proactive financial strategies are rewarded with subsequent finance-specific ownership advantages that can be most effectively exploited by undertaking FDIs. An explicit consideration of the exploitation of such advantages within the OLI framework will add explanatory value to the paradigm.

As proactive financial strategies we mention competitive sourcing of capital globally; cross-listing on foreign exchanges; providing accounting and disclosure transparency; maintaining strong commercial and investment banking relationships; maintaining a strong credit rating; negotiating financial subsidies and/or reduced taxation; reducing financial agency costs through FDI; and reducing operating and transaction exposure. Proactive finance-specific strategies are particularly important to MNCs resident in small, illiquid industrial or emerging capital markets. In addition to these proactive strategies we discuss financial reactive strategies that respond to foreign exchange rate market imperfections; misaligned stock markets; restrictions on the movement of funds; and opportunities to minimise taxation.

All the above financial strategies are presented within an OLI framework in order to explain whether they lead to ownership, location, or internalisation advantages. A survey of the existing literature on FDI and OLI suggests that finance-specific factors are recognised mostly as control variables or locational advantages to be dealt with by reactive financial strategies. Based on the OLI framework and case evidence we suggest for further testing eight hypotheses that link finance-specific factors and FDI within a proactive strategic framework.

First, we suggest that a firm is more likely to engage in FDI when it has access to competitively priced equity. Second, we hypothesise that a firm is more likely to engage in FDI when the firm is cross-listed in a prestigious capital market. Third, we suggest that a firm is more likely to engage in FDI when it is following globally recognised accounting and disclosure standards. Fourth, we hypothesise that a firm is more likely to undertake FDI after it has retained as advisors one of the prominent international banking institutions. Fifth, we suggest that a firm is more likely to engage in FDI when it enjoys a strong investment grade credit rating. Sixth, we hypothesise that a firm is more likely to engage in FDI when the firm is able to negotiate reduced taxation and/or to attract subsidies for financing it. Seventh, we suggest that a firm is more likely to engage in FDI when the firm is able to reduce financial monitoring costs through such an investment. Finally, we hypothesise that a firm is more likely to engage in FDI when this enables the firm to reduce financial and operating exposure. Successful testing of these hypotheses should lead to an enriched OLI paradigm.

Since financial capabilities and resources are important global competitive variables, the chapter suggests that small industrial and emerging market MNCs must be able to gain and retain their present access to foreign investors. The policy conclusion drawn for governments is not to overreact to economic crises by imposing controls that cut off critical access to global capital markets. Such policies would have serious repercussions for FDI by their own MNCs.

Notes

1 It should be noted that this chapter addresses the finance-specific factors only in relation to the MNC's outward FDI decision. Although domestic investment decisions are affected by the same finance-specific factors that affect outward FDI, the chapter does not cover investment in a country undertaken by firms resident in that country or as inward FDI. The chapter does not discuss the open question of whether or not FDI increases the share value of an MNC. However, it does suggest that a higher share value (lower cost of capital) could lead to increased FDI, especially in the case of making acquisitions paid for by shares of the MNC's stock.

2 Modén and Oxelheim (1997) concluded that a simultaneous cross-listing and new equity issue in the period 1981–93 created an 11 per cent cumulative abnormal return for Swedish firms during the first five days after the announcement of this dual strategy.

References

Aharoni, Y. (1966) *The Foreign Investment Decision Process,* Boston: Harvard Graduate School of Business Administration.

Aliber, R. (1970) 'A theory of foreign direct investment', in C.P. Kindleberger (ed.), *The International Corporation*, Cambridge, MA: MIT Press.

Boot, A. and Thankor, A. (2000) 'Can relationship banking survive competition', *Journal of Finance,* 55(2): 679–713.

Buckley, P. and Casson, M. (1976) *The Future of the Multinational Enterprise,* London: Macmillan Press.

Buckley, P. and Casson, M. (1998) 'Models of the multinational enterprise', *Journal of International Business Studies*, 29(2): 21–44.

Caves, R.E. (1971) 'International corporations: the industrial economics of foreign investment', *Economica*, 56: 279–93.

Caves, R.E. (1996) *Multinational Enterprise and Economic Analysis,* second edition, Cambridge: Cambridge University Press.

Chen, H. and Chen, T.J. (1998) 'Network linkages and location choice in foreign direct investment', *Journal of International Business Studies*, 29(3): 445–68.

Dunning, J.H. (1958) *American Investment in British Manufacturing Industry,* London: George Allen and Unwin.

Dunning, J.H. (1977) 'Trade, location of economic activity and the MNE: a search for an eclectic approach', in B. Ohlin, P.O. Hesselborn and P.M. Wijkman (eds), *The International Allocation of Economic Activity,* New York: Holmes and Meier.

Dunning, J.H. (1981) *International Production and the Multinational Enterprise,* London: Allen and Unwin.

Dunning, J.H. (1988) 'The eclectic paradigm of international production: a restatement and some possible extensions', *Journal of International Business Studies*, 19(1): 1–32.

Dunning, J.H. (1993) *Multinational Enterprises in the Global Economy,* Wokingham, Berks: Addison-Wesley.

Dunning, J.H. (1995) 'Reappraising the eclectic paradigm in an age of alliance capital', *Journal of International Business Studies*, 26(3): 461–91.

Dunning, J.H. (1997) *Alliance Capitalism and Global Business,* London: Routledge.

Dunning, J.H. (1998) 'Location and the multinational enterprise: a neglected factor', *Journal of International Business Studies,* 29(1): 45–66.

Dunning, J.H. (2000) 'The eclectic paradigm as an envelope for economic and business theories of MNE activity', *International Business Review,* 9: 163–90.

Dunning, J.H. and Dilyard, J.R. (1999) 'Towards a general paradigm of foreign direct and foreign portfolio investment', *Transnational Corporations,* 8(1) (April): 1–52.

Ekström, J. (1998) *Foreign Direct Investment by Large Swedish Firms – The Role of Economic Integration and Exchange Rates,* Lund: Lund University Press.

Foerster, S.R. and Karolyi, G.A. (1999) 'The effects of market segmentation and investor recognition on asset prices: evidence from foreign stock listing in the United States', *Journal of Finance,* 54(3) (June): 981–1013.

Forsgren, M. (1989) *Managing the Internationalization Process: The Swedish Case,* London: Routledge.

Gray, H.P. (1996) 'The eclectic paradigm: the next generation', *Transnational Corporations,* 5 (August): 51–65.

Gray, H.P. and Gray, J.M. (1981) 'The multinational bank, a financial MNC?', *Journal of Banking and Finance,* 5 (March): 33–63.

Glaum, M. (2000) 'Bridging the GAAP: the changing attitude of German managers towards AngloAmerican accounting and accounting harmonization', *Journal of International Financial Management and Accounting,* 11(1): 23–47.

Gruber, W., Mehta, D. and Vernon, R. (1967) 'The R&D factor in international trade and international investment of United States industries', *Journal of Political Economy,* February, 20–37.

Hennart, J.F. (1989) 'Can the new forms of investment substitute for the old forms? A transaction costs perspective', *Journal of International Business Studies,* 20 (Summer): 211–34.

Hirsch, S. (1967) *Location of Industries and International Competitiveness,* Oxford: Oxford University Press.

Hymer, S. (1976) *The International Operations of National Firms: A Study of Direct Foreign Investment,* Cambridge, MA: MIT Press (first published as a doctoral dissertation at MIT in 1960).

Itaki, M. (1991) 'A critical assessment of the eclectic theory of multinational enterprises', *Journal of International Business Studies,* 22(3): 446–60.

Ito, T. and Iwaisako, T. (1995) 'Explaining asset bubbles in Japan', *Monetary and Economic Studies,* 14(96): 143–93.

Johanson, J. and Vahlne, J.E. (1977) 'The internationalization process of the firm: a model of knowledge development and increasing foreign market commitments', *Journal of International Business Studies,* 8(1): 23–32.

Johanson, J. and Wiedersheim-Paul, F. (1975) 'The internationalization process of the firm – four Swedish cases', *The Journal of Management Studies,* 12(3): 305–33.

Johnson, S., Boone, P., Breach, A. and Friedman, E. (2000) 'Corporate governance in the Asian financial crisis', *Journal of Financial Economics,* 58(1–2): 141–86.

Kogut, B. and Zander, U. (1993) 'Knowledge of the firm and the evolutionary theory of the multinational corporation', *Journal of International Business Studies,* 24(4): 625–45.

Kogut, B. and Kulatilaka, N. (1994) 'Operating flexibility, global manufacturing and the option value of a multinational network', *Management Science,* 40: 123–39.

Krugman, P. and Venables, A. (eds) (1994) *The Location of Economic Activity: New Theories and Evidence,* London: Centre for Economic Policy Research.

Miller, D.P. (1999) 'The market reaction to international cross-listings: evidence from depositary receipts', *Journal of Financial Economics*, 51: 103–23.

Modén, K.M. and Oxelheim, L. (1997) 'Why issue equity abroad? – Corporate efforts and stock market responses', *Management International Review*, 37(3): 223–41.

Oregonian, (1998) 'Seizing opportunity, Tektronix buys plant in Asia at low price', June 23, http://oregonlive.advance.net/todaysnews/9806/st062314.html.

Oxelheim, L. (1993) 'Foreign direct investment and the liberalization of capital movements', in L. Oxelheim (ed.), *The Global Race for Foreign Direct Investments*, Berlin and New York: Springer Verlag.

Oxelheim, L. and Randøy, T. (2003) 'The impact of foreign membership on firm valuation', *Journal of Banking and Finance*, forthcoming.

Oxelheim, L., Randøy, T. and Stonehill, A. (2001), 'On the treatment of finance-specific factors within the OLI paradigm', *International Business Review*, 10(4): 381–98.

Oxelheim, L., Stonehill, A., Randøy, T., Vikkula, K., Dullum, K.B. and Modén, K.M. (1998) *Corporate Strategies to Internationalise the Cost of Capital*, Copenhagen: Copenhagen Business School Press.

Pedersen, T. and Petersen, B. (1998) 'Explaining gradually increasing resource commitment to a foreign market', *International Business Review*, 7(5): 483–501.

Rajan, R. and Zingales, L. (1998) 'Debt, folklore, and cross-country differences in financial structure', *Journal of Applied Corporate Finance*, 10(4): 102–7.

Randøy, T. and Dibrell, C. (2002) 'How Norwegian MNCs commit resources abroad: beyond choice of entry mode', *Management International Review*, 42: 119–40.

Rugman, A. (1980) 'Internalization as a general theory of foreign direct investment: a reappraisal of the literature', *Weltwirtschaftliches Archiv*, 116(2): 365–79.

Rugman, A. (1981) *Inside the Multinationals: The Economics of Internal Markets*, New York: Columbia University Press.

Saudagaran, S.M. (1988) 'An empirical study of selected factors influencing the decision to list on foreign stock exchanges', *Journal of International Business Studies*, Spring, 101–28.

Sharpe, S.A. (1990) 'Asymmetric information, bank lending and implicit contracts: a stylized model of customer relationship', *Journal of Finance*, 55: 1069–87.

Stevens, G. (1998) 'Exchange rate and foreign direct investment', *Journal of Policy Modelling*, 20(3): 393–403.

Stonehill, A. and Dullum, K.B. (1982) *Internationalizing the Cost of Capital in Theory and Practice: The Novo Experience and National Policy Implications*, Copenhagen: Nyt Nordisk Forlag Arnold Busck and London: Wiley.

Stulz, R.M. (1996) 'Does the cost of capital differ across countries? An agency perspective', *European Financial Management*, 2: 11–22.

Sundaram, A.K. and Logue, D.E. (1996) 'Valuation effects of foreign company listings on U.S. exchanges', *Journal of International Business Studies*, 27(1): 66–88.

Swenson, D.L. (1994) 'Impact of U.S. tax reform on foreign direct investment in the United States', *Journal of Public Economics*, 54: 243–66.

UNCTAD (2000) *World Investment Report*, New York.

Useem, M. (1998) 'Corporate leadership in a globalizing equity market', *Academy of Management Executive*, 12(4): 43–59.

Vernon, R. (1966) 'International investment and international trade in the product cycle', *Quarterly Journal of Economics*, May, 190–207.

11 The challenge of electronic markets for international business theory

John H. Dunning and Cliff Wymbs

Introduction

The study and research of international business and e-commerce has both many similarities and some differences. E-commerce touches many of the same disciplines as IB (Boddewyn and Iyer 1999), i.e. economics, political science, law, sociology and psychology, and functional or professional areas, i.e. management, marketing, finance, etc., which apply concepts, models and variables derived from these disciplines. We have observed how the network properties of the Internet and e-commerce have encouraged IB, but not necessarily the amount and internal relatedness of foreign direct investment associated with business models of the mid-1990s (Beck *et al.* 2000).

As with the difficulties cited by Vernon (1994) in truly understanding the structures, motivations and strategies of multinational enterprises, we are once again confronted with these problems with the emergence of global e-commerce companies, but now have much less history to guide us. Much of the earlier research in e-commerce focused on descriptive case studies, recounting what happened and why (Netscape, Yahoo, etc.), particularly in the US market, rather than taking the lead from the more careful theoretical analysis of Bartlett and Ghoshal (1993) in their evaluation of the major organisational changes then being introduced in such MNCs as ABB. We must attempt to introduce the IB discipline and eclectic inquiry process to analyse the new realities of globally networked businesses. This requires a basic understanding of the reality of business in the new economy and a great deal of imagination to develop creative theoretical solutions to unstructured and undetermined problems.

Since the 1960s, the eclectic paradigm (Dunning 1958, 1988, 1995, 2000) has sought to integrate the mainstream economic and behavioural theories to explain aspects of FDI and/or foreign-owned production. The paradigm offers an analytical framework for incorporating a number of context-specific and operationally testable theories, each of which seeks to explain a particular component of the internationalisation process (Dunning 2000). Expansion of the paradigmatic framework to e-commerce can help us formulate contextual and operationally testable theories. More specifically, this chapter examines the extent to which the tenets of the eclectic paradigm can be applied to e-commerce development and reviews many of the mainstream theories related to IB to establish this link.

The chapter will address two main issues:

1 *The specific impact of e-commerce on the framework of the eclectic paradigm of international production and some of the economic and behavioural theories that make up the paradigm.*

Much like the steam engine, the telegraph, the telephone, the railroads and the highway systems before it, the Internet is a facilitating technology that serves as a catalyst for new business combinations, permutations and mutations (Wymbs 2000a). Each of the above technologies was network-based and required a minimum number of users before its true economic benefit and a period of increasing returns could be realised (Bernstein 1998). Gates (1999) states that what is different about the Internet as a modality of service delivery is the speed at which its widespread use has followed its initial commercial introduction, namely, in approximately five years as compared with a decade or more earlier in the case of other communication technologies. This time compression will highlight some of the unique dislocations and discontinuities caused by or associated with e-commerce (Yoffie and Cusumano 1999; Iansitit and MacCormack 1997).

Although we will analyse it in more detail later, the basic tenets and predictions of the eclectic paradigm (Dunning 1988, 1993, 2000) appear to hold with respect to the Internet. Firms will seek to leverage Internet technology to sustain or augment rare, inimitable and immutable ownership (O) advantages both from a scale and scope perspective. They will seek efficient organisational structures and internalise asset-creating and value-added activities using electronic commerce when markets fail. In addition, firms will seek to invest in locations around the world where they can either best exploit their core competences or add to them.

The information technology revolution, however, is like an iceberg, the largest and most interesting part of which is below the paradigmatic water line. With regard to firm-, industry- and country-specific issues, key questions include: How will Internet technology affect the boundaries of firms, and how will firms mutate, combine, dissolve and spontaneously regenerate the resources and capabilities to take advantage of a new range of information and transaction-related economies? How may common cross-industry processes replace industries as the most appropriate classification of groups of firms? How may governments, both within countries and across countries, attempt to simultaneously encourage e-commerce development, yet limit some of its less desirable effects, including the sub-optimal behaviour of firms experiencing increasing returns on a global scale? We believe that the answers to these questions can be productively explored by using the methodology and contents of the eclectic paradigm.

2 *How contextual variables (industry, country and firm characteristics) affect OLI theory in an e-commerce world.*

Fundamental to exploring this issue is the need to answer the following basic question: Is it correct to think of e-commerce as a trajectory shift in our analysis of the determinants of international business (IB) activity?

The answer is both 'Yes' and 'No'. Technological advances are helping to transform our lives by inventing new, undreamed of products and producing them in new, undreamed of ways (Dicken 1998; Tomkins 1999; Reedy *et al.* 2000). Richard Lipsey (1997), among others, believes that the world is undergoing a deep structural adjustment in response to the advent of dramatically new information and communications technologies. Unlike structuralist scholars, such as Freeman and Perez (1988), who combine technology, structure and economic performance in one overarching concept, the so-called techno-economic paradigm, Lipsey and Bekar (1995) treat the three components separately and distinctly. We agree, and believe that the contemporary debate with respect to the impact of the Internet today is not so much about technology per se (we all expect a geometric increasing in computing power, transmission capacities and fast packet switching), but rather how the facilitating infrastructure will channel and nourish e-commerce growth; and how, and to what extent, the superior economic performance will flow to firms embracing new, increasing returns and business strategies, that both redefine existing markets and create new ones.

Steve Kobrin (1999) expands on Lipsey and Bekar's propositions by asserting that cyberspace is a marketplace unlike that of any other in history. Potentially, no physical product crosses geographic borders, no paper currency changes hands and there is no tangible record of the transaction (Kobrin 1999). Policy-makers, technologists and managers must confront the political and economic implications of digitisation. There appears to be an emerging asymmetry between economics and politics (Strange 1997). The former is becoming global rather than trans-border, and is being organised via the use of electronic networks, while the latter is overwhelmingly local and geographic.

The intangible aspect of cyberspace means that a transaction can no longer be mapped into two-dimensional space. Borders and jurisdictions are irrelevant, rather than ambiguous. The unique power of national governments to tax, punish and require participation is based on the idea of a national territory. Kobrin (1999) asserts that the basic disconnect between geographic space and cyberspace raises fundamental questions about the concept of national economic jurisdiction and/or control.

From an individual firm perspective, it is important to understand its implications on the three main modes of internationalisation, i.e. trade, foreign direct investment and international strategic alliances (Wymbs 1997). Few would argue with the proposition that the new information economy is reducing most forms of spatial transaction and coordination costs, and is thereby facilitating both more, and less costly, cross-border commerce. However, as Ronald Coase (1999) has observed, it is by no means clear which organisational entity is best equipped to own, control or access e-commerce. On the one hand, lower transaction costs would appear to favour a market – though not necessarily arm's-length market – solution; on the other, lower intra-firm coordinating costs and the potential for exploiting firm-specific scale and scope economies would seem to favour a hierarchical solution. This particular trade-off of Internet advantages is one of the most challenging questions now exercising the minds of organisational and IB scholars.

An in-depth analysis of these two basic issues follows.

The impact of electronic commerce on IB theory

As a context setter, this section first provides a brief overview of the eclectic paradigm. It then systematically analyses and applies each of the OLI components of the paradigm, and their related theories of IB, to e-commerce development.

The eclectic paradigm

In brief, the eclectic paradigm – also known as the OLI paradigm – states that the extent, pattern and form of IB activity will depend on the juxtaposition behaviour of three sets of advantages. The first is the competitive advantage- or ownership- (O) specific advantages of firms engaging in or contemplating value-added activities outside their national boundaries. The second is the locational (L) attractions of particular regions or countries in which firms might either create or augment these O advantages, or add value to them. The third advantage is the extent to which firms possessing O advantage will choose to coordinate these advantages with the L advantages of foreign countries through internalising (I) the former's cross-border markets for intermediate products, rather than engaging in arm's-length transactions or contractual agreements with foreign firms.

The paradigm was first put forward at a Nobel Symposium in Sweden in 1976 (Dunning 1976), and over the past 25 years it has been frequently modified and extended in the light of scholarly criticism and the changing world economic landscape.[1] In 1995, the OLI paradigm, originally developed during an era of hierarchical capitalism, was extended to incorporate alliance-related modalities. In this contribution, we propose another update to the paradigm to reflect the emerging e-commerce world. For example, Figure 11.1 compares the three temporal snapshots of the paradigm, while the discussion of the e-commerce parameters follows.

1. Ownership-specific advantages
(of enterprises of one nationality (or affiliates of same) over those of another)

Hierarchical advantages
c. 1993

- Property right and/or intangible asset advantages (O_a).
- Advantages of common governance, i.e. of organising O_a with complementary assets (O_t).
  - (i) Those that branch plants of established enterprises may enjoy over *de novo* firms.
  - (ii) Which specifically arise because of multinationality. Multinationality enhances operating flexibility, more favoured access and/or better knowledge of international markets, ability to take advantage of geographic differences in factor markets and the ability to diversify or reduce risk.

Alliance advantages
c. 1995

- Vertical alliances
  - (i) Backward access to R&D and suppliers.
  - (ii) Forward access to industrial customers and new markets.
- Horizontal alliances
  - (i) Access to complementary technologies and innovative capacity.
- Networks of similar firms
  - (i) Reduced transaction and coordination costs arising from better dissemination and interpretation of knowledge and information.
  - (ii) Business districts.

  As per (i) plus spatial agglomerative economies.

E-commerce advantages
c. 2001

- O_a property right based on technology, e.g. Cisco Systems Internet switches.
- O_a property right based on standards, e.g. Microsoft's operating system.
- O_a property right based on preferred customer interface, e.g. AOL and Amazon.com.
- O_t advantage based on e-commerce experience, e.g. Amazon.com's horizontal integration into CDs.
- Vertical alliances, e.g. commission paid based on sales referral.
- Vertical alliances, e.g. creating business partners to handle auctions (FreeMarkets.com).
- Horizontal alliances, e.g. GM & Ford partnering to create a purchasing system.
- Business districts, e.g. Silicon Valley, Silicon Alley.

2. Internalisation incentive advantages
(i.e. to circumvent or exploit market failure)

Hierarchical advantages

- Avoidance of search and negotiating costs.
- To avoid costs of moral hazard, information asymmetries and adverse selection.
- To avoid costs of broken contracts and ensuing litigation.

Figure 11.1 An extension of the eclectic paradigm of international production.*

Alliance advantages

- Alliances or network-related advantages are those which prompt a 'voice' rather than 'exit' response to market failure; they also allow many of the advantages of internalisation without the inflexibility or the risk-related costs associated with it.
- The growing structural integration of the world economy is requiring firms to go outside their immediate boundaries to capture the complex realities of knowledge exchange in innovation.

E-commerce advantages

- Disintermediation: the ability to substitute information technology for various components of an existing vertical value chain, e.g. auto manufacturers taking orders over the Web, bypassing dealers and Chemdex.com replacing catalogues.
- Re-intermediation: becoming an agent for buyers, seeking the lowest electronic price.
- Outsourcing: reduced transaction costs.
- Specialisation: size of core functions can grow due to reduced coordination costs.

3. Location-specific variables
(these may favour home or host countries)

Hierarchical advantages

- Spatial distribution of natural and created resource endowments and markets.
- Economic systems and policies of governments.
- Input prices (land, labour and capital, components).

Alliance advantages

- The L-advantage of alliances arises from the presence of a portfolio of immobile local complementary assets, which, when organised within the framework of alliances and networks produce a productive and stimulating environment.

E-commerce advantages

- The L-advantage arises from:
  (i) State-of-art telecommunications infrastructure.
  (ii) Flat rate access.
  (iii) Government adopting a minimalist market policy.
  (iv) Agglomerative economies associated with financing and technology.

Figure 11.1 continued

* Figure adapted from Table 1 in (Dunning 1995) 'Reappraising the eclectic paradigm in age of alliance capitalism'.

Specifically, in this chapter we ask the question: To what extent can Internet related IB activities be accommodated within the framework of the paradigm? After briefly indicating some of the modifications which seem to be required to the OLI configuration of firms resulting from the advent of e-commerce, we then proceed to consider how some of the specific theories which reference the eclectic paradigm need to be reconfigured.

Theories explaining O-specific advantages of firms

We initially address ownership-specific advantages of the firm as they relate to the eclectic paradigm and then discuss four important related theories of the firm, i.e. the resource-based view, evolutionary theory, organisational theory and economic strategic theory.

The eclectic paradigm O advantage

The eclectic paradigm clearly identifies two types of ownership advantages: the first pertains to the resources (asset) structure of the firm, which relates to property rights and/or intangible asset advantage (Oa), while the second pertains to the advantage of common governance, that is, of organising Oa with complementary assets (Ot), i.e. transaction cost minimising advantage

Oa advantages: We believe that, while e-commerce is changing the context and scope of the core competences of firms, such advantages are, if anything, becoming a more important discriminator in their strategic positioning and commercial success. At the same time, the Internet is creating quite a varied set of such opportunities, from Amazon.com establishing a relational community of over 22 million customers (Helft 2000), to Chemdex.com putting hundreds of thousands of speciality chemicals on line (Muehlbauer 2000), to GE creating an auction supply network (Margherion 1998), to DoubleClick using Internet usage data to create databases that target specific customer segments for clients (DoubleClick 2000). Each venture leverages network attributes of marketspace (which may be domestic or international) and attempts to exploit first mover and increasing returns advantages to preclude or inhibit market entry by competitors or potential competitors (Rayport and Sviokla 1998). Interesting relationships are also being forged between large retailers such as K-Mart and Wal-Mart who, initially, did not embrace the Internet, and by leading marketspace providers like Yahoo and AOL (*Standard* 1999). In many ways, these alliances are similar to options taken by MNCs in the 1980s in an attempt to expand their territorial boundaries (Kogut 1985). What is different today is that firms are expanding in cyberspace rather than in the marketplace, and are more likely to use the market to access resources and capabilities rather than to extend their ownership of these same resources and capabilities (*Standard* 1999; Rifkin 2000).

Ot advantages: At the same time, the organisational advantages of firms are becoming increasingly important. They are associated with the ability of firms to learn the business of electronic commerce and how its various components can be replicated and/or coordinated in diverse industries. For example, in the late 1990s, Priceline.com developed an auction model that initially worked well for the low-cost procurement of the excess capacity of airlines (*Economist* 2000a). It then leveraged the information system and management practices to establish itself in the related hotel industry. However, its next move was in a product area that was far removed from travel, creating an auction market for grocery items (Priceline 2000). Unlike the historical and functional linkages between the airline and hotel industries, this new linkage is based on the relatedness of processes (auctions) in ubiquitous marketspace (the Internet).

Another important Ot advantage is the ability to work efficiently and harmoniously with other enterprises, e.g. competitors, suppliers and customers, to innovate enterprises more productively or speedily, and/or more effectively utilise existing production and marketing opportunities. In addition to providing its proprietary on-line auction service functions, e.g. searchable databases, billing and collection, etc., e-Bay's value web includes electronic linkages with its customers, competitors such as Yahoo who route customers to e-Bay's site, but also provides a competing auction site, complementors such as banks that provide credit card verification and suppliers that provide software, hardware and network services (Cartwright and Oliver 2000). The Internet dramatically increases the amount of specialisation of value-adding activity that can economically take place in creating customer services. It also reduces the importance of the physical location of any particular value activity, and blurs the corporate lines between competitors, co-providers, and strategic partners for any particular service.

Resource-based view

The resource-based theory suggests that a firm's competitive advantages are internally generated rather than being determined by the industry of which they are part (Capron and Hulland 1999). It assumes that each firm has rare, sustainable and imperfectly imitable resources and that the firm uses these resources to obtain differential advantage (Barney 1991). As a general theory, the resource-based view (Wernerfielt 1984, 1995; Conner and Prahalad 1996) holds that firms are concerned with identifying and evaluating those assets that afford them a sustainable competitive advantage. However, unique, rare and not imitable assets identified in the current literature are those primarily based on physical resources and on the capabilities of firms to organise these effectively. In the electronic marketspace, knowledge creation and innovation are replacing physical processes as value-adding activities (Cartwright and Oliver 2000).

The competences associated with e-commerce are likely to include such intangibles as specialisation, speed and the ability to harness and deploy critical assets and to network efficiently. In particular, the experience gained by first-mover firms learning to operate in this new, information-driven and increasing-returns environment is likely to become a key component of their future competitive advantages. Because there are relatively low barriers to entry, the creating of sustainable and non-imitable advantages become serious problems for many start-up firms. However, a robust acquisition market for under-performing firms has developed, and when their capabilities are combined with their new owner's resources, a knowledge synergy can take place which helps to generate a new set of unique assets, e.g. Amazon.com buying PlanetAll so that Amazon can marry its customer preference data for 22.5 million customers with PlanetAll's gift suggestion software (Gimein and Lash 1998). Also, because of easy exit costs, the same team can quickly re-form around new ideas with new funding (Laseter *et al.* 2000). The creative applications of human capital, organisational learning and the network of relationships internal and external to the organisation are becoming the new rare and non-imitable assets of an electronic economy (Cartwright and Oliver 2000).

Evolutionary theory

Evolutionary theory's basic proposition relates to path-dependency and the accumulated competitive advantages of firms. More particularly, the theory asserts that the more efficient firms are in creating, accessing and managing these advantages, the more likely they will have the capability to engage in asset-exploiting and asset-augmenting FDI (Nelson and Winter 1982; Nelson 1991; Cantwell 1989, 1994; Teece *et al.* 1997). Not surprisingly, IBM has leveraged its computer and network skills and GE has leveraged its procurement network to become significant e-commerce players.

There are several components of e-commerce that are evolutionary; and most particularly, the building of a network infrastructure to provide high-speed, low-cost digital connectivity to all its users or potential users, and of a customer coverage that permits this infrastructure to be efficiently and economically exploited. Firms have been electronically evolving by replacing expensive private procurement systems like EDI with the Internet. Also, firms that have implemented Intranets usually evolve and develop gateways to allow their internal user community to communicate with external users via an Extranet. The Intranet/Extranet migration significantly increases communication efficiency with little incremental communication cost. Once, however, one goes beyond these substitution effects, the next stages of e-commerce are likely to increasingly focus on using electronic information to target customers, having customers create virtual communities and having customers use intelligent agents to deliver product information. Entrepreneur literature (Timmons 1990; Sahlman 1996) indicates that these revolutionary changes are more likely to come from smaller firms not wed to ingrained organisational processes and staid corporate cultures. But as these new Internet business models and processes become more widely established, they will create their own trajectories and path-dependent competitive processes (Hill 1997). Also, the Internet allows firms to better coordinate related asset-creating activities and exploit dynamic internalisation advantages.

Organisational (management related) theories

In the early years of the present millennium, managers will be required to devise appropriate organisational structures to harness and coordinate assets, processes and routines around the world. Information technology and the Internet will dramatically facilitate this process (Prahalad and Doz 1987; Bartlett and Ghoshal 1989, 1993; Doz *et al.* 1997). We would expect the value-added attributes of firms to undergo a fundamental change, with many non-core functions being increasingly contracted out (GE 1999). Similarly, Coase (1999) believes that this process will result in greater Smithian type specialisation and an enlarged size of specialist firms. When the most successful Internet start-up firms begin to earn profits, this will put increasing pressure on their competitors, in their search for both financial and human assets, to seek more leverage for their intangible assets, and to outsource their less profitable value activities. Indeed, a case has been made that, in the future, physical assets will be reduced to the status of commodities, and that the only assets that will be able to make monopolistic profits will be intellectual and relational (Evans and Wurster 1997). This suggests that organisational management structures will need to be flexible enough to simultaneously manage both Smithian growth resulting from the division of labour specialisation and Schumpeterian growth resulting from innovation of business processes.

Economic theory – follow my leader, 'tit for tat' strategy

This model based on Knickerbocker (1973) and Graham (1990, 1998) appears to be particularly relevant in an e-commerce environment. This is best illustrated by the behaviour of two large US Internet portal suppliers, Yahoo and Lycos, who appear to be chasing each other around the globe – with India being their most recent target country (Manzar 1999). As stated previously, many of the new companies entering cyberspace are facing increasing returns on their O-specific assets (Arthur 1996). In fact, the main reason for the high initial public offering prices for these firms is an expectation that they will be able to corner a specialised segment of the consumer market before others do so (Desmet *et al.* 2000). This mini-monopoly will then either be exploited directly by the founding firm or sold to a larger player, such as Amazon.com, who is looking to become the anchor tenant in the biggest mall on earth (Gimein and Lash 1998). Given the current monitoring and control by national governments, monopolies do not sit well from a public policy perspective. Therefore, the best these first movers can hope for is to leverage the increasing returns of their businesses, and to share their economic rents among a few global market players.

Theories explaining L-specific advantages of firms

The L-specific advantages of the firm as they relate to the eclectic paradigm are addressed first and this is followed by several important partial theories of the firm that relate to complementary assets, government-induced incentives and knowledge-enhancing dynamics.

The eclectic paradigm's L-advantage

The competitive advantage of countries or regions relates to the ability of locations (regions, countries or sub-regions) to offer the immobile assets necessary for the mobile assets of domestic and foreign firms to be used more efficiently – both to add value to those assets and to create (through innovation or tapping into indigenous capabilities) new assets. The locational (L) component of the eclectic paradigm is complex and must be addressed along several dimensions. The first deals with the separation of goods with high digital content from those without. While it is true that every product has to be produced and consumed at a specific point of space, the physical, intellectual and information path that it takes to get there is becoming increasingly complicated (Kobrin 1999). For high digital content products, such as CDs and software, there has been a replacement of the physical products with 'virtual' ones. Even for goods without a high digital content, there appears to be a split developing between their tangible and intangible value chains. While, for example the physical components of a car, e.g. the engine, transmission, etc., still have locational space, the information properties relating to these components can appear only in cyberspace. Databases that contain product specific design information can be sited in any country that has a server on the Internet and can be accessed by anyone with a computer and network access. What gives the information value is its connectivity to other parts of the network, rather than where it is stored. Physical access points to the information that can be identified, but the value associated with the traditional process of car design is the synergy of the parts working together (Henson 2000).

The second locational dimension depends on whether the firm is an 'e-commerce firm' or an 'existing firm using e-commerce'. E-commerce firms are network-based and, as such, experience, in most cases, increasing rather than decreasing returns when they produce additional or sell existing products in new markets. Many of these firms seek horizontal integration, e.g. Amazon entering into auctions, toy and CD markets, and geographic expansion, e.g. Yahoo is currently (2000) seeking to develop its international operations by accessing new resources and capabilities where it makes sense, and using its leverage to gain market share where it does not (Larson 2000); and, by so doing, to spread their high initial fixed costs over greater revenue streams and gain economies of scale and scope. Unlike old economy firms, these e-commerce firms enter foreign markets instantaneously when their services are posted on the Internet. Of course, each firm can expend resources to culturally adapt its site to a particular region; however, in almost all cases, the cost of serving foreign markets is dramatically lower than those associated with the traditional product expansions of the 1990s.

Locational choice decisions by existing firms using e-commerce centre mostly on the opportunities for horizontal or vertical integration that affect the economics of FDI and trade. The Internet reduces both internal co-ordination (favouring FDI) and market transaction costs (favouring trade) so the relative rate of cost reduction becomes important and this is likely to be firm and industry specific. The cultural predisposition of firms within a particular location also affects the above trade-off. For example, Toyota, which is part of the GM/Ford/Daimler-Benz auction network is choosing to put out to bid only non-essential supply chain components, while the other auto makers are including most of their essential auto components. Japanese companies in general, and Toyota specifically, have greater cultural bonds with their suppliers that militate against pure market solutions (Dunning 1993). However, if GM/Ford/Daimler-Benz experience significant cost reductions in their supply chains because they have put critical components out to auction, Toyota may have to forsake cultural bonds out of global competitive necessity.

The third locational dimension pertains to domestic government policies toward the basic telecommunication and operating legal environment. Clearly, the quality of the telecommunication infrastructure is the driver of the information revolution (Bond 1998); and such a structure, once established, tends to be location-bound. Low-cost, flat-rate access stimulates Internet traffic, while e-commerce knowledge intensive centres *à la* Porter (1990) are developing, first around the US and slowly across the world. The legal environment is a key locational determinant of e-commerce growth. For example, the US model that uses advertising to subsidise much of the cost of new Internet services would not be as effective in the European Union (EU) because e-commerce consumers have the right not only to know about any personal data files kept by companies, but to see, change and delete them, and to collect damages in cases of data abuses (*Standard* 2000a). Also, tax laws, as they relate to immediate realisation of personal income when stock options are granted (as opposed to when they are exercised as is the custom in the US), could dramatically affect location choice of a firm (Sprenger 2000).

Theories related to complementary assets

Home and host complementary and location-bound assets comprising the telecommunication infrastructure of countries are key to helping e-commerce firms add to and exploit their O advantages (Applegate 1995). Countries that have low flat-rated telephone access, such as the US, record a much higher Internet penetration than those that have high variably priced access. Ubiquitous, countrywide Internet connectivity and a high penetration of personal computers are also key complementary assets to the use of business to consumer retail e-commerce applications. With regard to specific technologies, the presence of an upgradeable fibre optic infrastructure, both within and between countries, facilitates the exploitation of high-speed transmission of video-related e-commerce. For on-line delivery of digital products, the contemporary communication infrastructure is the counterpart of the roads and rails of a past generation, while the information bits are equivalent to the products carried by the trucks and trains. The critical difference, however, is that the bits do not go through customs as they cross borders.

Theories related to government-induced incentives

To a large extent, national governments provide the basic rules of the game for the conduct of service transactions, though supra-national entities, e.g., WTO and the European Union, are likely to play a more important role in the future. To date, most governments have allowed the existing body of commercial law – including that of intellectual property rights – to apply to electronic commerce (Arrow 1999). This is likely to become increasingly strained, particularly in the area of taxes, privacy, pornography, uniform contracts and intellectual property rights (Kobrin 1999). However, Peterson (1986) cautions that government intervention could be a heavy-handed solution to a relatively minor problem. Most policies set by governments are national (some are regional like the EU), but could dramatically affect the ability of indigenous firms to compete on a global basis (Kobrin 1999). However, government policy can on occasion enhance the competitiveness of indigenous firms' competitiveness. For example, the establishment of the European Union GSM cellular standard allowed firms in the region to lead the development of M-commerce, cellular-based Internet business.

There appears to be significant opportunity for the creation of private/public partnerships. One such example is the authorising by the Chinese government to be a key information site for companies looking to do business in China. Another example of a company working with a government agency is e-Bay consulting with the Better Business Bureau to limit fraud and copyright infringements on the Internet. As commerce on the Internet grows, business will increasingly identify areas where it will seek out government assistance to help create order in the virtual e-commerce world (Lessig 1999).

Knowledge-enhancing (dynamic) theories of location

In the last decade, the attractiveness of particular locations has been increasingly explained by their ability to help investing firms upgrade their knowledge and learning experiences. Such asset-augmenting theories of FDI, as, for example, put forth by Dunning (2000), Kogut and Zander (1994), Florida (1995), Kuemmerle (1999) and Porter (1994, 1998) are particularly applicable to our understanding of the location of e-commerce-related activities. At least four areas of the US appear to be distinguishing themselves as Internet Knowledge centres, namely Silicon Valley, MIT media lab in Boston, New York's Silicon Alley and Microsoft's Seattle (Florida 1998; IDA 1999); significant e-commerce development is taking place in and about London, and the Scandinavian countries, particularly Finland, are developing e-commerce products and services.

The US entrepreneurial culture and deep-venture capital market are also greatly facilitating additional investment in these areas (Dunning and Wymbs 1999). Though physical in location, each knowledge centre's value-added is composed almost entirely of intangible resources, e.g. people with ideas, organisational modalities of firms that readily respond to market needs, and when necessary, can rapidly reform on new ideas, funding mechanisms that encourage firm growth and entrepreneur profits and universities that provide positive externalities associated with knowledge spillover. However, as real estate rents for physical space in these knowledge locations dramatically increase, many start-ups are choosing to locate in spoke communities around these knowledge centres and use their previous relationships and the Internet to serve as virtual bridges.

Risk diversification theories

Risk diversification – a motive for foreign direct investment, identified and analysed by such scholars as Vernon (1973, 1983), Rugman (1979) and Kogut (1985) – appears to be increasing on a strategic product basis, but decreasing on a locational basis. Firms are now hedging between electronic versus bricks and mortar applications, e.g. the *New York Times* and the *Wall Street Journal* have electronic additions, Merrill Lynch launched Internet trading, while Barnes & Noble has an on-line book selling operation (Henry 1999). Access to these electronic applications requires only a phone line and a computer, and can occur 24 hours a day, 7 days a week and 365 days a year on a global basis. Also, the recent market shake-out has permitted established firms to execute a relatively low-cost product diversification strategy by buying assets of failing dot.coms for cents on the dollar. With regard to physical location, the new knowledge economy is unlike the old economy because it is based on intangible assets that are mobile, rather than tangible assets that are fixed. With information being stored in many places throughout networks, there is less need to geographically diversify a firm's assets to reduce transaction, translation and asset exposure risk.

Theories explaining I-specific advantages of firms

The I-specific advantages of the firm as they relate to the eclectic paradigm are first addressed followed by several important partial theories of the firm.

The eclectic paradigm's I-advantage

With regard to internalisation (I) theory, e-commerce is likely to have the greatest short-term effects by reducing the transaction and coordinating costs of economic activity, particularly in the business sector. In the past, large businesses have spent billions of dollars using private electronic data interchange (EDI) networks to lower transaction costs (Henry 1999). Today, the Internet can be used to construct EDI-like networks at a fraction of the price, and in so doing, opens up these economies to virtually all business customers. Increasing access to markets dramatically increases competition, lowers prices and enhances innovation. This is particularly true where producers have been able to set up auctions to get competitors in real time to bid against one another (Turban *et al.* 2000). The range of projects is quite broad, and embraces everything from municipal bond underwriting to the sale of aircraft and vehicle parts to the creation and supply of electronic money (Cohen 2000).

At this point, it may be instructive to look at how the Internet has changed the main drivers of transaction cost economics as, for example, identified by (Buckley and Casson 1976, 1985; Hennart 1982; Rugman 1981; Williamson 1985). Simply stated, transaction cost economics asserts that the firm is an avoider of market costs that result from exchange. It will internalise the markets for those intermediate products that it perceives will derive greater benefit from this modality than from an arm's-length transaction. The three most often-cited reasons for internalisation are information asymmetry, bounded rationality and asset specificity. Information asymmetry is predicated on the fact that one of the parties in any transaction lacks information relative to the other. Clearly, the Internet network dramatically increases the amount of information available to all parties for informed market decisions. Bounded rationality deals with a human inability to process all information in making a decision. When used correctly, the Internet permits relevant information to be obtained, distilled and transferred in a timely fashion. As cited above, the creation of auctions is a market solution to the bounded rationality problem. Asset specificity pertains to having to commit a large sum of capital to produce a specialised product. To the extent that the Internet permits increased specialisation of activities along the value chain, it may help reduce asset specificity required for each component. In fact, much of the value-adding processes associated even with the production of physical products is not capital intensive, e.g. marketing, R&D, inter-firm coordination. For many leading firms in industries, the capital intensive components have been outsourced, e.g. auto parts, PC production, sneakers, etc.

On the other hand, asset specificity might increase if firms focused more on their core competences, and if these were interrelated with those of other firms. However, the assets that drive these core competences will likely be intangible (people and relational) resources. Because these assets are more mobile than physical assets, they are less likely to cause asset specificity problems.

Simultaneously, information technology is reducing the coordinating costs within and between organisations, thereby permitting much larger firms to evolve. In consequence, firms using e-commerce tend to be more focused on their core competences, and to outsource a higher proportion of their non-core activities. There appears to be a fundamental reordering of firms using the Internet world, based on whether market-specific transaction costs are decreasing as fast as firm-specific coordination benefits are increasing.

Orthodox internalisation theory

Orthodox internalisation theory, which is essentially concerned with maximising the economic rent of a *given* set of O specific advantages, explains why firms coordinate these advantages with the L advantages of countries through internal fiat rather than using external markets (Caves 1996; Anderson and Gatignon 1986; Buckley and Casson 1976, 1985; Hennart 1982, 1989). We now consider a number of issues arising from e-commerce in question form, which, we aver, orthodox internalisation theory needs to address.

1 *What are the differences between e-commerce and non e-commerce, and how do they affect the propensity of firms to internalise the cross-border market for Internet related services?*

The willingness and ability of firms to internalise markets depend on the type of intermediate products and end products being supplied. In supplying products directly to the final consumer, companies like Amazon.com or Priceline.com find that there is a tremendous potential to horizontally expand and internalise heretofore different markets (Taylor 2000). This they do, for example, by creating a customer database, developing a business model or recipe, and attempting to replicate this model in as many markets as are appropriate. Priceline.com gets bids on airline tickets as well as groceries while Amazon.com has expanded into toys, auctions, and CDs (Watson *et al.* 1998).

The business-to-business (B2B) market is different, in that it affords firms an ability to internalise their operations more efficiently by use of such means as global e-mail, e-mail attachments, CAD and intranets (IBM 1999). However, many of these communication tools also increase the firm's ability to outsource business functions and better control them (Kotabe *et al.* 1998), resulting in immediate enhanced bottom-line financial performance (Hunt 1997).

2 *What are electronic market failures?*

The reasons why traditional markets fail are well identified in the literature and we have already suggested above that electronic markets may reduce some of these failures. At the same time, electronic markets generate their own set of imperfections, namely, (a) market failures incurred by consumers when shopping on the net and (b) market failures incurred by firms when selling or purchasing skills and resources.

Margherion (1998) has identified six of these consumer failures as they relate respectively to issues of security, privacy, indecency, intellectual property, contracts and consumer protection. The first three failures deal with the medium itself and the last three with its potential results. Security, e.g. with respect to piracy, hacking, etc., is often quoted as the number one impediment limiting e-commerce (Pitt *et al.* 1999). To mitigate this problem, credit card companies have created industry standards that use sophisticated algorithms to encrypt and decrypt credit card numbers. The privacy and indecency issues are already present in communication technologies. However, these are dramatically escalated, since with a keystroke, the Internet permits access to indecent material (Henry 1999). Intellectual property rights, contracts and consumer protection directly relate to the global dimension of the Internet media, and open up a new challenge with respect to the boundaries and content of both national and international law. These uncertainties create impediments to what is purported to be a frictionless process. The last item also pertains to the application of law to firms attempting to extract monopolistic profits in global marketspace (Kobrin 1999). Many of the above actions relate to firms working with institutions to circumvent market imperfections so that they may more fully recover returns from their ownership advantages/capabilities.

The market for the purchase of skills and resources by firms has greater imperfections in an electronic market than in a physical one. As stated previously, the key value-adding resources in an electronic world are intangible resources that, by their very nature, are hard to define and thus make informed assessments of value difficult. The technology explosion associated with the Internet has precluded firms from internally pursuing many good opportunities. In many instances, firms have used alliances as knowledge extensions to keep their options open in new emerging areas. Alliances are also being used as a means to get first-hand information and make better decisions on acquisitions. However, the increased acquisition of start-up firms by established players in particular new foreign markets entering the e-commerce area, e.g. AOL's acquisitions of access providers in Europe, could lead to further market failures mainly related to limited competition. Even the sharing of information by alliance partners could lead to collusive behaviour, e.g. government agencies in the US are currently evaluating the airlines' planned response to Priceline.com and the auto parts alliance.

3 *Can we distinguish between* static *and* dynamic *electronic and other kinds of market failures?*

Traditional endemic market failures relate mainly to the presence of static transaction costs, such as those relating to opportunism, bounded rationality, asymmetry of information, externalities and the vertical integration of markets (Buckley and Casson 1976, 1985; Williamson 1985; Hennart 1982, 1989; Peterson 1986). Clearly, the Internet, through its ability to access information, can substantially reduce some of these imperfections. However, in a dynamic, rapidly changing environment, this increases the complexity of management choices and leads to a series of innovation-related market failures (OECD 1999). Information on the Internet also exhibits the characteristics of a public good, in the sense that one person's consumption of a particular product does not limit that of another. As stated previously, the increasing returns component of the Internet leads to significant economies of scale for the first mover (Arthur 1996). The Internet also allows firms to better coordinate related asset-creating activities and exploit dynamic internalisation advantages.

Market power/efficiency/knowledge acquisition theories

The acquisition or access to all kinds of information (Kogut and Zander 1994; Wesson 1993) is appearing as a major rationale for engaging in cross-border e-commerce (Margherion 1998). e-Bay's purchase of the Butterfield & Butterfield's high-end auction operations in Germany and Cisco Systems, the largest Internet switch manufacturer, using acquisitions as its main R&D vehicle are two such instances (Wymbs 2000b; Rabinovitz *et al.* 2000). Examples of acquisitions, the sole purpose of which are to reduce competition and increase market power include Proflowers.com acquiring Flowerfarm.com and luxury-goods retailer Ashford.com snapping up Jasmin.com, a perfume site (Rabinovitz *et al.* 2000). Amazon and e-Bay expressed acquisition strategy targets investment in companies that will spread their operating efficiency and franchise in new horizontal categories and in multiple regions throughout the world (ibid. 2000). Invariably, acquisitions increase firm revenues, a key financial metric in the evaluation of dot.com companies, and also provide a way of acquiring unique resource bundles (Capron and Hulland 1999; Rabinovitz *et al.* 2000).

Real options theory

As firms create innovative strategies and new information-centric business models in an increasingly uncertain world, they are, in effect, developing a portfolio of real options (Kogut and Kulatilaka 1994). These learning options are similar to financial options and, as such, increase in value with rising uncertainty (Copeland and Keenan 1998). This partially explains the high market capitalisation of Web-based service firms, and the vast number of start-up service firms. It also explicates why established service firms are creating Web-based spin-offs, outsourcing service functions and engaging in alliances among service firms. Consistent with real options theories, Web-based alliances are different than traditional ones in four main ways.

First, they involve a much larger and varied group of companies, e.g. e-Bay's value web includes over six different companies that vary from credit card processing to a competitive auction site referring customers to e-Bay; second, they rely on more informal business relationships, e.g. each part of the value web provides a unique, coordinated service integrated through information rather than ownership; third, they require leadership by one or two companies to define standards for all Web members and create incentives that attract more companies to it (OECD 1998); and fourth, their market value is based on creating new marketspace rather than on models predicting net present value (NPV) returns (Kim and Mauborgne 1999). However, fundamental to these alliances and real options is a desire to move from a marketspace of diminishing returns to one of increasing returns.

To summarise the points made in the previous paragraphs, Figure 11.1 identifies some of the modifications to the OLI configuration depicting IB activity that e-commerce requires to be made. We now turn to examine how some of the mainstream and contextually related IB theories may need to be modified to incorporate the characteristics of e-commerce.

E-commerce and context-specific variables

The eclectic paradigm identifies three sets of contextual variables likely to affect the extent, pattern and form of MNC activity. These are (a) the types of activities engaged in by the firms; (b) the countries or regions of origin of investing firms; (c) certain characteristics specific to individual firms – other than their nationality of ownership – e.g. their size, products and innovative strategies. Let us consider each of these in turn in so far as they may help us explain the extent to which the emergence of e-commerce requires modification to existing theories.

Activity specific factors

Before we address the effect of e-commerce on business activity, it is instructive to set the context of analysis by providing the overall size of the emerging Internet economy. In June 1999, a University of Texas project made a first attempt to quantify the Internet-related sector of the US economy (Thompson 1999). The results of the study suggest that the US Internet Economy generated $301 billion in revenues and employed 1.2 million people in 1998. Working with the International Data Corporation, the University of Texas forecasted the worldwide Internet economy to be $1.8 trillion in 2003 (Thompson 1999). A Commerce Department study put the Internet economy at 8 per cent of the US gross domestic product (GDP) for 1999, while a more conservative estimate by Goldman Sachs, which, *inter alia*, takes the view that a $30 book sold on Amazon.com should not count as an Internet economy transaction, put the Internet economy at

5 per cent of GDP (Ledbetter 1999). A Boston Consulting group study found 1998 B2B e-commerce transactions amounting to $671 billion, and predicts it will grow to over $2 trillion in 2003 (Paperfree 2000). The above numbers provide a range of estimates; however, it is clear that e-commerce spending is large today (3–8 per cent of the US GDP) and will likely double in percentage terms by 2003. We now look at the economic effect of the Internet on the activities of MNCs, on potential MNCs, within traditional industries and on e-commerce providers.

Traditional – minor impact

Traditional goods and services industries that have a relatively simple customer interface and little information and/or cultural asymmetry between buyers and sellers are likely to be the least affected by e-commerce. At McDonalds, for example, although much of the ordering function can be automated, the customer is still getting a standard product (a hamburger) with a standard customer interface (a salesperson). Of course, McDonalds' back office processes associated with physical commodity procedures and the supply function will use intranets and extranets (OECD 1998). A similar minor impact is likely to be observed initially at the customer/store interface for purchasing building supplies at a store like Home Depot; however, even here, e-commerce models associated with grocery purchases are likely to be implemented.

E-commerce is likely to remain excluded from the actual manufacturing, assembling and delivery of most goods, e.g. autos ($350 bln), energy ($230 bln), but will play a key support role (Thompson 1999). However, if these vertically integrated industries choose to outsource key business segments, e.g. manufacturing, then the Internet could play an increased coordinating function. There will likely be slower Internet penetration of efficiency-enhancing measures in the public sector, e.g. primary and secondary education, government agencies, etc., than in the private sector. With regard to horizontally integrated activities, the more culture prone or idiosyncratic an activity is, i.e. language- or contact-dependent, the less likely it is that an international e-commerce solution will be provided (Knight 1999).

Traditional – major impact

The information content, context and preferred delivery modality of traditional products and services determine, in a large part, how the Internet affects them. This is discussed first, followed by an assessment of the effect Internet will have on the overall category.

Those industries most likely to be affected by information content are those which supply products requiring extensive buyer–seller information exchange, e.g. the sound of a CD that is trialed on Amazon.com; those which require a large number of separate transactions, e.g. Detroit's Big Three automakers creating an auction supplier network to process $240 billion in annual purchases and cut ordering processing fees by 90 per cent (Dalton 2000); and those that require the production and consumption of the product simultaneously, e.g. reading the on-line version of the *Wall Street Journal* (OECD 1999). One unique feature of e-commerce information-gathering is that it may occur temporally and/or be geographically separate from the actual business transaction. Separately, consumers may research an automobile cost and features on AutoByTel.com, but purchase it at a dealer a week later.

The industries where the information context is sufficient for the customer to make a purchase decision, e.g. the size, colour and price of a dress from an on-line vendor like Lands End, are most affected. When one source of information must be combined with other information, e.g. a doctor's signature for ordering from Drugstore.com, the impact is dramatically reduced (Hof and McWilliams 1998).

The last part of the information variable is infrastructure that affects the modality of information delivery, i.e. electronically, print, in person. Goods and services industries that involve timely, stand-alone, electronic information are most likely to be replaced by Internet services (Hamel and Sampler 1998), e.g. stock quotations (E-Trade), speciality chemical catalogues (Chemdex.com), parts ordering on the CommerceOne site being constructed for the auto industry. Dell and Cisco Systems are taking their ordering systems and most of their customer service functions on-line while FedEx focuses only on the customer service tracking component on-line.

It is estimated by Thompson (1999) that these traditional goods and services industries that made up approximately 95 per cent of the US economy in the late 1990s are likely to account for 86 to 90 per cent of the economy in 2003. The vast majority of the $1.5 trillion dollar Internet spending in 2003 will remain between businesses for traditional cost-saving (supply chain management) and revenue-targeting applications (Lawrence 2000). In fact, the International Data Corporation predicts that the percentage of B2B will increase from 72 per cent in 1999 to over 86 per cent in 2003 of total Internet spending (ibid.).

E-commerce providers

More than 200 Internet firms went public in 1999 and, in so doing raised about $20 billion (Ledbetter 1999). This was more than double the number which went public in the previous four years. Clearly, any product or function of a product that can be delivered digitally (newspapers, books, CDs, software), or whose major processes can be linked digitally, is a candidate to become an e-commerce industry. Some deliveries will have to wait for infrastructure improvements, such as video on demand and expanded wireless, but it is expected that advances in high-speed infrastructure technology will make them viable in the very near future. For example, an exciting new range of products about to emerge is that of Internet home appliances, e.g. refrigerators electronically monitoring the use of milk and reordering its bar code when supplies are low via a wireless link to the store.

Access companies are branching out into content, such as the AOL merger with Time-Warner, and portal companies are increasing their customer base, such as Yahoo possibly merging with e-Bay (Lewis 2000). E-commerce-related industries will be created that redefine customer information collection and use. As more transactions occur on the Internet, companies such as BroadVision and DoubleClick are assembling increasingly sophisticated customer profiles, and using artificial intelligence algorithms to target customers for advertising. To allay public concern and to preclude government solutions to privacy market failure, many data collection companies are supporting industry standards with respect to the information they will collect on Internet customers (Gork 2000).

Another area where e-commerce is likely to excel is the creation of on-line auctions. Indeed, as the London *Economist* (2000c) has recently observed, the Internet provides a perfect medium for aggregating buyers and sellers from all around the world (*Economist* 2000c). This application can be both consumer-to-consumer (e-Bay) and business-to-business (the Big Three auto parts network). Its basic property is that it creates an efficient market by congregating informed buyers and sellers in cyberspace rather than at a specific physical location. The increased number of participants makes the auctions more efficient, while the auction companies attempt to establish rules of conduct.

Cross-industry effect

More generally, the advent of e-commerce is causing us to question the appropriability of the industry as a unit of analysis. A reconfigured specialisation of Internet firms is creating a whole new set of strategic business groups (SBGs), e.g. security on the net, customer Internet tracking, shopping agents (services that electronically compare prices for consumers), routers (fast packet switches), etc. The boundaries of existing industries are being reconfigured. The US automobile industry for example, which was previously vertically integrated, has now evolved an active auction market for parts procurement, while auto consumers are accessing information about product, prices and dealers cost on the Internet (and are gaining an advantage in price negotiations). Priceline.com is de-branding the hotel and airline sectors by allowing customers to bid for lodging and air transport, but not permitting them to question the specific provider. Amazon.com and Chemdex.com are disintermediating the retail book distribution and chemical catalogue businesses, etc. (OECD 1999). Customer interface processes, WEB portals have the potential to replace product groupings as a more meaningful way of segmenting businesses (Malone and Laubacher 1998). Amazon.com's business model is based on competing for customers across a broad set of unrelated products rather than on the traditional industry model which started with a particular product class and becomes more focused until a specific product is identified for purchase (*Economist* 2000b).

Country- or region-specific factors

There is little doubt that the Internet reduces spatially related transaction costs and makes international market transactions more efficient. Increased trade is likely to bring new pressures on hierarchal and alliance modalities, particularly by forcing them to innovate and become more efficient. However, in support of increases in alliances and FDI, Hart (1996) found that the Internet has become an effective medium for bringing together people around the world as business partners – people who probably would otherwise never meet. It is likely that in most industries all three (trade, FDI, alliances) will coexist; however, one will likely dominate based on the unique characteristics of the industry in question, and of its locational needs.

The assessment of trade, FDI and alliance trade-offs using a market failure lens provides additional insights. As product markets become more imperfect, market failure due to poor information is increased. Production will be increasingly outsourced and driven to areas of low factor costs, thereby increasing trade. However, because of the increased time pressure associated with the Internet economy, existing long-distance trade patterns, for example, between the US and Southeast Asia for computer parts, may be shortened to trade patterns between the US and Mexico. FDI either through acquisition or greenfield will both stimulate and mitigate market failures. E-commerce ventures by US firms in foreign countries associated with vertical integration will either expand their supply chain or retailing, and will likely have positive externalities on local firm e-commerce activities due to spillover effects.

On a country specific basis, the dominant home country for e-commerce has been, and will likely remain in the near future, the United States. In 1999, it accounted for two-thirds of the world market (Lawrence 2000). *Inter alia*, the US's entrepreneurial climate, *laissez-faire* government policy, low and flat-rated telecommunication access structure, high quality tech chip and communication research, extensive PC penetration and extensive venture capital market have all facilitated the e-commerce explosion (OECD 1999). The spectacular advances in telecommunication technology have been a major factor leading to increases in US productivity and stock prices that have fuelled further growth in e-commerce (Henry 1999). However, by 2003, the US e-commerce share is projected by International Data Corporation to decrease to 54 per cent with Western Europe increasing 16 percentage points to 33 per cent and Asia accounting for the remaining 14 per cent (Lawrence 2000). The major reasons for the projected percentage decline include: (1) European venture capitalists are finally realising that they need to get up to speed on the Web; (2) the Finns, who pioneered the mobile phone revolution, are now way ahead in the next trend: viz., wireless Web; (3) the European Union is telling US companies not to collect personal data via the Internet on its people; (4) e-commerce and on-line advertising are ripe to bloom in China and Japan; (5) Spain's largest Internet provider has introduced a free service in Brazil, and this will dramatically change the country's e-commerce landscape (*Standard* 2000b).

Firm-specific factors

We shall discuss two main groups of factors specific to influencing individual firms. The first are those identified by most theories embodied in the eclectic paradigm which treat the firm as a stand-alone entity. The second are those which view a firm OLI configuration as being affected by network functions and complexities of the networks in which it is involved; then it harnesses and consolidates network advantages with its own core competences.

The firm as a stand-alone entity

Received economic and managerial theories of IB activity[2] view the firm as a seeker after natural resources, markets, efficiency and strategic assets.

Resource-seeking investment relates to Internet activities that are, for the most part, created rather than natural, and intangible rather than tangible. With the exception of low-wage programmers in India, resource-seeking investment from Internet specialist firms will focus on acquiring marketing expertise in culturally diverse countries through alliances and joint ventures, e.g. AOL, Lyco and AOL in India (Manzar 1999; Rabinovitz *et al.* 2000). For firms wishing to save operating costs, the Internet permits a much tighter linking of manufacturers' supply chains, thereby facilitating foreign subcontracting FDI, e.g. US clothing manufacturers in Vietnam (Baxter *et al.* 1998). One can make a case for the reduction of Internet-related resource-seeking FDI by assuming that firms can more easily affect customer buying behaviour in foreign nations from their home servers. If so, this may eliminate the need for a local partner who is familiar with the local market. Moreover, the Internet is likely to have a homogenising effect across countries that also would reduce the need for a local partner (Wymbs 1997).

Market-seeking FDI by Internet companies is likely to favour a location that offers agglomerative economies, e.g. US information centres like Silicon Valley, New York and Boston. These areas have entrepreneurs who have built billion-dollar businesses and are seeking to recycle themselves. In fact, Jim Clark of Silicon Valley has already built three billion-dollar businesses, Silicon Graphic, Netscape and Healtheon (Lewis 1999). Also, the US capital market is well schooled in funding e-commerce start-ups, which is not the case around the world. With regard to firms using e-commerce, the move on-line by Asian suppliers has profound implications for the way business is conducted worldwide: developing Asia is already home to a high percentage of the world's manufacturing capacity, and increased virtual links with the West will benefit Asian exporters, e.g. the Web allows the obscure computer peripherals manufacturer in Taipei to catch the eye of a large computer manufacturer in Texas (Jacob 2000).

Efficiency seeking investment, driven by Smithian effects, has gravitated to Silicon Valley, London, Finland, New York and Boston (Florida 1995, 1998). However, the distributed properties of the Internet will result in other development centres possibly springing up in Singapore and Hong Kong, both of which are rapidly becoming knowledge-intensive co-ordination and management centres for Southeast Asian business (Enright 2000). Cultures and government telecommunication policies may stimulate or retard the formation of efficient production centres in continental Europe and Japan. For example, historically, American regulators have viewed the telephone as a basic necessity and developed pricing structures to encourage its maximum penetration. In most other countries, the telephone has been provided by a state enterprise who viewed it more as a luxury good, and priced access on a high, usage-sensitive (per minute) basis. However, in March 2000, AltaVista and NTL began flat-rate access in the UK and this is likely to encourage e-commerce usage in that country. Interestingly, the exact opposite is true for the cellular telephone service, which American regulators treated as a luxury, and, until recently, they permitted a duopoly market structure that resulted in relatively high prices (Aufderheide 1999). By contrast, cellular service, particularly in Europe and Japan, has been looked at as a cost-effective alternative to high-priced traditional service. Not surprisingly, because most computers today are connected via traditional telephone and cable service, the US has developed a tremendous lead in Internet usage and applications. However, the Europeans, particularly the Scandinavian countries, and Japan are viewed as the leaders in wireless applications connecting to the Internet.

Strategic asset investment designed to protect or augment firm-specific advantages will be from foreign e-commerce firms investing in the United States to tap into its established information knowledge nodes like Silicon Valley and centres of funding like New York City. For example in March 1999, the Singapore AMO, a competitor of CommerceOne, moved its headquarters to Silicon Valley. AMO's chairman, T.K. Wong said, 'This (Silicon Valley) is where the action is. If you're not in Silicon Valley, you are playing a defensive strategy – waiting in Asia for some American company to come and acquire you' (Hamilton 2000). In the short term, the majority of the strategic asset-augmenting investment will take place primarily in two areas. The first one is by leading-edge technology firms, e.g. Cisco

Systems, who, due to their high market capitalisations, are able to buy high quality R&D in the market through acquisition rather than develop it themselves. The second area involves e-commerce businesses that are looking to expand into new markets. As European and Asian e-commerce start-ups emerge, US cash-rich firms will actively seek to acquire them. Clearly, the increasing returns nature of the network-centric business encourages acquisitions, reducing the number of competitors and increasing first-mover advantage (Rayport and Sviokla 1998).

The firm as part of a network of interrelated activities

E-commerce, along with globalisation and the emergence of knowledge-based and alliance capitalism, is requiring scholars to reappraise their concept of the firm as a stand-alone entity. In particular, the Internet, as its very name implies, is not only associated with more and deeper networking activities – both of a horizontal and vertical kind – but is upgrading the firm-specific advantages arising from such extra-firm linkages.[3] E-commerce affords the participating firms greater, more efficient, speedier and more cost-effective access to resources and customers and a different set of ownership, location and organisational capabilities than non-members. Hamel (1999) views networks to be driving the new techno-economic business paradigm and believes that a firm's strategy, i.e. response to e-commerce, is its key to survival. Strategy is particularly important because the network-centric Internet world is transforming many previous business strengths, such as broad market coverage and significant bricks and mortar assets into liabilities. It is forcing traditional firms to evaluate how new kinds of information and new modes of delivery will affect their businesses and whether they need to create Internet specific responses. If they do, then they have to manage the very difficult problems of cannibalisation and channel conflict between two worlds (*Economist* 2000d).

This emerging paradigm for certain electronic products, e.g. news, information, CDs, software, is likely to substantially remove distance and geographic barriers from the equation. From a strategic perspective, electronic products take on the characteristics of a global product because they can be produced in one location and distributed around the world. However, from the marketing perspective, these products can be tailored to the needs of a particular country or a particular region, and, as a result, have the look and feel of a multi-domestic product. In effect, the Internet permits a company to exploit previously untapped opportunities of wide reach (size of audience), great richness (customised products) and strong affiliation (personalised responses to customers' interests) (Evans and Wurster 1998).

A second strategic implication of the network-centric Internet world is that it separates the information value chain from the production value chain. The electronic linking of supply chain production processes electronically has been going on for a while (KPGM 1999). What appears to be new is the emergence of information businesses that collect, package and sell information on consumers' buying behaviour and habits and the emergence of agents acting on behalf of consumer groups. The former allow firms to better target customers and to create a one-to-one marketing experience, while the latter offer consumers value-added services (*Economist* 2000a).

More specifically, the application of the core competence of firms has to be segmented by the following four applications (*Economist* 2000c): business-to-business; business-to-consumer (B2C); consumer-to-consumer (C2C); consumer-to-business (C2B). For B2B, large firms that can dictate terms to suppliers are likely to experience the greatest benefits of price reductions associated with e-commerce. For example, GE, via their transaction processing network (TPN), GM and Ford have extracted billions of dollars of cost savings from their suppliers (Margherion 1998).

With regard to B2C, the most impressive innovations are coming from *smaller firms* who do not have existing customer interface baggage. They are using information technology to attack separate pieces of the value chain. Each new approach to better serving customers or to making a distribution process more cost effective is brought to the well-established venture capital market for review. Each time one of these companies obtains funding, it represents an attack on an inefficient dimension of an existing company's value chain. Few established firms have created procedures to beat back or even neutralise these aggressive start-ups.

C2C activity is associated with the proliferation of auctions such as e-Bay, while the C2B is related to the power of the Internet to drive transactions the other way around, i.e. would-be passengers bidding for airline tickets on PriceLine.com (*Economist* 2000c).

Conclusions and future research

This chapter has attempted to reorient the received theories and the eclectic paradigm of IB activity to take account of the growth of e-commerce. Though all the dimensions are not fully fleshed out, we believe that the chapter provides a foundation to begin debate and assessment of defensible trajectories for the IB field in an information-centric world.

International business theories that seek to address the competitive advantages of MNCs appear to be applicable. The criteria of 'what a core competence is' will change, e.g. maintaining buying profiles and preferences on 22.5 million customers like Amazon.com, but the necessity of core competences to achieve and advance sustainable competitive advantages is still applicable. Internalisation to increase efficiency and reduce related transaction cost appears valid in a network-centric world, e.g. Nike, WalMart and the auto companies are using the Internet to seek out and work with suppliers (*inter alia,* the Internet aids the exchange of data regarding design, development, specification). Even in a virtual world, location still matters. In fact, the Internet makes it all the more important for governments – local and national – to ensure their countries have (or can provide) all the necessary Internet infrastructure (robust telecommunication networks and commerce-friendly legal, financial and tax environments) to attract and retain 'e-commerce firms' and 'firms using e-commerce'. Though the Internet is labelled as a distance-insensitive medium and can efficiently create links among firms around the world, it is somewhat ironic that e-commerce firms are concentrated in relatively small areas, e.g. Silicon Alley and Silicon Valley, that can clearly be traced to computer, communication and media knowledge centres of the mid-1990s.

We believe that the Internet has provided, and will continue to provide greater access to information and opportunities to build relational assets both within and between firms. Exploring the many dimensions of how relational networks associated with OLI advantages will change the competitive dynamics among firms appears to be fertile ground for future research.

Notes

1 See, e.g., Dunning (1988, 1995, 1998). For a recent exposition of the contents of the paradigm, and how it relates to a number of mainstream theories of international business, see Dunning (2000, 2001).
2 As reviewed by Dunning (2000)
3 Some of these are described in recent papers by Chen (2000) and Chen and Chen (1998).

References

Anderson, E. and Gatignon, H. (1986) 'Modes of foreign entry: transaction costs and propositions', *Journal of International Business Studies*, 17: 1–26.

Applegate, L. (1995) 'Electronic commerce: trends and opportunity', *Harvard Business School Case No 9-196-006*, 6 October.

Arrow, K. (1999) *Amici Curiae Brief*, DOJ vs. Microsoft Case.

Arthur, W. (1996) 'Increasing returns in the new world of business', *Harvard Business Review*, www.hbsp.harvard.edu/products/hbr/julang96/96401.html.

Aufderheide, P. (1999) *Communication Policy and the Public Interest,* New York: Guilford Press.

Barney, J.B. (1991) 'Firm resources and sustained competitive advantage', *Journal of Management*, 17: 99–120.

Bartlett, C.G. and Ghoshal, S. (1989) *Managing Across Borders: The Transnational Solution,* Boston, MA: Harvard Business School Press.

Bartlett, C.G. and Ghoshal, S. (1993) 'Beyond the M-form: towards a managerial theory of the firm', *Strategic Management Journal*, 14(1): 23–46.

Baxter, A., Koudal, P. and Mathur, S. (1998) *Global Report: 1998 Vision in Manufacturing,* survey by Deloitte and Touche and Deloitte Consulting, New York.

Beck, J., Morrison, A. and Bouquet, C. (2000) 'Netchising: the next global wave', *e-Commerce and Global Business Forum*, Santa Cruz, mimeo.

Bernstein, P. (1998) 'Are networks driving the new economy?', *Harvard Business Review*, November–December, Reprint 98602.

Boddewyn, J.J. and Iyer, G. (1999) 'International business research: beyond déjà vu', *International Management Review*, 39: 161–84.

Bond, J. (1998) 'The divers of the information revolution – cost, computing power and convergence', *Private Sector*, World Bank, Washington, DC, 27–30.

Buckley, P.J. and Casson, M.C. (1976) *The Future of the Multinational Enterprise*, London: Macmillan.

Buckley, P.J. and Casson, M.C. (1985) *The Economic Theory of the Multinational Enterprise*, London: Macmillan.

Cantwell, J.A. (1989) *Technological Innovation and Multinational Corporations*, Oxford: Basil Blackwell.

Cantwell, J.A. (ed.) (1994) 'Transnational corporations and innovatory activities', *United Nations Library on Transnational Corporations*, 17, London: Routledge.

Capron, L. and Hulland, J. (1999) 'Redeployment of brands, sales forces and general market management expense following horizontal acquisitions: a resource-based view', *Journal of Marketing*, 63(2): 41–54.

Cartwright, S. and Oliver, R. (2000) 'Untangling the value web', *Journal of Business Strategy*, 20: 22–7.

Caves, R.E. (1996) *Multinational Firms and Economic Analysis, 2nd edition,* Cambridge: Cambridge University Press.

Chen, T.J. (2000) 'Network resources for internalization', *The Case of Taiwan's Electronics Firms*, Taipai: National Taiwan University.

Chen, T.J. and Chen, H. (1998) 'Network linkages and location choice in foreign direct investment', *Journal of International Business Studies*, 29: 445–69.

Coase, R. (1999) 'Talking about tomorrow – Ronald Coase', T. Petzinger, Jr. *Wall Street Journal*, 12/31/99, 36.

Cohen, B. (2000) 'Marketing money: currency policy in the globalized world' in A. Prakash and J. Hart (ed.), *Coping with Globalization*, London and New York: Routledge.

Conner, K.R. and Prahalad, C.K. (1996) 'A resource based theory of the firm knowledge versus opportunism', *Organizational Science*, (5): 477–501.

Copeland, T.E. and Keenan, P.T. (1998) 'Making real options', *McKinsey Quarterly*, 3: 128–41.

Dalton, G. (2000) 'The suppliers' demand', *The Standard*, 6 March.

Desmet, D., Francis, T., Hu, A. and Koller, T. (2000) 'Valueing dot-coms', *McKinsey Quarterly*, 1.

Dicken, P. (1998) *Global Shift,* London: Guilford Press.

DoubleClick (2000) October 25, 2000.

Doz, Y.L., Asakawa, K., Santos, J.F.P. and Williamson, P.J. (1997) 'The Metanational Corporation, Fontainebleau', France: *INSEAD Working Paper* 97/60/SM.

Dunning, J.H. (1958) *American Investment in British Manufacturing Industry*, London: George Allen and Unwin, Reprinted (1976) New York: Arno Press.

Dunning, J.H. (1976) *US Industry in Britain,* London: Wilton House.

Dunning, J.H. (1988) *Explaining International Production*, London: Unwin Hyman.

Dunning, J.H. (1993) *Multinational Enterprises and the Global Economy*, Wokingham, Berkshire: Addison Wesley.

Dunning, J.H. (1995) 'Reappraising the eclectic paradigm in the age of alliance capitalism', *Journal of International Business Studies*, 26: 461–91.

Dunning, J.H. (1998) 'Location and the multinational enterprise: a neglected factor', *Journal of International Business Studies*, 29(1): 45–66.

Dunning, J.H. (2000) 'The eclectic paradigm as an envelope for economic and business theories of international business activity', *International Business Review*, 9: 163–90.

Dunning, J.H. (2001) 'The eclectic, or OLI, paradigm of international production past, present and future', Reading and Rutgers Universities, mimeo.

Dunning, J.H. and Wymbs, C. (1999) 'The geographical sourcing of technology-based assets by multinational enterprises' in D. Archibugi, J. Howells, and J. Michie (eds), *Innovation Policy in a Global Economy,* Cambridge: Cambridge University Press.

Economist, The (2000a) 'In the great Web bazaar', 2/26/00–3/3/00: 40–4.

Economist, The (2000b) 'Amazon's amazing ambition', 2/26/00–-3/3/00: 24.

Economist, The (2000c) 'First America, then the world', 2/26/00–-3/3/00: 49–53.

Economist, The (2000d) 'Something old, something new', 2/26/00–3/3/00: 15–16.

Enright, M. (2000) 'Globalization, regionalization and the knowledge-based economy in Hong Kong', in J. Dunning, *Globalization, Regionalization and The Knowledge-Based Economy,* New York: Oxford University Press.

Evans, P. and Wurster, T. (1998) 'Strategy and the new economics of information', *Harvard Business Review*, September–October, 70–82.

Evans, P. and Wurster, T. (1997) 'Strategy and the new economics of information', *Harvard Business Review,* September–October, 70–82.

Florida, R. (1995) 'Toward the learning region', *Futures* 27(5): 527–36.

Florida, R. (1998) 'Calibrating the learning region', in J. DeLaMothe and G. Paquet (eds), *Local and Regional Systems of Innovation,* Boston, MA: Kluwer Academic Publishers.

Freeman, C. and Perez, C. (1988) 'Structural crisis of adjustment, business cycles and investment behaviour', in G. Dosi, C. Freeman, R. Nelson, G. Silverberg and L. Soete (eds), *Technical Change and Economic Theory,* London: Pinter.

Gates, W. (1999) *Business @ The Speed of Thought,* New York: Warner Books.

GE (General Electric) (1999) 'Building virtual global trading communities through electronic commerce outsourcing', White Paper.

Gimein, M. and Lash, A. (1998) 'Amazon moves to build E-commerce empire', *The Standard,* 7 August.

Gork (2000) 'DoubleClick cries uncle', *The Standard, Media Gork,* 3/3/00.

Graham, E.M. (1990) 'Exchange of threats between multinational firms as an infinitely repeated non-cooperative game', *International Trade Journal,* 4(3): 259–77.

Graham, E.M. (1998) 'Market structure and the multinational enterprise: a game-theoretic approach', *Journal of International Business Studies*, 29(1): 67–84.

Hamel, G. (1999) 'The Strategos manifesto', http:www.strategosnet.com/about/manifesto.html.

Hamel, G. and Sampler, J. (1998) 'The E-Corporation', *Fortune*, 7 December.

Hamilton, A. (2000) 'Asia: from east to west', *The Standard*, 7 February.

Hart, A. (1996) 'Howdy Partner', *Internet World*, 7(11), http://www.iworld.com.

Helft, M. (2000) 'Amazon.com: more than a merchant', *The Standard*, 17 January.

Hennart, J.F. (1982) *A Theory of Multinational Enterprise*, Ann Arbor, MI: University of Michigan Press.

Hennart, J.F. (1989) 'The transaction cost theory of the multinational enterprise', in C.N. Pitelis and R. Sugden (eds), *The Nature of the Transnational Firm*, London: Routledge.

Henry, D. (1999) 'The emerging digital economy II', US Department of Commerce, http:www.ecommerce.gov.

Henson, C. (2000) 'David.com vs. Goliath', *Best's Review*, 101(1) (May): 57–60.

Hill, C. (1997) 'Establishing a standard: competitive strategy and technological standards in winner-take-all industries', *The Academy of Management Executive*, 11: 7–25.

Hof, R. and McWilliams, G. (1998) 'The "click here" economy', *Business Week*, 6/22/98, http:www.businessweek.com(1998/25itspec98).htm.

Hunt, S. (1997) 'Resource-advantage theory: an evolutionary theory of competitive firm behaviors?', *Journal of Economic Issues*, 31(1): 59–77.

Iansiti, M. and MacCormack, A. (1997) 'Developing products on Internet time', *Harvard Business Review*, September–October, 108–17.

IBM (1999) 'IBM targets US$10–US$15bn e-business sales in 1999', *New Straits Times Press*, 6 April, 18.

IDA (1999) New York City Industrial Development Agency, New York.

Jacob, R. (2000) 'Buyers and sellers flock to online Asian bazaar', *Financial Times*, 3/29/00: 10.

Kim, E.C. and Mauborgne, R. (1999) 'Importance of an alternative path', *Financial Times*, 6/10/99: 24.

Knickerbocker, F.T. (1973) *Oligopolistic Reaction and the Multinational Enterprise*, Cambridge, MA: Harvard University Press.

Knight, G. (1999) 'International services marketing: review of research, 1980–98', *Journal of Services Marketing*, 13(4/5): 346–60.

Kobrin, S. (1999) 'You can't declare cyberspace national territory', in D. Tapscott (ed.), *BluePrint to the Digital Economy*, Cambridge, MA: Harvard University Press.

Kogut, B. (1985) 'Designing global strategies: profiting from operational flexibility', *Sloan Management Review*, 26 (Fall): 27–38.

Kogut, B. and Kulatilaka, N. (1994) 'Operational flexibility, global manufacturing and the option value of a multinational network', *Management Science*, 40(1): 123–39.

Kogut, B. and Zander, U. (1994) 'Knowledge of the firm and the evolutionary theory of the multinational corporation', *Journal of International Business Studies*, 24(4): 625–46.

Kotabe, M., Murray, J. and Javalgi, R. (1998) 'Global sourcing of services and market performance: an empirical investigation', *Journal of International Marketing*, December: 10–31.

KPGM (1999) *Supply Chain Solutions*, New York.

Kuemmerle, W. (1999) 'The drivers of foreign direct investment into research and development: an empirical investigation', *Journal of International Business Studies*, 30(1): 1–24.

Larson, P. (2000) 'Yahoo! Looks to expand abroad', *Financial Times*, 4/6/00: 20.

Laseter, T., Houston, P., Wright, J. and Park, J. (2000) 'Amazon your industry. Extracting value from the value chain', *Strategy and Business*, 94–105.

Lawrence, S. (2000) 'The net world in numbers', *The Standard*, 8 February.

Ledbetter, J. (1999) 'The Internet economy gets real', *The Standard*, 20 December.

Lessig, L. (1999) *Code and Other Laws of Cyberspace*, New York: Basic Books.

Lewis, M. (1999) *The New New Thing: A Silicon Valley Story*, New York: W.W. Norton & Co.

Lewis, W. (2000) 'Yahoo! And e-Bay restart merger talks', *Financial Times*, 3/24/00: 1.

Lipsey, R.G. (1997) 'Globalization and national government policies: an economist's view', in J. Dunning (ed.), *Governments, Globalization and International Business*, Oxford: Oxford University Press.

Lipsey, R.G. and Bekar, C. (1995) 'A structuralist view of technical change and economic growth' in P.J. Courchene (ed.), *Technology, Information & Public Policy*, Kingston, Ontario: John Deutsch Institute.

Malone, T. and Laubacher, R. (1998) 'The dawn of the E-lance economy', *Harvard Business Review*, September–October, Reprint 98508.

Manzar, O. (1999) 'Yahoo, Lycos to launch India portal', *The Standard*, 18 November.

Margherion, L. (1998) 'The Emerging Digital Economy', US Department of Commerce, 1-50.

Muehlbauer, J. (2000) 'Does Chemdex need a mom', *Forbes*, 10 February.

Nelson, R.R. (1991) 'The role of firm differences in an evolutionary theory of technical advance', *Science and Public Policy*, 18(6): 347–52.

Nelson, R.R. and Winter, S. (1982) *An Evolutionary Theory of Economic Change*, Cambridge, MA: Belknap Press.

OECD (1998) 'The Economic and Social Impact of Electronic Commerce' .

OECD (1999) 'E-commerce', http:www.oecd.org/subject/e_commerce/summary.htm.

Paperfree (2000) 'Multiple messaging standards', *Electronic Commerce World*: 12.

Peterson, H. (1986) *Business and Governments*, New York: Harper & Row.

Pitt, L., Berthon, P. and Watson, R. (1999) 'Cyberservices: taming service marketing problems with the WWW', *Business Horizons*, 42(1): 11–18.

Porter, M.E. (1990) *Competitive Advantage of Nations*, New York: Free Press.

Porter, M.E. (1994) 'The role of location in competition', *Journal of Economics of Business*, 1(1): 35–9.

Porter, M.E. (1998) 'Location, clusters and the "new" micro-economics of competition', *Journal of Business Economics*, 33(1): 7–13.

Prahalad, C.K. and Doz, Y.L. (1987) *The Multinational Mission: Balancing Local Demands and Global Vision*, New York: Free Press.

Priceline (2000) 'Now you can name your own price for groceries!', *Priceline WebHouse Club*, February: 29.

Rabinovitz, J., Dalton, G. and Helft, M. (2000) 'The urge to merge', *The Standard*, 17 January.

Rayport, J. and Sviokla, J. (1998) *Winning in Electronic Commerce*, Tape, New York: Harvard Business School Publishing.

Reedy, J., Schullo, S. and Zimmerman, K. (2000) *Electronic Marketing*, New York: Harcourt College Publishing.

Rugman, A.M. (1979) *International Diversification and the Multinational Enterprise*, Lexington, MA: Lexington Books.

Rugman, A.M. (1981) *Inside the Multinationals and The Economics of Internal Markets*, London: Croom Helm.

Sahlman, W.A. (1996) *Some Thoughts on Business Plans, The Entrepreneurial Venture*, Cambridge, MA: Harvard Business School Press.

Sprenger, P. (2000) 'U.K. government backhands dot-coms', *The Standard*, 24 March.

Standard, The (1999) 'March on the marts', 16 December.

Standard, The (2000a) 'Rare European win in e-commerce', 29 February.

Standard, The (2000b) 'The Internet economy goes global', 7 February.

Strange, S. (1997) 'An International Political Economy Perspective', in J. Dunning (ed.), *Governments, Globalization and International Business*, Oxford: Oxford University Press.

Taylor, P. (2000) 'How the Internet will reshape worldwide business activity', *Financial Times*, 4 July: 1–2.

Teece, D.J., Pisano, G. and Shuen, J. (1997) 'Dynamic capabilities and strategic management', *Strategic Management Journal*, 18(7): 509–33.

Thompson. M.J. (1999) 'Net economy pegged at $301 billion', *The Standard*, June 10.

Timmons, J. (1990) *New Venture Creation*, Boston, MA: Irwin.

Tomkins, R. (1999) 'Reinventing the dinosaur', *Financial Times*, 8 June: 13.

Turban, E., Lee, J., King, D. and Chung, H. (2000) *E-Commerce*, Upper Saddle River, NJ: Prentice Hall.

Vernon, R. (1973) *Sovereignty at Bay*, Harmondsworth: Penguin.

Vernon, R. (1983) 'Organizational and institutional responses to international risk', in R.J. Herring (ed.), *Managing International Risk*, Cambridge: Cambridge University Press.

Vernon, R. (1994) 'Research on transnational corporations', *Transnational Corporations*, 3: 137–58.

Watson, R., Akelsen, S. and Pitt, L. (1998) 'Attractors: building mountains in the flat landscape of the world wide web', *California Management Review*, 40(2) (Winter): 36–56.

Wernerfelt, B. (1984) 'A resource-based view of the firm', *Strategic Management Journal*, 5(2), 17: 1–80.

Wernerfelt, B. (1995) 'The resource-based view of the firm: ten years after', *Strategic Management Journal*, 16: 171–4.

Wesson, T.J. (1993) 'An alternative motivation for foreign direct investment', Cambridge, MA: Ph.D. dissertation, Harvard University Press.

Williamson, O.E. (1985) *The Economic Institutions of Capitalism*, New York: Free Press.

Wymbs, C. (1997) 'Internet: its effect on international trade, foreign direct investment and strategic alliance formation', International Telecommunications Society and International Council for Computer Communications Conference in Calgary, Canada, June, mimeo.

Wymbs, C. (2000a) 'How electronic commerce is fundamentally changing the service industries,' *Journal of Services Marketing*, 14(6): 463–78.

Wymbs, C. (2000b) 'The impact of the information revolution of the global corporation', in J. DeLaMothe and G. Paquet (ed.), *Information, Innovation, and Impacts*, Boston, MA: Kluwer Academic Publishers.

Yoffie, D. and Cusumano, M. (1999) 'Building a company on Internet time: lessons from Netscape', *California Management Review*, 41(3) (Spring): 8–28.

12 The recent globalisation of the Italian banking sector

An interpretation based on the eclectic paradigm

Lucia Piscitello

Introduction

Over the last decades or so, a number of events have occurred that fostered unprecedented opportunities and challenges to the whole world economy. The pre-eminent driving force behind these events has been a series of systemic technological and political changes ranging from liberalisation of trade policies, privatisation processes and globalisation of markets (Dunning 1995; Miller and Parkhe 1998; Berger *et al.* 1999; Mallampally and Zimny 2000). Service sectors and, particularly, financial and banking sectors, have experienced a season in which both the products offered and the regulatory framework have undergone significant changes. The range of products available to multinational banks (MNBs) has expanded significantly, and at the same time many of the domestic banking markets, within which the MNBs also compete, have undergone significant regulatory reforms (Dewatripont and Tirole 1994; Williams 1997). In fact, at the end of the 1990s the world FDI outward stock in the tertiary sector accounted for 54.7 per cent of the total, and the inward stock for 50.3 per cent. Financial services constituted 27.3 and 16 per cent of the total of services, respectively (UNCTAD 2001).

Numerous studies on the internationalisation of banking and financial services (Dunning 1993) have flourished. Most of the existing literature refers to the US (e.g. Grubel 1977; Goldberg and Saunders 1980, 1981a, 1981b; Gray and Gray 1981; Ball and Tschoegl 1982; Kindleberger 1983; Yannopoulos 1983; Aliber 1984; Nigh *et al.* 1986; Hultman and McGee 1989; Goldberg and Johnson 1990; Goldberg and Grosse 1994; Miller and Parkhe 1998; Berger and DeYoung 2001). Some studies have been put forward for the UK (Jones 1992; Moshirian and Van der Laan 1998), France (Marois and Abdessemed 1997), Germany (Buch and Lapp 1998; Moshirian and Van der Laan 1998; Buch 1999), Japan (Aggarwal 1993; Yamori 1998), and the Nordic European countries (Vastrup 1983; Hellman 1984; Laakso 1984; Flatraaker and Husevåg 1991; Jacobsen and Tschoegl 1997, 1999; Engwall and Wallenstål 1998; Engwall *et al.* 2001). Some evidence has been also proposed with reference to the Italian case but it is mostly limited to some explorative and qualitative studies (see Marchi and Panizza 1993; Nieri 1991, 1993, 1994; Crecchia 1996; Di Quirico 1999; Stanciu 2000).

Both the theoretical and the empirical debate about the international strategies of MNBs is lively, and it mainly focuses upon four issues:

1 The determinants of MNBs' expansion abroad. Most of the papers have dealt with this issue at an aggregate level, focusing on the determinants of the size of the foreign bank presence in a given host country (Tschoegl 1982, 1988; Cho 1985, 1986; Hultman and McGee 1989; Goldberg and Johnson 1990; Grosse and Goldberg 1991; Heinkel and Levi 1992; DeYoung and Nolle 1996);
2 The entry mode choice of banks on foreign markets. Studies dealing with this issue have normally undertaken micro approaches investigating the bank's motivation to adopt different organisational forms[1] (Ball and Tschoegl 1982; Heinkel and Levi 1992; Ursacki and Vertinsky 1992; Gabrowski *et al.* 1993; Blandón 1998);
3 The survival of MNBs in foreign and unfamiliar contexts (e.g. Baldwin and Rafiquzzaman 1995; Zaheer and Mosakowski 1997);
4 The impact of multinationality on banks' productivity and performance (Berger *et al.* 2000a, 2000b; Berger and DeYoung 2001).

The present chapter combines approach (1) and (2) as it addresses the issue of the determinants of the multinational banks' expansion upon foreign markets at a micro individual level, thus allowing for the organisational form chosen by the bank. Specifically, we concentrate on the two organisation forms that refer to the bank's corporate internal growth.[2] Ranked in order of increasing foreign involvement, these forms are: representative offices (employed by multinational banks essentially to build up contacts and relationships with foreign markets), and branches (which have instead the full authority to represent and commit the bank).

Empirical evidence is provided with reference to the Italian case in the decade 1989–99, a period in which important phenomena (such as the upsurge of M&As and equity agreements among banks in the European Union, and the advances in the liberalisation and privatisation processes in Central and Eastern Europe countries, which opened new opportunities for international growth) have changed the competitive scenario and influenced the internationalisation strategies of banks. Specifically, accepting the remit given by the editors of the present book, our aim is to evaluate how the eclectic paradigm is applicable to this particular context, and how it has responded to recent developments and changes.

An eclectic interpretative framework

The international literature on multinational banking mainly relies on the theory of the multinational company (MNC), which explains why MNBs offer transaction services via direct presence rather than by using the open market (Grubel 1977; Gray and Gray 1981; Sabi 1988; Williams 1997). The wider international banking literature relates geographic expansion of banks to the possibility to move towards a more efficient structure thanks to efficiency improvements (or X-efficiency), which may accrue in the form of lower costs or higher revenues (DeYoung and Nolle 1996). Indeed, geographic expansion allows banks to grow upon new markets when domestic growth is limited (the latter, for example, justified globalisation of Japanese banks in the 1980s, see Aggarwal 1993); to diversify the risks in different regional economic environments thus reducing the variation in the organisations' earnings (Aggarwal and Durnford 1989; Berger and DeYoung 2001); to exploit internationally high domestic market to book ratios (Aliber 1984) as the easy availability of capital seems to provide bankers with a significant competitive edge in global markets (Marr *et al.* 1989). Nonetheless, these theories are only partial and they can only explain some kinds of foreign direct investment.

The present chapter relies instead on the eclectic paradigm approach (Dunning 1977) which points to a methodology and to a generic set of variables which contain the ingredients necessary for any satisfactory explanation of all aspects of the phenomenon.[3] The paradigm views the foreign direct investment decision to be a combination of Ownership, Internalisation and Location advantages and, specifically, it asserts that, at any given moment of time, such a decision will be determined by the configuration of three sets of forces (see Chapter 2):

1 The competitive advantages which firms of one nationality possess over those of another nationality in supplying any particular market or set of markets (*ownership advantages*).
2 The extent to which firms perceive it to be in their best interests to internalise the markets for the generation and/or the use of these assets.
3 The extent to which firms choose to locate these activities outside their national boundaries.

Ownership advantages

Ownership advantages are crucial as they allow the foreign bank to overcome the advantages possessed by the domestic banks due to incumbency, and then to compete effectively with them.[4] However, ownership advantages must be thought of also in relation to the international competition mainly from other multinational companies (banks). Today MNCs are generally competing with one another in international markets, they are usually not in the earliest stage of internationalisation and their investment is not all of a local market-oriented kind (see Chapter 1).

In a study prepared for the UNCTC in the late 1980s and published on *The Service Industries Journal*, Dunning (1989) identified five competitive or ownership advantages which are particularly relevant in the case of service firms: (a) quality consistency, reputation and product differentiation; (b) economies of scope; (c) economies of scale and specialisation; (d) access to, control of and ability to effectively process and disseminate information; and (e) favoured access to inputs and/or markets.

Specifically, with reference to multinational banks, ownership competitive advantages are found to arise essentially from their ability to differentiate their banking products (Cho 1986). Such ability stems from the availability of financial and human resources, the possession of superior and unique banking techniques, extensive banking experience, skilled personnel, organisational skills, and a large domestic capital and deposit base (Lees 1974; Cho 1985; Yannopoulos 1983; Aliber 1984).

The majority of studies indicate that larger firms are likely to prefer high levels of equity ownership and commitment (Gomes-Casseres 1985). Smaller firms, which lack the resources or expertise needed to venture into foreign markets, have a different (lower) attitude with regards to risk, thus preferring slimmer organisational forms. Indeed, smaller companies have fewer means of reducing the uncertainty connected with FDI, and they are therefore particularly exposed to the risks inherent in it. Consequently, they tend to prefer internationalisation strategies based on prudence and minimisation of risk, and will therefore prefer representative offices, which allow both set-up costs and the risks involved in FDI to be reduced. This remains the accepted view and will accordingly be stated as the hypothesis. However, it should be pointed out that other studies suggest the opposite, namely that these assumptions may not apply to several service sectors, where the advantages of size may equally well derive from a global network of partnerships and alliances (Gatignon and Anderson 1988; Contractor and Kundu 1998). Nonetheless, as we believe in the 'schumpeterian' relationship between size and the abilities to innovate and differentiate products, the endowment of skills and capabilities, we propose to test the first viewpoint:

H1: We expect a positive relationship between the bank size, its possession of unique resources and capabilities, and its decision to go abroad.

In general, FDI represents greater risk than domestic investments, as it implies entry into a complex environment, and the company must face many unfamiliar factors that cannot always be managed on the basis of experience gained in the country of origin. The lack of international experience may cause the investor to take inappropriate decisions or lead to errors in managing relations with the workforce, customers, competitors and local authorities. As a result, the novice investor tends systematically to overstate risks and understate expected returns of foreign operations (Davidson 1980; Caves 1982). Once the first experience of FDI has been gained, the newly multinational company sets in motion a cumulative learning process in 'going abroad'. The perception of uncertainty decreases and the company acquires growing faith in its own ability to manage foreign operations and to assess the risks and the expected economic returns of FDI correctly (Johanson and Vahlne 1977). As a result, the propensity to accept greater commitments tends to increase while experience in dealing with international operations is accumulated (Anderson and Gatignon 1986). Our hypothesis is therefore:

H2: Rising levels of international experience are positively associated with the bank's decision to go abroad with higher commitment level.

Empirical evidence provided in the literature tends in the main to support these hypotheses with reference to FDI undertaken in both manufacturing (Stopford and Wells 1972; Davidson 1980; Gomes-Casseres 1985; Gatignon and Anderson 1988; Kogut and Singh 1988), in the banking sector (Ball and Tschoegl 1982; Ursacky and Vertinsky 1992; Blandón 1998), and in other service sectors as well (see, for example, Contractor and Kundu 1998, for the hotel sector). The dimensional scale and the relevant endowment of resources, international experience, knowledge and information about foreign markets have been shown to be important factors in explaining banks' international growth (Normann 1984; Goshal 1987; Enderwick 1989; Nayyar 1990; Erramilli and Rao 1990; Erramilli 1991). Ball and Tschoegl (1982) found that foreign direct investment in banking is significantly influenced by the size of the bank and its multinational experience. More recently, Ursacki and Vertinsky (1992), investigating foreign banking expansion in Japan and Korea, confirmed that the size of the bank and its experience in operating in a multinational environment constitute significant determinants of the banks' expansion through branches, while only the experience of the bank operating in a multinational environment reveals as an important factor in explaining foreign banking expansion through representative office (Blandón 1998).

Internalisation advantages

Internalisation advantages explain why multinational firms offer cross-border transaction services via a direct presence rather than by using some other form of international involvement (e.g. licensing to produce the service in a foreign country). As a broad generalisation, the transaction costs of using the market, compared with the hierarchical route of exchanging intermediate services, are likely to be higher than in case of goods (Dunning 1993). Indeed, most services contain a larger element of customer tailoring than do goods and, as information or knowledge related to service activities may be inexpensive to replicate, the possibility of abuse or dissipation of that knowledge is a real threat to the firm possessing it. Additionally, many services are impossible or difficult to trade over space, thus explaining the rapid growth of hierarchical forms in this sector.

Internalisation discussed in the case of multinational banking relates to the product and factor market imperfections and to the economies of internal operations (Gray and Gray 1981). Specifically, internalisation is largely sourced from the role of information, which has a crucial role in banking (Yannopoulos 1983; Tschoegl 1987). The bank–client relationship consists primarily of flows of information and information inputs are difficult to obtain at arm's length due to the failure in the market for information. Dunning (1981) refers to this as cognitive imperfection, which arises whenever information about a product or service is not readily available or is costly to acquire. Information capital provides opportunities for multinational banks to internalise a firm-specific advantage (Miller and Parkhe 1998). Indeed, as observed by Casson (1990), personal contact constitutes one of the main advantages for banks. Preserving established accounts by opening foreign offices becomes a means for protecting knowledge and information networks, thus achieving internalisation advantages. This idea, originally associated to that of a defensive strategy,[5] is also and better known as the 'follow the customer' approach (Kindleberger 1983). Our hypothesis is:

H3: The presence of bank's overseas offices in a country will be positively related to FDI already undertaken by national firms in that country.

Empirical evidence for the hypothesis has already been provided by both trends followed by FDI in the 1960s and 1970s (UNCTC 1987, 1988, 1989), and research studies on multinational banks' activities (Pringle 1966; Dahl 1967; Brimmer and Dahl 1975; Grubel 1977; Fieleke 1977; Metais 1979; Terrel 1979; Goldberg and Sanders 1981a, 1981b; Vastrup 1983). In particular, Nigh *et al.* (1986), Goldberg and Johnson (1990) and Miller and Parkhe (1998) found a positive correlation between US banks' foreign activities and manufacturing activities set abroad by US firms in the periods 1976–82, 1972–95 and 1987–95, respectively. Similarly, Goldberg and Saunders (1981a, 1981b), Hultman and McGee (1989) and Goldberg and Grosse (1994) confirmed the same result with reference to the foreign participation in the US market in the period 1973–86. Yamori (1998) corroborated the evidence with a study on Japanese banks in the first years of the 1990s showing that Japanese banks can exploit their superior knowledge of and contacts with their home-based multinational corporations. Interestingly, a study by Seth *et al.* (1998) using data on the financial sources of affiliates of foreign firms in the US as well as data on the activities of foreign banks in the US, found strong evidence that banks from Japan and the UK did not follow their customers, while the reverse was true for the other countries.

Additionally, the establishment of any initiative by Italian firms in a foreign country also implies flows of goods and services across national borders. These flows necessitate the conversion of the currency of the importing country into the currency of the exporting country, which is facilitated by the presence of banking operations in loco (Gray and Gray

1981), also due to the intangibility and inseparability of banking and financial services, which limit the possibility of trading them across boundaries (Dunning 1989). Accordingly, empirical studies suggest a positive relationship between the presence of banks in a foreign country and the extent of commercial relationships between the two countries involved (Goldberg and Johnson 1990; Miller and Parkhe 1998; Yamori 1998; Buch 1999), as a further evidence for the 'follow the customer' hypothesis. Hence:

H4: The presence of overseas offices of banks in a country will be positively associated with commercial relationships with that country.

This hypothesis has been confirmed by Goldberg and Johnson (1990), and Grosse and Goldberg (1991), who found that bilateral trade contributes to the expansion of foreign banks in the US. Likewise, Miller and Parkhe (1998) and Yamori (1998) found that bilateral trade influenced the behaviour abroad of the US and the Japanese banks, respectively; Galiatsos and Papapetrou (1995) reached similar conclusions for foreign banks in Greece, and Buch (1999) for German banks' internationalisation choices.

Location advantages

Location advantages are regarded as another necessary condition for internationalisation (Dunning 1981, 1993). Many of the factors affecting the location of goods-producing activities help explain the location of services but two additional variables seem to be particularly relevant in the siting of the production of services: (a) the first is the extent to which the service is tradable. Many services are, in fact, location-bound, thus requiring face-to-face contacts between buyers and sellers. Such services can be provided to consumers either by a foreign firm via FDI or by indigenous firm under licence to the foreign producer; (b) the second factor, especially relevant to the location of services, is the regulatory environment of host countries.

With specific reference to multinational banking, those advantages encompass a variety of factors (Gray and Gray 1981; Yannopoulos 1983; Nigh *et al.* 1986) including differences in regulatory structures, the geographical dispersion of the bank's client base, the desire of investors to separate currency from political risk, the entry into growing markets, and the access to a skilled pool of labour. Indeed, banks are found to be attracted to foreign markets to exploit favourable (or less restrictive) foreign banking environment and/or to take advantage of local banking opportunities in foreign countries[6] (Brimmer and Dahl 1975; Kelly 1977; Fieleke 1977; Terrell 1979; Khouri 1979, 1980; Goldberg and Saunders 1980, 1981a, 1981b; Rugman 1981; Aliber 1984; Cho 1985). Therefore:

H5: Banks' FDI is likely to be directed towards more industrialised and high-income countries.

However, the existence of locational advantages in a given country exploitable even by foreign banks is reflected in the characteristics of the local banking market in terms of size, development and efficiency (e.g. Sabi 1988; Amel and Liang 1997; Buch and Lapp 1998; Miller and Parkhe 1998; Buch 1999). Therefore:

H6: Banks' FDI is likely to be directed where business opportunities in the financial sector are higher.

Finally, several studies have already shown that multinational banks' locational choices in a country are strongly attracted by the presence of an important international financial centre in that country (Brealey and Kaplanis 1996; Buch and Lapp 1998; Buch 1999). That constitutes our final hypothesis:

H7: Banks are more likely to undertake activities in a foreign country if the country hosts an important international financial centre.

The empirical analysis

The data concern the set of the Italian banks that established at least one operational unit abroad in the 1990s. Specifically, the data set has been built from the information provided by the Association of the Italian Banks (ABI), the annual reports and the websites of the Italian banks. The information gathered relates to the firm-specific characteristics of both the Italian parent banks and their foreign initiatives set abroad through internal growth (i.e. branches and representative offices).

Table 12.1 reports the evolution of Italian multinational banks' direct presence in foreign countries in the period 1989–99. The number of banks with a foreign branch and/or a representative office rises from 55 to 67 as well as the relevant number of branches (from 85 to 95); the number of representative offices tends to remain stable over the decade (from 177 to 176). However, focusing on the second half of the decade, it emerges that the number of representative offices dropped from 202 to 176, this reflecting the restructuring processes that followed the recent mergers occurred amongst Italian banks over the period.

Table 12.2 shows the geographical breakdown of the direct presence of Italian banks in foreign countries in the period 1989–99. The figures reveal that Italian banks' foreign activities (branches and representative offices) are mainly placed in Europe and in the main international financial centres. Specifically, the presence in Western Europe recorded a noteworthy increase in terms of branches, from 33 to 49, due to the increased interest towards UK,

Table 12.1 Multinational Italian banks, branches and representative offices, 1989–1999

Year	*No. multi-national banks*	*No. banks with foreign branches (only)*	*No. banks with foreign r. offices (only)*	*No. banks with branches and r. offices*	*No. branches*	*No. representative offices*
1989	55	0	40	15	85	177
1990	56	2	40	14	95	184
1991	55	2	36	17	102	195
1992	56	4	35	17	105	184
1993	58	3	35	20	108	179
1994	59	6	35	18	103	186
1995	63	10	36	17	106	202
1996	54	7	29	18	98	182
1997	68	10	41	17	100	192
1998	66	7	41	18	97	191
1999	67	9	39	19	95	176

Sources: Our elaboration on data provided by ABI – Association of Italian Banks.

Luxembourg and Spain, while the total number of representative offices reflects the several restructuring processes experienced by Italian banks in the second half of the 1990s (from 81 representative offices in 1995, to 71 in 1999).

Asian countries are also strongly favoured by Italian banks that are present there through an ever-increasing number of both branches (from 14 to 18) and representative offices (from 38 to 53). The presence in the US shows instead a noteworthy decrease, both in North America, where the number of branches and representative offices declines from 24 to 15, and from 17 to 13, respectively; and in Latin America, where from the early 1990s Italian banks do not present any branch office.

In order to corroborate the interpretative framework based on the eclectic paradigm and to validate the relevant hypotheses advanced, we developed an econometric model. Specifically, the model refers to the 98 Italian banks that were present in foreign countries with at least a branch and/or a representative office in the period 1989–99. The observation unit is the extent of the presence of bank i in country j, and the model specification is the following:

$$BANK_{ij} = a + bOWNER_i + cINTERNAL_{ij} + dLOCAT_j + e_{ij}$$

i=1, … , 98 Italian banks j=1, …, 46 countries

The dependent variable $BANK_{ij}$ assumes two distinct specifications according to each organisational form considered. Therefore:

$BANK_{ij} = BRANCHES_{ij}$, that is the total number of the bank's i foreign branches in country j;

$BANK_{ij} = OFFICES_{ij}$, that is the total number of the bank's i representative offices in country j.

Table 12.2 Geographical breakdown of overseas activity (direct presence through branches and representative offices) of Italian banks, 1989–1999

	1989		*1992*		*1995*		*1999*	
	branch	*repr. office*	*branch*	*repr. office*	*branch*	*repr. office*	*branch*	*repr. office*
Western Europe	**33**	**82**	**50**	**85**	**57**	**81**	**49**	**71**
France	4	10	6	8	6	8	4	4
Germany	8	25	10	22	9	16	6	7
UK	12	24	15	24	15	25	17	34
Luxembourg	1	–	3	2	9	1	9	1
Netherland	1	2	1	2	1	2	1	1
Spain	6	2	12	5	13	8	9	5
Switzerland	–	6	–	4	–	2	–	2
Eastern Europe	**–**	**11**	**–**	**12**	**–**	**17**	**1**	**15**
Russian Fed.	–	8	–	8	–	8	–	7
Poland	–	1	–	1	–	2	–	3
Czech Rep.	–	–	–	2	–	3	–	3
North America	**24**	**17**	**24**	**16**	**18**	**17**	**15**	**13**
Canada	–	–	–	2	–	1	–	–
USA	24	17	24	14	18	16	15	13
Latin America	**2**	**12**	**–**	**11**	**–**	**10**	**–**	**12**
Argentina	2	4	–	3	–	3	–	4
Brazil	–	2	–	3	–	3	–	4
Asia	**14**	**38**	**15**	**45**	**16**	**60**	**18**	**53**
China	–	7	–	9	–	24	2	21
Japan	3	6	4	7	4	3	3	4
Hong Kong	7	20	7	21	7	23	6	14
Singapore	4	2	4	2	5	2	7	1
Near East	**7**	**4**	**7**	**4**	**7**	**5**	**5**	**5**
Turkey	2	1	2	1	2	2	2	2
Lebanon	4	1	4	1	4	1	2	1
Africa	**1**	**6**	**1**	**5**	**–**	**5**	**–**	**4**
Caribbean Islands	**4**	**–**	**8**	**–**	**8**	**–**	**7**	**–**
Cayman Islands	4	–	7	–	7	–	6	–
Australia	**–**	**7**	**–**	**6**	**–**	**7**	**–**	**3**
Total	**85**	**177**	**105**	**184**	**106**	**202**	**95**	**176**

Sources: Our elaboration on data provided by ABI – Association of Italian Banks.

The model has been run for four periods: 1989, 1992, 1995 and 1999. As the dependent variable is a limited count variable, where the large majority of observations is zero (see Table 12.3), a simple ordinary least squares regression analysis would yield biased results. For this reason, the model has been estimated through negative binomial regression models, which are the most appropriate for count data.[7]

The independent variables

According to the hypotheses previously developed, the variables suggested by the international literature and the availability of data, the independent variables relate to ownership-, internalisation- and location-specific advantages as follows.

Ownership advantages (OWNERi)

Size: The ability of a bank to differentiate products and services, as well as the availability of 'exclusive' resources has been proxied by its dimensional scale, in line with the traditional Schumpeterian hypothesis on the positive relationship between size, resources and innovativeness. Specifically, the size of the Italian bank has been measured by the number of the parent's domestic employees (SIZE1), of the counters in Italy (SIZE2); by the bank's financial outcome (SIZE3), the bank's deposits (SIZE4) and funds (SIZE5).[8]

International experience: Variables relating to the parent company's experience in managing operations abroad aim to capture both the cumulative aspect – the variable AGE is the number of years the parent company had been present in that country – and the geographical dimension, that is, the bank's international experience related to the accumulation of experience in different countries. Specifically: N_COUNTRY is the number of foreign countries where the bank is already operating; SIM_COUNTRY is the number of foreign countries culturally similar[9] to Italy, where the bank is already present. Additionally, in order to take into account the experience of previous restructuring processes undertaken by the bank, the variable RESTRUCT measures the number of operations dismissed by the bank until the moment considered.

According to the hypotheses advanced in the earlier sections, the expected sign for these variables is positive.

Internalisation advantages (INTERNALij)

Internalisation advantages mainly relate to the 'follow the customer' hypothesis. Therefore, the proxy considered measures the extent of foreign direct investment undertaken by Italian firms in the same home country. Specifically, the presence in foreign countries has been measured by the actual number of plants or employees, which allows a rather precise

Table 12.3 Number of branches and representative offices of Italian banks

No. branches/ r. offices	*Branches*								*Representative offices*							
	No. banks				*%*				*No. banks*				*%*			
	'89	*'92*	*'95*	*'99*	*'89*	*'92*	*'95*	*'99*	*'89*	*'92*	*'95*	*'99*	*'89*	*'92*	*'95*	*'99*
0	40	35	36	39	73	62	57	59	0	4	10	9	0	7	16	13
1	2	6	12	15	4	11	19	23	23	19	17	27	42	34	26	41
2	3	4	4	5	5	7	6	8	15	17	6	7	27	30	10	10
3	1	2	3	0	2	4	5	0	2	4	13	13	4	7	21	20
4	1	0	0	1	2	0	0	1	1	0	3	2	2	0	5	3
5	1	1	1	1	2	2	2	1	8	1	4	2	15	2	6	3
>6 and <9	4	5	4	3	7	9	6	4	3	7	6	4	5	13	10	6
>10	3	3	3	3	5	5	5	4	3	4	4	3	5	7	6	4

estimate of the phenomenon.[10] The variable FDI is the number of manufacturing affiliates of Italian firms in each country in the antecedent year. However, the variable FDI_EMP, measured by the total number of employees in the mentioned manufacturing affiliates, allows for the dimensional scale of these foreign initiatives by Italian firms. Although considering only manufacturing affiliates could be constraining as it excludes commercial affiliates of manufacturing firms and affiliates of services firms, nonetheless the problem is smoothed by the two following facts: (1) FDI in service sectors are often pulled by manufacturing MNCs (which indeed generate the demand for services); and (2) the presence of commercial activities is partly captured by the proxy (TRADE) used for commercial relationship between the countries involved (see below). Data are obtained from the database Reprint, which censuses outward and inward foreign direct investment in Italy from 1986. The database has been developed at Politecnico di Milano in cooperation with the National Council for Economics and Labour (CNEL).[11]

Along with the hypothesis put forward in the second section, the expected correlation between the presence abroad of Italian banks and the presence of manufacturing affiliates of Italian firms in the same country is positive.

However, in order to take into account the extent of commercial relationships between the two countries involved (suggested as a further side of the 'follow the customer' hypothesis), the variable TRADE is the sum of exports and imports of goods between Italy and each country considered. Likewise, EXP and IMP measure exports and imports, respectively.[12] Data come from the Italian Institute for the International Trade (ICE), and the expected signs are positive.

Location advantages (LOCATj)

Wealth and size of the local market: The wealth of the foreign market has been measured by GDP and GDP_POP, which are the national GDP and GDP per capita in the period considered. The variable MKT (a measure of the country population) is the proxy for the size of the local market. Data come from the International Monetary Fund (1999), and the expected impact on the local presence of Italian banks is positive.

Characteristics of the local banking market: As far as the characteristics of the local banking market (Demirgüç–Kunt and Levine 1996; Miller and Parkhe 1998), we considered: MKT_SIZE (measured by the total deposits) as a proxy for the size of the banking market; MKT_DEV (measured as the ratio of the total claims of deposit money banks to the national GDP) as a proxy for the development of the banking system; and SPREAD (measured by the difference between bank lending and borrowing rate) as a proxy for the banking efficiency. Finally, we considered the exchange rate between the Italian lira and the local currency (EXCH). The source for all these variables is International Monetary Fund (1999).

Presence of an international financial centre: According with Davis and Lewis (1982), Brealey and Kaplanis (1996), Buch and Lapp (1998) and Buch (1999), the variable FIN_CEN is a dummy which equals one for the following countries: Bahamas, Cayman Islands, Hong Kong, Indonesia, Japan, Luxembourg, Singapore, Switzerland, United Kingdom and USA. As the hypothesis is that the presence of an important international financial centre in a country represents a crucial attractive factor for multinational banks, the expected sign is positive.

Country political and economic risk: Finally, as some of the empirical studies so far proposed in the literature suggested that FDI by banks can be negatively influenced by the target country's risk (e.g. Hultman and McGee 1989; Yamori 1998), we included the variable RISK, obtained from the Institutional Investors Credit Rating Index (IICRI)[13] for each country where the foreign operations of Italian banks are located. However, because the IICRI index is on an inverted scale of highest risk=0 to lower risk=100, our variable RISK is (100 – IICRI).

Empirical findings

Results of the econometric estimates (both the most general model in which the whole set of independent variables are estimated – Models A –, and the best specification of the model, in which only the variables significant at least at $p<0.10$ are left – Models B –) are reported in Tables 12.4 and 12.5. Specifically, Table 12.4 reports the estimates of the equation for the presence of Italian banks in foreign countries through branch offices in the four periods considered. Likewise, estimates for presence through representative offices are reported in Table 12.5.

The results obtained generally provide strong support for the relevance of the eclectic paradigm as a theoretical explanation for the international growth patterns followed by the Italian banks in the 1990s, and to the hypotheses developed as far as the level of services and commitment undertaken on foreign countries.

Ownership advantages are indeed crucially important for a bank wishing to expand abroad through internal growth, and that holds both in case of branches and representative offices. First of all, the dimensional scale of the bank, that is the availability of resources and skills, shows a positive and significant coefficient, especially at the beginning of the period (SIZE1 and SIZE3 are jointly significant at $p<0.01$ in the first half of the decade, while they become alternative later on). Similarly, the bank's international experience, that is the cumulated degree of familiarity with foreign countries, shows an increasing relevance throughout the period. Indeed, both the temporal dimension (AGE) and the geographical one (N_COUNTRY) show positive and significant coefficients along the decade, and the effect increases towards the end (both the variables become significantly different from zero, at $p<0.01$ in 1999, in the best specifications of both the models).

Table 12.4 Econometric estimates (dependent variable=BRANCHES)

	1989		*1992*		*1995*		*1999*	
	Model A	*Model B*	*Model A*	*Model B*	*Model A*	*Model B*	*Model A*	*Model B*
CONSTANT	−9.39	−7.73	−8.45	−7.58	−10.59	−6.93	−6.45	−6..95
	(−7.55)***	(−15.53)***	(−9.63)***	(−16.14)***	(−9.12)***	(−18.45)***	(−4.79)***	(−17.96)***
SIZE1	0.0002	0.0002	0.0001	0.0001	0.0001	0.0001		
	(8.24)***	(8.29)***	(3.29)***	(8.29)***	(3.50)***	(4.25)***		
SIZE3	1.56E−06	1.71E−06	3.07E−06				6.80E−07	6.10E−07
	(2.50)**	(2.82)***	(1.96)**				(2.48)**	(2.52)**
AGE	0.03	0.05	0.06	0.05	0.04	0.05	0.08	0.07
	(1.81)*	(2.58)***	(5.38)***	(5.38)***	(3.57)***	(6.38)***	(6.91)***	(7.98)***
N_COUNTRY			0.06		0.07	0.05	0.11	0.099
			(2.05)**		(2.08)**	(2.05)**	(5.18))***	(5.28)***
RESTRUCT	0.01	0.02	0.01	0.01	0.03	0.007	0.009	0.009
FDI	(4.41)***	(5.64)***	(5.21)***	(4.79)***	(6.48)***	(3.62)***	(3.14)***	(5.38)***
FDI_EMP								
TRADE				3.95E−05		2.55E−05	2.50E−05	
EXP				(2.84)***		(2.65)***	(1.52)	
GDP_POP	4.37E−08				7.35E−08			
	(0.66)				(1.65)*			
EXCH	0.0005							
	(0.14)							
SPREAD							−0.11	
							(−1.12)	
MKT_DEV					0.0006			
					(1.05)			
FIN_CEN	1.47	2.17	2.10	2.31	1.36	2.27	2.38	2.40
	(2.90)***	(6.07)***	(6.07)***	(7.02)***	(2.74)***	(8.10)***	(4.11)***	(7.98)***
RISK	−0.03		−0.007		−0.04		0.01	
	(−1.53)		(−0.57)		(−2.64)***		(0.56)	
Included obs.	2694	3387	3000	3157	2432	3268	2739	3652
Log likelihood	−128.66	−139.72	−139.66	−173.66	−114.44	−227.52	−149.07	−205.27
LR statistic	266.14***	284.69***	373.23***	387.89***	252.76***	388.43***	327.24***	362.56***
LR index (pseudo R^2)	0.51	0.50	0.58	0.53	0.52	0.46	0.52	0.47

Notes: Number in brackets are z–ratios. ***significant at $p<0.01$; **significant at $p<0.05$; *significant at $p<0.10$.

Table 12.5 Econometric estimates (dependent variable=OFFICES)

	1989		*1992*		*1995*		*1999*	
	Model A	*Model B*	*Model A*	*Model B*	*Model A*	*Model B*	*Model A*	*Model B*
CONSTANT	−5.96 (−8.38)***	−5.82 (−9.12)***	−5.93 (−16.67)***	−5.70 (−13.22)***	−4.70 (−13.57)***	−4.55 (−24.70)***	−5.11 (−18.39)***	−5.33 (−23.82)***
BRANCHES	−1.53 (−3.16)***	−1.41 (−3.02)***	−2.51 (−3.64)***	−2.46 (−4.73)***	−2.1 5(−4.52)***	−2..51 (5.70)***	−2.8 1(−5.75)***	−3.00 (−6.13)***
SIZE1	6.89E−05 (1.48)	9.3E−05 (4.24)***	0.0001 (5.48)***	0.0001 (5.96)***	7.76E−05 (5.74)***	7.92E−05 (6.35)***		
SIZE3	1.67E−06 (2.26)**	2.22E−06 (3.65)***	2.92E−06 (2.40)**	3.26E−06 (3.30)***	2.04E−07 (0.97)		5.57E−07 (3.08)***	5.99E−07 (3.67)***
AGE	0.12 (8.22)***	0.13 (9.05)***	0.09 (6.78)***	0.12 (9.85)***	0.10 (10.28)***	0.10 (10.51)***	0.11 (10.10)***	0.11 (10.70)***
N_COUNTRY	0.04 (1.04)						0.07 (4.50)***	0.08 (5.73)***
RESTRUCT	−1.32 (−1.02)		−1.13 (−1.83)*	−2.14 (−3.57)***	−0.25 (−1.14)		−0.52 (−2.26)**	−0.54 (−2.30)**
FDI		0.003 (1.70)*	0.01 (3.59)***	0.004 (2.82)***	0.005 (3.48)***	0.004 (2.16)**	0.004 (2.10)**	0.005 (2.89)***
FDI_EMP	1.56E−05 (3.03)***							
TRADE			1.56E−05 (3.07)***			6.52E−06 (1.79)*		
POP	0.0009 (1.69)*	0.0009 (1.77)*	0.0009 (1.01)		0.001 (5.69)***	0.002 (6.48)***	0.002 (6.32)***	0.002 (6.73)***
EXCH	0.0004 (2.11)**		0.0005 (1.22)**				0.0002 (1.59)	0.0002 (2.13)**
SPREAD	0.07 (2.55)**	0.06 (2.36)**					0.01 (0.52)	
MKT_SIZE			0.0009 (1.22)					
FIN_CEN	1.22 (4.19)***	1.27 (5.05)***	0.84 (1.89)*	1.71 (6.85)***	1.33 (5.54)***	1.38 (7.17)***	1.65 (5.75)***	1.66 (6.38)***
RISK	−0.01§ (−1.43)	−0.02 (−2.05)**		−0.01 (−1.82)*	−0.006 (−1.03)			
Included obs.	1923	1924	2279	3039	2839	3191	2738	3568
Log likelihood	−327.06	−335.29	−258.45	−409.43	−516.76	−543.63	−399.53	−455.07
LR statistic	314.93***	298.60***	299.98***	415.38***	347.49***	375.84	403.88***	485.11***
LR index (pseudo R^2)	0.32	0.31	0.37	0.34	0.25	0.26	0.34	0.35

Notes: Numbers in brackets are z–ratios. ***significant at $p<0.01$; **significant at $p<0.05$; *significant at $p<0.10$.

However, the model for representative offices includes also a dummy variable, BRANCHES, controlling for whether the Italian bank already owns a branch in the same foreign country. If this is the case, the bank is less likely to open there a representative office (indeed the variable BRANCHES is always significant at $p<0.01$). The presence through representative offices is also sensitive to the restructuring occurring within the parent bank in the past, that is to the number of foreign operations the bank has already closed (RESTRUCT is indeed negative in 1992 and 1999). In other words, if the Italian bank has started a restructuring process it is less likely to open new representative offices abroad.

Empirical findings strongly confirm that the Italian banks tend to save their pre-existing relationships with their clients, following them to foreign countries, especially through the establishment of branch offices. Indeed, the latter provide the whole set of services and complementary assets Italian firms might need to support their global dimension. The variable FDI shows a positive and significant sign throughout the period, and the same holds for the proxies for commercial relationships between Italy and each foreign country, from 1992 on. Only for those clients which undertook the bigger initiatives abroad, and especially at the beginning of the period, Italian banks located representative offices in addition to branches (FDI_EMP is significant at $p<0.01$, see Table 12.5).

Location-specific aspects traditionally used in studies for firms' locational choices (namely, wealth and size of the target country), as well as the characteristics of the local banking system, do not influence the presence of Italian banks on foreign countries when it is obtained through branch offices; while they influence – but only marginally – the location of representative offices. However, as expected, the latter often represent only explorative attempts and are therefore intentionally located by the parent bank in more promising countries.

Finally, the presence of an important international financial centre significantly explains the presence of Italian banks abroad throughout the period (FIN_CEN is always significant at $p<0.01$).

Conclusions

This study provides further empirical support for the belief that the eclectic paradigm offers a useful framework for explaining foreign activities undertaken by multinational banks. Indeed, econometric estimates corroborate the theory that the choice of a multinational bank to grow internationally through internal growth and direct presence in a foreign country, significantly depends on the following:

1 Ownership advantages related to the bank's dimensional scale, and its international experience already cumulated on foreign countries.

2 The need to follow its clients who have already expanded their activities onto foreign markets, in order to save their pre-existing relationships. Indeed, personal contact constitutes one of the main advantages for banks and preserving established accounts by opening foreign offices become a means for protecting knowledge and information networks, thus achieving internalisation advantages. In other words, banks go abroad to service their clients who have preceded them abroad. It is worth observing that while many banks initially followed their clients (mostly goods-producing companies) to the countries in which they set up their factories, today it is the ability of banks to offer global services which appeals to their clients who themselves are global corporations (Dunning 1993).

3 The access to location advantages represented by positive externalities offered by the most important international financial centres, while the traditional location variables do not seem to significantly influence MNBs' location choices.

These findings seem to offer some support to the evidence about the recent MNCs attitude to source key competitive advantages from abroad. Namely, MNCs' are increasingly undertaking foreign direct investment not only as a means of exploiting the existing ownership specific or competitive advantages, but also as a vehicles to augment them or source them from a foreign location (see McKaig-Berliner and Dunning 2001).

In fact, a complementarity seems to emerge between the MNB's ownership advantages and its ability to consolidate and extend these advantages through the international network placed both where existing customers have already extended their activities, and where they can enjoy (industry-specific) spillovers and (specialisation) externalities. For this reason, as well described in the introductory Chapter 1, ownership and location advantages have begun to develop cumulatively together.

However, as far as internalisation advantages are concerned, it has been highlighted that hierarchical control and full internalisation is no longer always a first-best option to MNCs (see Chapter 1). A series of events over the last two decades has led scholars to suggest that the world is moving to embrace a new process of globalisation and the associated advent of alliance capitalism. That has also ushered changes in the way MNBs undertake cross-border activities, shifting the emphasis away from hierarchies towards a richer variety of organisational models. In fact, external growth and non-equity alliances are becoming more important forms of international economic involvement. Nonetheless, very few empirical works have been so far put forward (see, for example, Jacobsen and Tschoegl 1999), which investigates Nordic banks' international alliances and consortia). Much work has still to be done, and the challenge to the eclectic paradigm as an interpretative framework is still open. That having been said, it is worth reminding ourselves that Dunning himself in a paper published in *JIBS* in 1995 already acknowledged that the eclectic paradigm actually requires some reappraisal in order to

> consider more explicitly the competitive advantages arising from the way firms organise their inter-firm transactions, the growing inter-dependencies of many intermediate product markets, and the widening of the portfolio of the assets of districts, regions and countries to embrace the external economies of interdependent activities.
>
> (Dunning 1995: 461)

Therefore, we cannot but agree with the prognosis advanced by the Editors of the present volume about the good state of health of the eclectic paradigm.

Notes

1 The following organisational forms are generally available to the bank: representative office, branch and subsidiary (Blandón 1998). The representative office is the cheapest overseas banking organisational forms, and it usually consists in a small commercial office designed to help the parent bank and its customers in their financial and commercial activities. A foreign branch constitutes a higher level of commitment to the market, as it actively participates in the host-country banking system. Foreign subsidiaries instead are incorporated to the host banking system. Miller and Parkhe (1998) observe that organisational complexity decreases from *subsidiaries*, which maintain separate legal entities and resemble host-country banks, to *representative offices*, which are instead primarily used for exploratory purposes. *Branches* lie in the middle, since they perform all of the traditional banking functions but their financial policies are mandated by the parent bank.

2 External growth refers mainly to foreign subsidiaries and joint ventures. Nonetheless, as they involve different actors and do not relate to a single bank decision, we will not consider them. For a complete description of all the organisational forms employable by a multinational bank, see Houpt (1999) and Bain *et al.* (1999).

3 It is worth observing that other authors have applied Dunning's paradigm to international banking (e.g. Gray and Gray 1981; Cho 1985, 1986; Sagari 1992; Blandón 1998) for the investigation of the bank's decision to invest abroad.

4 The same idea is also known as 'liability of foreignness' and relates to the fact that foreign firms are disadvantaged *vis-à-vis* domestic firms. For an application to the global banking case, see Miller and Parkhe (2002), and Berger *et al.* (2000a).

5 Brimmer and Dahl (1975) defined as 'defensive expansion' the reaction of US banks to specific regulations which forced them to expand offshore. Banks respond to the expansion of their clients abroad to defend their client–bank relationship. In fact, if the banks do not accompany their client abroad, their client will establish a banking relationship that could expand to supplant any domestic banking relationship.

6 It is worth observing that the same hypotheses has been tested for domestic new entrants in local banking markets, which are attracted by high profits, market size and market growth (Amel and Liang 1997).

7 The other possible model normally used for count data, the Poisson model, presents a major drawback related to the fact that the conditional mean is assumed to be equal to the conditional variance, so that any cross-sectional heterogeneity is ruled out. The negative binomial model provides a generalisation that permits us to solve the problem, by introducing an individual unobserved effect into the conditional mean (Greene 1997). The author is grateful to Adrian Tschoegl for this valuable suggestion.

8 It is worth observing that these variables related to dimensional aspects of the Italian banks have been also calculated as logarithm but the significance in the estimates has been systematically lower. The source of data is Mediobanca.

9 Specifically, the whole set of countries considered have been grouped into ten clusters, following, with appropriate modifications, the classification proposed by Ronen and Shenkar (1985) and subsequently taken up by Gatignon and Anderson (1988). It is also worth observing that a proxy for the cultural distance between Italy and each foreign country has been tested as an independent variable. Nonetheless, as it never came out significant (as already found in Contractor and Kundu 1998, for the hotel service sector), we decided not to include it among the independent variables.

10 Traditionally instead, FDI has been measured by FDI flows and stocks, as recorded in the balance of payments. Nonetheless, we are convinced that they might present some limits and biases which undermine their significance for detailed analyses at country and industry level. First, they follow a *first country, first sector* criterion. For instance, if an Italian parent firm sets up a financial holding in the Netherlands which acquires a manufacturing firm in the US, the Italian balance of payments records an Italian FDI in the Netherlands in the financial sector, and does not record any FDI in the US. Moreover, only part of the capital invested abroad by multinational enterprises is financed though movements recorded in the balance of payments. Finally, further problems relate to the fact that FDI are recorded at their historical value.

11 For further details, see Cominotti *et al.* (2001).

12 According to other empirical studies (e.g. Buch 1999) we also tested other variables (e.g. the ratio (IMPORT–EXPORT)/(IMPORT+EXPORT). Nonetheless, we reported only the most significant.

13 The IICRI index is widely known among financial operators and is calculated twice a year on the basis of information supplied by the world's leading banks, which are asked to assess the economic situation and the risk of insolvency in all countries apart from their own.

References

Aggarwal, R. (1993) 'The global expansion of Japanese financial service firms: role of domestic economic and regulatory policies', in Y. Aharoni, *Coalitions and Competition: The Globalization of Professional Business Services*, London: Routledge.

Aggarwal, R. and Durnford, J. (1989) 'Market assessment of international banking activity: a study of US bank holding companies', *Quarterly Review of Economics and Business*, 29: 58–67.

Aliber, R.Z. (1984) 'International banking: a survey', *Journal of Money, Credit and Banking*, 16: 661–712.

Amel, D.F. and Liang, J.N. (1997) 'Determinants of entry and profits in local banking markets', *Review of Industrial Organization*, 12: 59–78.

Anderson, E. and Gatignon, H. (1986) 'Modes of foreign entry: a transaction cost analysis and propositions', *Journal of International Business Studies*, 17: 1–26.

Bain, E.J., Fung, J. and Harper, I. (1999), 'Multinational banking: historical, empirical and case perspectives', Melbourne Business School Working Paper No. 5.

Baldwin, J.R. and Rafiquzzaman, M. (1995) 'Selection versus evolutionary adaptation: learning and post-entry performance', *International Journal of Industrial Organization*, 13: 501–22.

Ball, C. and Tschoegl, A. (1982) 'The decision to establish a foreign bank branch or subsidiary: an application of binary classification procedures', *Journal of Financial and Quantitative Analysis*, 17: 411–24.

Berger, A.N. and DeYoung, R. (2001) 'The effects of geographic expansion on bank efficiency', *Journal of Financial Services Research*, 19(2/3).

Berger, A.N., Demsetz, R.S. and Strahan, P.E. (1999) 'The consolidation of the financial services industry: causes, consequences, and implication for the future', *Journal of Banking and Finance*, 23: 135–94.

Berger, A.N., Bonime, S.D., Goldberg, L.G. and White, L.J. (2000b) 'The dynamics of market entry: the effects of mergers and acquisitions on entry in the banking industry', Wharton School Center for Financial Institutions Working Paper, no. 00–12, University of Pennsylvania.

Berger, A.N., DeYoung, R., Genay, H. and Udell, G.F. (2000a) 'Globalization of financial institutions: evidence from cross-border banking performance', Brookings-Wharton Papers, 3.

Blandón, J.G. (1998) 'The choice of the form of representation in multinational banking: evidence from Spain', Economics Working Papers from Department of Economics and Business, Universitat Pompeu Fabra.

Brealey, R.A. and Kaplanis, E.C. (1996) 'The determination of foreign banking location', *Journal of International Money and Finance*, 15: 577–97.

Brimmer, A. and Dahl, F. (1975) 'Growth of American international banking: implications for public policy', *Journal of Finance*, 30: 341–63.

Buch, C.M. (1999) 'Why do banks go abroad? Evidence from German data', Kiel Working Papers No. 948: Kiel Institute of World Economics.

Buch, C.M. and Lapp, S. (1998) 'The Euro – no big bang for European financial markets, *Konjunkturpolitik*, 47: 11–78.

Casson, M. (1990) 'Evolution of multinational banks: a theoretical perspective', in G. Jones (ed.), *Banks as Multinationals*, London: Routledge.

Caves, R.E. (1982) *Multinational Enterprise and Economic Analysis*, Cambridge: Cambridge University Press.

Cho, K.R. (1985) *Multinational Banks: Their Identities and Determinants*, Michigan: UMI Research Press.

Cho, K.R. (1986) 'Determinants of international banks', *Management International Review*, 26: 10–23.

Cominotti, R., Mariotti, S. and Mutinelli, M. (2001) *Italia Multinazionale*, Roma: Cnel.

Contractor, F. and Kundu, S.K. (1998) 'Modal choice in a world of alliances: analysing organizational forms in the international hotel sector', *Journal of International Business Studies*, 29: 325–58.

Crecchia, C. (1996) *Banche italiane e mercati esteri: il processo di internazionalizzazione del sistema bancario italiano*, Bologna: Il Mulino.

Dahl, F. (1967) 'International operations of US banks: growth and public policy implications', *Law and Contemporary Problems*, 33: 100–30.

Davidson, W.H. (1980) 'The location of foreign investment activity', *Journal of International Business Studies*, 11: 9–22.

Davis, K. and Lewis, M. (1982) 'Foreign Banks and the Financial System', Australian Financial System Inquiry, Commissioned Studies and Selected Papers, Part I. Canberra: Australian Government Publishing Service.

DeYoung, R. and Nolle, D.E. (1996) 'Foreign-owned banks in the United States: earning market share or buying it?', *Journal of Money, Credit and Banking*, 28: 622–36.

Demirgüç-Kunt, A. and Levine, R. (1996) 'Stock market development and financial intermediation', *The World Bank Economic*, 10: 291–321.

Dewatripont, M. and Tirole, J. (1994) *The Prudential Regulation of Banks*, Cambridge: MIT Press.

Di Quirico, R. (1999) 'The initial phases of Italian banks' expansion abroad, 1900–1931', *Financial History Review*, 6: 7–24.

Dunning, J.H. (1977) 'Trade, location of economic activity and the MNE: a search for an eclectic approach', in B. Ohlin, P. Hesselborn and P. Wijkman (eds), *The International Allocation of Economic Activity*, London: Macmillan.

Dunning, J.H. (1981) *International Production and the Multinational Enterprise*, London: Allen and Unwin.

Dunning, J.H. (1989) 'Multinational enterprises and the growth of services: some conceptual and theoretical issues', *The Services Industries Journal*, 9: 5–39.

Dunning, J.H. (1993) 'The globalisation of service activities', in J.H. Dunning (ed.), *The Globalisation of Business*, London: Routledge.

Dunning, J.H. (1995) 'Reappraising the eclectic paradigm in an age of alliance capitalism', *Journal of International Business Studies*, 26: 461–91.

Enderwick, P. (1989) 'Some economies of service-sector multinational enterprises', in P. Enderwick (ed.), *Multinational Service Firms*, London: Routledge.

Engwall, L. and Wallenstål, M. (1998) 'Tit for tat in small steps: the internationalisation of Swedish banks', *Scandinavian Journal of Management*, 4: 147–55.

Engwall, L., Marquardt, R., Pedersen, T. and Tschoegl, A.E. (2001) 'Foreign bank penetration of newly opened markets in the Nordic countries', *Journal of International Financial Markets, Institutions and Money*, 11: 53–63.

Erramilli, M.K. (1991) 'The experience factor in foreign market entry behaviour of service firms', *Journal of International Business Studies*, 22: 479–501.

Erramilli, M.K. and Rao, C.P. (1990) 'Choice of foreign entry modes by service firms: role of market knowledge', *Management International Review*, 30: 135–50.

Fieleke, N.S. (1977) 'The growth of US banking abroad: an analytical survey', in *Key Issues in International Banking:* Federal Reserve Bank of Boston.

Flatraaker, D.-I. and Husevåg, V. (1991) 'Norwegian banks abroad', *Norges Bank Economic Bulletin*, March: 181–95.

Gabrowski R.N., Rangan N. and Rezvanian, R. (1993), 'Organizational forms in banking: an empirical investigation of cost efficiency', *Journal of Banking and Finance*, 17: 531–8.

Galiatsos, K. and Papapetrou, E. (1995) 'The determinants of foreign banking activity in Greece', in J. Doukas and L. Lang (eds), *Research in International Business and Finance*, Greenwich, CT: JAI Press.

Gatignon, H. and Anderson, E. (1988) 'The multinational corporation's degree of control over foreign subsidiaries: an empirical test of a transaction cost explanation', *Journal of Law, Economics and Organisation*, 4: 305–36.

Goldberg, L.G. and Grosse, R. (1994) 'Location choice of foreign banks in the United States', *Journal of Economics and Business*, 46: 367–79.

Goldberg, L.G. and Johnson, D. (1990) 'The determinants of US banking activity abroad', *Journal of International Money and Finance*, 9: 123–37.

Goldberg, L.G. and Sanders, A. (1980) 'The causes of US banking expansion overseas: the case of Great Britain', *Journal of Money, Credit and Banking*, 12: 630–43.

Goldberg, L.G. and Sanders, A. (1981a) 'The determinants of foreign banking activity in the United States', *Journal of Banking and Finance*, 5: 17–32.

Goldberg, L.G. and Sanders, A. (1981b) 'The growth of organisational forms of foreign banks in the United States', *Journal of Money, Credit and Banking*, 13: 365–74.

Gomes-Casseres, B. (1985) *Multinational Ownership Strategies*, University of Missouri: UMI Press.

Goshal, S. (1987) 'Global strategy: an organising framework', *Strategic Management Journal*, 11: 425–40.

Gray, H. and Gray, J. (1981) 'The multinational bank: a financial MNC?', *Journal of Banking and Finance*, 5: 33–63.

Greene, W.H. (1997) *Econometric Analysis*, New York: Prentice Hall.

Grosse, R. and Goldberg, L.G. (1991) 'Foreign bank activity in the United States: an analysis by country of origin', *Journal of Banking and Finance*, 15: 1093–112.

Grubel, H.G. (1977) 'A theory of international banking', *Banca Nazionale del Lavoro Quarterly Review*, 123: 349–64.

Heinkel, R.L. and Levi, M.D. (1992) 'The structure of international banking', *Journal of International Money and Finance*, 11: 251–72.

Hellman, P. (1984) 'The internationalisation of Finnish financial service companies', *International Business Review*, 5: 191–208.

Houpt, J.V. (1999) 'International activities of US banks and in US banking markets', *Federal Reserve Bulletin*, September: 599–615.

Hultman, C.W. and McGee, L.R. (1989) 'Factors affecting the foreign banking presence in the US', *Journal of Banking and Finance*, 13: 383–96.

International Monetary Fund (1999) *International Financial Statistics*, 52, January.

Jacobsen, S.F. and Tschoegl, A.E. (1997) 'The internationalisation of the Norwegian banks', Wharton Financial Institutions Center Working Paper: 97–38.

Jacobsen, S.F. and Tschoegl, A.E. (1999) 'The Norwegian banks in the Nordic consortia: a case of international strategic alliances in banking', *Industrial and Corporate Change*, 8: 137–63.

Johanson, J. and Vahlne, J.E. (1977) 'The internationalization process of the firm: a model of knowledge development and increasing foreign market commitments', *Journal of International Business Studies*, 8: 22–32.

Jones, G. (1992) 'The legacy of the past: British multinational banking strategies since the nineteenth century', in P. Buckley and M. Casson (eds), *Multinational Enterprises in the World Economy*, Aldershot: Edward Elgar.

Kelly, J. (1977) *Bankers and Borders: The Case of American Banks in Britain*, Cambridge, MA: Ballinger.

Khouri, S. (1979) 'International banking: a special look at foreign banks in the US', *Journal of International Business Studies*, 10: 36–52.

Khouri, S. (1980) *Dynamics of International Banking*, New York: Praeger Publishers.

Kindleberger, C.P. (1983) 'International banks as leaders or followers of international business', *Journal of Banking and Finance*, 7: 583–95.

Kogut, B. and Singh, H. (1988) 'The effect of national culture on the choice of entry model', *Journal of International Business Studies*, 19: 411–32.

Laakso, J. (1984) 'The internationalisation of Finnish banking operations', *Bank of Finland Monthly Bulletin*, 58: 26–31.

Lees, F. (1974) *International Banking and Finance*, London and New York: Macmillan.

Mallampally, P. and Zimny, Z. (2000) 'Foreign direct investment in services: trends and patterns', in Y. Aharoni (ed.), *Globalisation of Service*, London: Routledge.

Marchi, A. and Panizza, U. (1993) 'Il credito e i servizi finanziari', in R. Cominotti and S. Mariotti (eds), *Italia multinazionale nei servizi. Radiografia degli investimenti diretti esteri in Italia*, Milano: EtasLibri.

Marois, B. and Abdessemed, T. (1997) 'The determinants of foreign direct investment of the banking industry: the French case', paper presented at the EIBA 23rd Annual Conference, Stuttgard, 14–16 December.

Marr, M.W., Rogowski, R.W. and Trimble, J.L. (1989) 'The competitive effects of US and Japanese commercial bank participation in Eurobond underwriting', *Financial Management*, 18: 47–54.

McKaig-Berliner, A.L. and Dunning, J. (2001) 'The geographical source of competitiveness of professional service firms', paper presented at the EIBA 27th Annual Conference, Paris, 13–15 December.

Metais, J. (1979) 'The multinationalisation of the large commercial banks: an industrial economic approach', *Revue Economique*, 3: 487–517.

Miller, S.R. and Parkhe, A. (1998) 'Patterns in the expansion of US banks' foreign operations', *Journal of International Business Studies*, 29: 359–90.

Miller, S.R. and Parkhe, A. (2002) 'Is there a liability of foreignness in global banking? An empirical test of U.S. banks' X-efficiency', *Strategic Management Journal*, 23: 55–75.

Moshirian, F. and Van der Laan, A. (1998) 'Trade in financial services and the determinants of banks' foreign assets', *Journal of Multinational Financial Management*, 8: 23–38.

Nayyar, P.R. (1990) 'Information asymmetries: a source of competitive advantage for diversified service firms', *Strategic Management Journal*, 11: 13–19.

Nieri, L. (1991) *L'espansione all'estero delle banche italiane: modalità di presenza e caratteristiche dell'attività svolta, Banca Impresa Società*, 3: 379–97.

Nieri, L. (1993) 'Motivazioni strategiche e caratteristiche operative delle filiali di banche italiane a Londra', *Banche e banchieri*, 7: 521–37.

Nieri, L. (1994) *L'espansione all'estero delle banche italiane: aspetti strutturali e problematiche gestionali*, Bologna: Il Mulino.

Nigh, D., Cho, K.R. and Krishan, S. (1986) 'The role of location-related factors in U.S. banking involvement abroad: an empirical examination', *Journal of International Business Studies*, 17: 59–72.

Normann, R. (1984) *Service Management: Strategy and Leadership in Service Business*, Chichester: Wiley.

Pringle, R. (1966) 'Why American banks go overseas', *The Banker*, November, 770–85.

Ronen, S. and Shenkar, O. (1985) 'Clustering countries on attitudinal dimensions: a review and synthesis', *Academy of Management Review*, 10: 435–54.

Rugman, A. (1981) *Inside the Multinationals*, London: Croom Helm.

Sabi, M. (1988) 'An application of the theory of foreign direct investment to multinational banking in LDCs', *Journal of International Business Studies*, 19: 433–47.

Sagari, S.B. (1992) 'United States foreign direct investment in the banking industry', *Transnational Corporations*, 1: 93–124.
Seth, R., Nolla, D.E. and Mohanty, S. (1998) 'Do banks follow their customers abroad?', *Financial Markets, Institutions and Instruments*, 7: 1–25.
Stanciu, L. (2000) 'Italian multinational banking in interwar East Central Europe', *Financial History Review*, 6: 45–66.
Stopford, J.M. and Wells, L.T. (1972) *Managing the Multinational Enterprise*, New York: Basic Books.
Terrel, H. (1979) 'US banks in Japan and Japanese banks in the United States: an empirical comparison', *Economic Review*, Federal Reserve Bank of San Francisco, 18–30.
Tschoegl, A.E. (1982) 'Foreign direct investment in banking in Japan and California', in A. Rugman (ed.), *New Perspectives on the Multinational Enterprise: Theory and Applications*, London: Croom Helm.
Tschoegl, A.E. (1987) 'International retail banking as a strategy: an assessment', *Journal of International Business Studies*, 18: 67–88.
Tschoegl, A.E. (1988) 'Foreign banks in Japan', *Bank of Japan Monetary and Economic Studies*, 6: 93–118.
UNCTAD (2001) *World Investment Report. Promoting Linkages*, Geneva: United Nations.
UNCTC (1987) *Foreign Direct Investment, the Service Sectors and International Banking*, New York: United Nations.
UNCTC (1988) *Transnational Corporations in World Development*, New York: United Nations.
UNCTC (1989) *Foreign Direct Investment and Transnational Corporations in Services*, New York: United Nations.
Ursacki, T. and Vertinsky, I. (1992) 'Choice of entry timing and scale by foreign banks in Japan and Korea', *Journal of Banking and Finance*, 16: 405–21.
Vastrup, C. (1983) 'Economic motives for foreign banking: the Danish case', *Kredit und Kapital*, 16: 117–25.
Williams, B. (1997) 'Positive theories of multinational banking: eclectic theory versus internalisation theory', *Journal of Economic Surveys*, 11: 171–211.
Yamori, N. (1998) 'A note on location choice of multinational banks: the case of Japanese financial institutions', *Journal of Banking and Finance*, 22: 109–20.
Yannopoulos, G. (1983) 'The growth of transnational banking', in M. Casson (ed.), *The Growth of International Business*, London: George Allen and Unwin.
Zaheer, S. and Mosakowski, E. (1997) 'The dynamics of the liability of foreignness: a global study of survival in financial services', *Strategic Management Journal*, 18: 439–64.

13 A critical reflection and some conclusions on OLI

*Lorraine Eden**

Introduction

When I was in kindergarten, my teacher taught me that the way to learn about a new object was to ask five basic questions: Who or what, when, where, why and how? I have always believed that the OLI (or eclectic) paradigm was about a subset of those questions. The 'who or what' is, of course, the multinational enterprise (MNE). While John Dunning's writings often start with a definition of the MNE, the focus of the eclectic paradigm from its beginnings up to the present day has been O (why?), L (where?) and I (how?).

For me, the OLI paradigm is best understood by reading three out of the dozens of books that John Dunning has authored over the past 25 years: *International Production and the Multinational Enterprise* (1981b), *Explaining International Production* (1988a) and *Multinational Enterprises and the Global Economy* (1993a). Each book pulls together and synthesises Dunning's writings from the previous period.[1]

In the concluding chapter to this book on the eclectic paradigm, what I would like to do is provide an overview of the development of the OLI paradigm as seen through my 'lens', and link the book chapters to its development. My thesis is that OLI should best be seen as a *way of looking* at the phenomenon of multinational enterprises and their activities. OLI addresses three of the five kindergarten questions – the why, where and how of MNE activities. Each of these questions can be addressed at a different level: macro (big picture, country), meso (mid picture, industry) or micro (firm, top management team). As examples of the why-where-how questions I list some of the research questions that have or currently do engage international business scholars in Figure 13.1. I argue that the OLI paradigm has been most successful at the macro and meso levels, and less so at the micro level, but I do not see that as a weakness. Paradigms do not

* *Acknowledgements*: I would like to thank John Cantwell, Rajneesh Narula, Peter Gray and Dan Li for their helpful comments. Any errors or omissions are my own responsibility.

- **Macro (big picture, country level) questions:**
  - O: Why do MNEs exist? Why are MNEs a successful organisational form?
  - L: Why are some countries a home, and others principally a host, to MNEs? How has the pattern of international investment across countries changed over time? How does regional integration affect the pattern of FDI and international production?
  - I: How do market imperfections affect the pattern of international production? How are transactions costs related to the firm's optimal mode of entry into foreign countries?

- **Meso (industry level) questions:**
  - O: Why are some industries dominated by MNEs and others are not?
  - L: What are the changing patterns of international production in the semiconductor (or other) industry? Why do certain industries cluster geographically together?
  - I: Why are certain modes of entry more predominant in one industry than another?

- **Micro (firm, top management team) questions:**
  - O: Why do firms differ? How do these differences affect their performance? Why do some firms choose to become MNEs (engage in international production) and others do not? How do O advantages relate to differences in firm performance? How does the management of knowledge flows within the MNE network affect its performance? Why is one firm better able to manage its O assets than another?
  - L: How does an MNE's locational choices affect its performance?
  - I: How are modes of entry choices related to firm performance?

Figure 13.1 Macro, meso and micro research questions in international business.

have to explain everything, nor does the 'big tent' of OLI have to shelter all theories under its umbrella.

This chapter is organised as follows. First, I address the issue of what is OLI – a paradigm, theory or model? I argue that OLI has moved from a theory to a paradigm, the pre-eminent one in the international business (IB) field. Second, I look at the development of OLI from its beginnings in the 1970s to its latest evolution, highlighting major changes and new directions. As the IB field grew and developed, and as today's global economy emerged, the OLI paradigm faced new challenges. It evolved in response to these challenges similar to the way that a 'tent' needs to grow to accommodate more 'children' underneath it. I argue that OLI has gone through four stages: Mark I (moving from a theory to a paradigm) and three challenges: Mark II (deepening the paradigm), Mark III (OLI and alliance capitalism) and Mark IV (OLI and strategic management). Insights from the chapters in this book are incorporated into each of these different stages. Lastly, I conclude with some thoughts about new extensions and challenges to the OLI paradigm.

What is OLI?

OLI has been called, variously, a framework, paradigm, theory and model.[2] Which is it? Let us start with a *framework*, which can be defined as a basic, conceptual structure of ideas. Frameworks establish structures for thinking about ideas. Nested within the concept of a framework is a paradigm. *Paradigms*, according to *Webster's Dictionary*, are philosophical and theoretical frameworks within which theories, laws and generalisations, and the experiments performed to test them, are formulated. A paradigm is an archetype for modelling and solving problems. Inherent in a paradigm are basic assumptions about the nature of problems and how they are to be approached. In sum, paradigms are the way that we think about problems, a set of background assumptions against which theories are developed.

Nested within a paradigm are theories. A *theory* is an organised system of accepted knowledge that applies in a variety of circumstances to explain a specific set of phenomena; in other words, a theory is a general statement of cause and effect relationships between phenomena. Theories are explanations of observations or laws.[3] Theories normally consist of assumptions, causal laws ('A causes B') and/or hypotheses ('we expect A to cause B'), and explanations for the causal laws or hypotheses ('A causes B because A causes X, which causes Z, which causes B').[4]

Lastly, theories are modelled by using mathematical relationships to formalise the relationships between assumptions, causal laws or hypotheses, and explanations. A *model* is a description or analogy used to help visualise something (e.g. an atom) that cannot be directly observed. It includes a system of postulates, data and inferences presented as a mathematical description of an entity or state of affairs. Models describe observed behaviour but simplify by ignoring certain details.[5]

It is clear that Dunning originally saw OLI as a theory, in particular, as an *eclectic theory*, drawing together different strands of economic theories of international production. His early writings consistently refer to the 'eclectic theory of international production'. It is not until the late 1980s that Dunning adopted the term 'eclectic paradigm' and began to argue that other theories (e.g. internalisation) were partial explanations that focused on particular issues with respect to international production.[6] OLI was the theoretical framework under which theories could be developed about particular aspects of international production. Dunning makes this point quite clearly:

> The purpose of the eclectic paradigm is not to offer a full explanation of all kinds of international production but rather to point to a methodology and to a generic set of variables which contain the ingredients necessary for any satisfactory explanation of particular types of foreign value-added activity.
>
> (Dunning, this volume, Chapter 2: 29)

Tolentino, in her review of the evolution of OLI, also argues that OLI was originally a theory that evolved over time into a paradigm. She sees OLI now as:

> [A] general framework of analysis that explains the level and pattern of foreign value-added activities of firms, and/or of countries, and allows for the co-existence of complementary and alternative theories in the discipline of international economics in a logically consistent manner without being inextricably wedded to any one particular approach.
>
> (Tolentino 2001: 191)

Dunning (2000b) makes the strongest case for arguing that OLI is not only a paradigm, but also the reigning paradigm – or 'envelope' – for all economic theories of the MNE. He reviews an enormous literature of IB theories, grouping them under the O, L and I 'sub-paradigms'. The article concludes that OLI is the reigning paradigm of MNE activities because: (1) the value of OLI is greater than the sum of the theories that can be contained under the envelope (i.e. the whole is greater than the sum of the parts); (2) OLI continues to offer value-adding generic hypotheses about MNE activity despite the growing complexities in O advantages; (3) the paradigm continues to address significant problems; and (4) there are no other IB paradigms that are serious contenders to OLI. Thus, OLI has grown from a theory to a 'big tent'.

How did OLI metamorphasise from a theory into the reigning paradigm of international production? Were the changes internally or externally driven? How influential were external critics relative to changes in institutional realities and the emergence of the global economy? In the next section of this paper, I examine changes in OLI, pinpointing key turning points in the development of the paradigm.

OLI: from a theory to the reigning paradigm

Mark I: developing the OLI paradigm

The fundamental puzzle that started the three decade-old history of the OLI paradigm were questions such as: Why do firms invest overseas? What determines the amount and composition of international production? From the beginning, the eclectic paradigm has been preoccupied with explaining the origin, level, pattern and growth of firms' offshore activities.

The early development of the OLI paradigm came from Dunning's searches across different literatures for answers to these questions. Dunning (1973), for example, is an enormous literature review, focusing on lessons about MNEs and international production drawn from surveys and theories of capital flows, international trade, location, industrial organisation and market structure. He defines the MNE as 'an enterprise which operates and controls income-creating activities in more than one country' (Dunning 1973: 290). The paper proposes two new lines of research. The first was that scholars should shift their question from 'why international production?' to explaining the growth rate, geographic and industrial patterns of multinational enterprises. Second, researchers should focus on the distinctiveness of MNEs and their forms of market penetration, by country and industry.

In his 1976 Nobel Symposium lecture in Stockholm, subsequently published as Dunning (1977), the first version (Mark I) of the OLI paradigm had begun to take shape. Dunning (1977, 1979) outlined the components of the OLI paradigm as three conditions that determined whether or not a firm would engage in FDI.[8]

The first condition – O – answered the 'why go abroad' or 'how is it possible to go abroad' question. O advantages (primarily from possession of intangible assets) were characteristics of MNEs that gave them a net competitive advantage over other firms supplying particular foreign markets. O advantages were broken into three types:

- Type 1: advantages that do not arise from multinationality but are advantages that any firm may have over another producing in the same location, i.e. advantages stemming from size, monopoly power and better resource capability and usage. These enable the firm to achieve more technical or cost efficiency or more market power than another firm.
- Type 2: advantages from being part of a multi-plant enterprise, such as economies of scale in non-production overheads (e.g. centralised accounting) and access to internal resources at lower cost than on the external market (e.g. internal borrowing).
- Type 3: advantages that come specifically from multinationality, such as wider opportunities and the ability to exploit differences in factor endowments and markets across countries; such advantages increase along with the number of foreign countries in which the MNE has operations and the diversity of their economic environments.

Dunning recognised that type 1 advantages were potentially available to all firms, but type 2 and 3 advantages came from being part of a multinational group rather than a *de novo* enterprise (Dunning 1981b: 27).[9]

The second condition – I – answered the 'how' or 'by which route' question. I advantages meant that it was more beneficial to the firm to use its O advantages internally rather than lease or sell them in the external market. Dunning (1977, 1979) saw the incentives of firms to internalise activities as twofold: to either avoid the disadvantages (or capitalise on the advantages) of imperfections in external mechanisms (market or government) for allocating resources or capitalise on the advantages. Market imperfections were either structural or cognitive. Structural imperfections arose when there were barriers to competition, transaction costs were high, or the economies of interdependent activities could not be fully captured by the market. Cognitive imperfections arose whenever information about products was not easily available or was too costly to acquire. Government intervention in resource allocation (e.g. R&D subsidies and patents, differences in corporate income tax rates across countries) also provided reasons for internalising cross-border activities.

The third condition – L – answered the 'where' or 'why do firms produce in one country rather than in another' question. L advantages meant that it was more profitable for the firm to use its O advantages together with factor inputs outside the home country. Dunning (1979) simply listed L advantages (e.g. spatial distribution of inputs and markets, transport and communication costs, government intervention, psychic distance) without grouping them.

Mark II: deepening the OLI paradigm

Throughout the 1980s and 1990s, the eclectic theory was deepened in several ways, but at least four dominate, in my view: responses to criticisms by internalisation theorists, the investment development cycle, applications to different industries, and incorporating geography. The first was Dunning's response to counter-arguments that internalisation theory, not OLI, was *the* reigning explanation for international production. The second was Dunning's attempt to make OLI more dynamic in terms of explaining changing patterns of FDI over time as a function of economic development. The third broadened the focus of OLI from manufacturing to services. The fourth changed the focus of the L component from the country level upwards (regional integration) and downwards (clustering).

OLI and internalisation theory

The eclectic theory did not meet with universal acceptance. The strongest criticisms came from the proponents of internalisation theory (see, for example, Buckley 1981, 1983; Rugman 1980). They argued that market failures in international intermediate product markets were both necessary and sufficient to explain the existence of MNEs. Since Dunning's O advantages were neither necessary nor sufficient, internalisation theory should be seen as the *theory of the multinational enterprise.*

As a response, Dunning made the first of several accommodations to his critics. Dunning (1983) reorganised the O advantages into two groups: Type 1 advantages were relabelled as Oa (asset) advantages and types 2 and 3 were grouped and relabelled as Ot (transactional) advantages. While Oa advantages were unchanged, Ot advantages

> . . . mirror the capacity of MNE hierarchies, *vis-à-vis* external markets, to capture the transactional benefits (or lessen the transactional costs)

> arising from the common governance of a network of these assets, located in different countries.
>
> (Dunning 1998a: 42)

This was a direct attempt to incorporate transaction costs into O advantages. O advantages are summarised with the nice statement:

> . . . O advantages of MNEs stem from their exclusive possession and use of certain kinds of income-generating assets; their ability to co-ordinate separate value-added activities across national boundaries; and their capacity to reduce environmental and foreign exchange risks.
>
> (Dunning 1988a: 25)

The criticisms continued, however (see, for example, Buckley and Casson 1985; Casson 1987). Dunning (1988a) explicitly addresses the criticisms of the internalisation school. In a section entitled 'Criticisms of the Eclectic Paradigm', he asks, 'are competitive or ownership advantages necessary to explain international production?' (1988a: 42). His response is to differentiate between the capability and the willingness of MNEs to internalise markets. O advantages provide the *capability* to internalise markets whereas I advantages provide the *willingness*. He also criticises internalisation theory for assuming market imperfections are always exogenous to the firm. In Dunning's paradigm, market imperfections can also be endogenous because MNEs can erect barriers to entry and exploit their monopoly power in cross-border markets.

Making the macro level dynamic

From the beginning, Dunning was interested in explaining changes in the pattern of international production, addressing the research agenda he posed in his 1973 article. The answer was the *investment development path* (Dunning 1981a, reprinted in 1981b, Chapter 5). Dunning argued that the net FDI position of a country depended on its firms' OLI advantages. Countries in Stage 1, at the lowest level of economic development, have very little inward or outward FDI because their firms' O advantages are weak or non-existent and the country's L advantages are either weak or unexploited due to weak institutions and infrastructure. As per capita income rises, shifting the country into Stage 2, FDI is attracted into the country. By Stage 3, domestic firms have strengthened sufficiently to move abroad themselves; eventually by Stage 4, the country becomes a net outward investor. In later work, Narula (1993) added Stage 5,[10] where inflows and outflows moved into balance. The overall pattern, therefore, is one of shifting from a host country to balanced FDI inflows and outflows.

Dunning continued writing on this topic (see, for example, Dunning 1988: Chapter 5; Dunning and Narula 1996: 1–41), but the topic has not received the attention that his work on other subjects (e.g. alliance capitalism) has received. Perhaps this is because inward and outward flows of FDI have become much more equal for OECD member countries (where the bulk of FDI occurs) so that international business scholars accept the investment development path as straightforward. More likely, it is because micro – not macro – research questions are currently in fashion in the international business literature.

However, IB scholars who study developing countries and emerging market economies continue to explore the macroeconomic implications of the eclectic paradigm. The chapter by Ozawa and Castello (Chapter 4) provides a good example of how OLI can provide a theoretical foundation for 'MNC-cum-government driven endogenous growth'. The authors see MNEs and governments as co-drivers of economic growth, turning L disadvantages (growth constraints; savings, foreign exchange and human resource gaps; macro-organisational weaknesses) into new opportunities for O appropriation by MNEs. Thus, virtuous circles of economic development can emerge through the interactions of government policies and MNE activities.

From manufacturing to services and beyond

From the early days of OLI, Dunning argued that the paradigm applied not only to manufacturing but also to services and natural resource industries. Perhaps the first application of OLI at the meso level was to the international hotel industry (Dunning and McQueen 1981; Dunning 1988a). An important piece is Chapter 10 in Dunning (1993b) on the globalisation of service industries, which I often ask the students in my graduate MNEs seminar to read.

In this book, Piscitello (Chapter 12) provides a good example of applying OLI to the financial services sector. She use proxies for the O, L and I advantages of Italian banks to empirically test the locational patterns of their foreign branches and offices over the 1990s. O advantages are proxied by bank size and international experience, I advantages by the need to follow the customer (Italian manufacturing FDI in the host country), and L advantages by host country market size and risk, and industry characteristics. The author finds that O and I advantages and the presence (absence) of an international financial centre are the primary factors determining banking location.

Perhaps the most interesting industry study is Dunning and Wymbs (Chapter 11) and their application of OLI to electronic commerce. The authors argue that the Internet is a facilitating technology that acts as a catalyst for business networks, disrupting traditional borders and boundaries for firms, industries and governments. However, they believe that the basic tenets of OLI continue to hold even in a world of e-commerce.

Clusters and regional integration

A fourth broadening of the OLI paradigm was directed at the locational or country-specific advantage. Dunning's original formulation of the location advantage was a simple list of country-specific items that could induce international production (Dunning 1979).[11] Later, these items were grouped (Dunning 1988c, 1993a) into the ESP paradigm: Environment (resources, stage of economic development, cultural/historic background), System (institutional framework) and Policies (macro, macro, general and FDI policies).[12] His own efforts to the contrary, Dunning has continued to argue that insufficient attention has been paid to location by international business scholars. Dunning (1998) is a plea for IB scholars to engage in more research on the L component of the eclectic paradigm.

Throughout this period, Dunning's primary OLI focus was the *country* level. However, Dunning, from very early on, was interested in location at other levels of analysis. Dunning's work on location at the sub-national level goes back at least to 1981. Dunning (1981b: Chapter 9) asked whether FDI strengthened agglomeration tendencies within the United Kingdom, and explored the effects of locational subsidies on regional economic development patterns.

At the supra-national level, a continuing preoccupation has been regional integration. The first piece, to my knowledge, was Dunning (1988a: Chapter 11).[13] In this chapter, he explored the linkages between the integration of corporations (goal: improve profits and long-run competitiveness) and the integration of countries (goal: increase efficiency, resource usage and competitiveness). He argued that cross-border integration by MNEs facilitated, and created a demand for, cross-border integration of countries, and vice versa.[14]

In later work, Dunning (1993a: Chapter 17; 1993b: Chapter 15) explicitly examined the linkages between MNEs and regional integration, focusing on the European Union, the Canada–US Free Trade Agreement and the NAFTA. Dunning (1994), which compared and contrasted the roles MNEs have played in NAFTA and the European Community, was particularly well argued. However, perhaps his best-known work on regional integration was Dunning (1997), the huge, two-part literature review and analysis of FDI and European integration, pre- and post-EC 1992, in the *Journal of Common Market Studies*. Most recently, Dunning (2000a) incorporated both interests – sub-national and supra-national location – in an edited volume that focused on clustering and agglomeration.[15]

Mark III: OLI and alliance capitalism

As early as 1982, researchers had begun to notice two new trends in international production. The first was a shift by firms, from using one of the two polar modes of entry (markets or hierarchies) to intermediate modes such as equity joint ventures, subcontracting, and R&D alliances. The second, which was tied to the first trend, was the emergence of non-traditional MNEs from developing countries, in particular, East Asia. Non-traditional MNEs were much more likely to use joint ventures than wholly owned subsidiaries as an entry mode (Giddy and Young 1982). Dunning (1984) dealt explicitly with non-equity forms of 'international involvement' and their implications for OLI.

Perhaps the best piece marking the change was Chapter 13 in Dunning (1988a) on 'The New Style Multinationals – Circa the Late 1980s and Early 1990s', which began with the prophetic statement:

> There are gathering signs that the internationalisation of value-adding activities by enterprises has reached a new watershed in its evolution . . . the MNE is now increasingly assuming the role of an orchestrator of production and transactions within a cluster, or network, of cross-border internal and external relationships, which may or may not involve equity investment, but which are intended to serve its global interests.
>
> (Dunning 1988a: 327)

The creation of these 'new-style MNEs' cried out for an explanation, one that John Dunning clearly wanted to fit within the OLI paradigm. A first response was to broaden the definition of international production to include any 'value-adding activity owned or controlled, and organised by a firm (or group of firms) outside its (or their) national boundaries' (ibid.: 1). This enabled international production to include modes of entry such as joint ventures and strategic alliances.

Chapter 13 provided a nice discussion of three major drivers of the new-style MNEs (the introduction of information technologies; the increased importance of non-market forces especially government policies for international production; and the rise of international alliances). Porter's value chain was used to illustrate the range of value-adding activities that can occur within the MNE network.[16] A discussion of networks followed. After reviewing alliances and networks, Dunning argued that 'the conceptual and analytical structure of the paradigm remains largely unimpaired', but worried that 'its operational usefulness decreases as the complexity of the variables making up the OLI configuration increases' (ibid.: 342). He concluded with the observation that MNEs should now be seen as 'organizers and co-ordinators of clusters of cross-border producing and transacting activities; to which a latitudinal dimension, which embraces a series of co-operative inter-firm relationships is added' (ibid.: 345).

In 1993, Dunning's major opus, *Multinationals and the Global Economy*, was published (Dunning 1993a). Inter-firm relations were examined in detail in Chapter 9 with a full discussion of the variety of entry modes. Technology-based alliances were also analysed in Dunning (1993b: Chapter 8).

The definitive piece was Dunning (1995) on 'Reappraising the Eclectic Paradigm in the Age of Alliance Capitalism', where he separated old-style hierarchical capitalism from new-style alliance capitalism and compared their O, L and I advantages. He argued that hierarchical capitalism was akin to an 'exit' strategy (replacing the external market with the internal hierarchy) whereas alliance capitalism was similar to a 'voice' strategy since firms adopt cooperative solutions within the market. In conclusion, Dunning argued that firms and governments needed to focus more on innovation and to recognise that firms go abroad to acquire O advantages as well as to exploit them. Alliances ('voice') should also be seen as an efficiency-improving mode of entry and a way to enhance competitiveness.

Mark IV: OLI and strategic management

OLI plus S?

The third major challenge to the OLI paradigm appeared in the mid-1980s; however, this challenge was a theoretical one, not empirical. International business scholarship had been dominated by economists throughout the 1970s and 1980s; with the emergence of strategic management as a separate discipline, strategy scholars began asking questions and developing theories that were drawn from disciplines such as sociology, psychology and labour relations. Initially, strategy research focused on domestic firms, but scholars such as Michael Porter, Sumatra Ghoshal, Christopher Bartlett and Michael Hitt began to develop a new subfield in international/global strategic management, linking strategy with international business. Could OLI expand to incorporate this new discipline?

Dunning's first attempt to incorporate strategy into the OLI paradigm (Dunning 1993b: Chapter 4) compares and contrasts OLI with strategy. Perhaps the most telling statement is:

> From the economist's perspective, strategy related variables are most often treated as part of the 'unexplained' (or unexplainable) variables, whereas they are the main subject of interest to the business analyst.
>
> (Dunning 1993b: 83)

Recognising that strategy is missing from the eclectic paradigm, Dunning's solution was to add it on at the end as a 'dynamic add-on', creating OLIS. He recognised that OLI could induce strategy, but that strategy could also affect OLI. He argued that firms with different OLI configurations would adopt different S. As a result, he concluded that S should be added to OLI as a fourth component. An appendix at the end of the chapter detailed the components of O, L, I and S (where S includes 10 topics such as technology, sourcing, HRM, marketing).

Personally, I have never liked this attempt to weld strategy on at the end of OLI. It does not work. Going back to my kindergarten questions at the beginning of this essay, strategic management, as a discipline, addresses all the 'Who, what, where, when, why and how?' research questions. Strategy questions, however, are micro questions at the level of the individual firm, or even go inside the firm to focus on the entrepreneur, top management team or board of directors. Strategy therefore cuts across all the kindergarten questions, but at the micro level. The OLI paradigm, on the other hand, has been focused on the why, where and how research questions, primarily at the macro and meso levels. If we think of the kindergarten questions as a matrix with who/what, where, why, how and when as columns and macro, meso and micro as rows, OLI fits primarily into the why-where-how columns and macro-meso rows; whereas strategy, as a discipline, crosses all the columns but primarily at the micro row level. Therefore, for me, adding S as the last 'column' in OLI does not and cannot fully integrate strategy into the eclectic paradigm.

Along a similar vein, Guisinger's revision of the OLI paradigm to OLMA (Guisinger, this volume, Chapter 6) attempts to shift OLI down to the micro level of the firm by replacing I with M (mode of entry) and adding A (adaptation of businesses processes to the environment) to the end, where A is basically S (strategy formulation and implementation). As such, OLMA suffers from the same problems as OLIS.

A more promising route to accommodating strategic management within OLI would follow the same approach that Dunning adopted with Porter's diamond of competitive advantage, where he incorporated international dimensions (defined as multinational business activity) into each of the diamond factors (Dunning 1992). Dunning (1988a: Chapter 12) began this process of linking OLI to strategy by reviewing eight different disciplinary approaches to international production – ranging from strategic management to law and economic history. In each case, he showed how insights about international production from a discipline could be included under the O, L and I 'umbrella'. He concluded that the eclectic paradigm was 'robust' and 'offer[ed] a powerful tool of analysis' for understanding MNE activities (1988: 325).

In this volume, Devinney, Midgley and Venaik (Chapter 8) also provide an example of how strategy and the OLI paradigm can be integrated. The authors argue that the eclectic paradigm pays insufficient attention to what goes on inside the firm; that is, to the role played by managers and the dynamic evolution of the MNE. Recognising that O, L and I affect S and that S also affects O, L and I, they develop a theory where OLI is both exogenous and endogenous. The initial environment (O, L and I) determines the set of all possible strategic orientations for the firm. The MNE's existing structure acts as a constraint, determining the technologically feasible set of strategic orientations open to the MNE. Managers' beliefs about the feasible set then determine the strategies they choose. As the

MNE's managers implement their chosen strategy, this in turn affects the firm's OLI advantages. Because certain strategies dominate others in terms of profitability and market contestability, MNEs make adjustments over time to move to the optimal position. Thus, strategic considerations and managers' beliefs are incorporated into the OLI paradigm. In my view, the intertwining of strategy and OLI in this chapter has potential, but needs to be 'fleshed out' through application to particular strategic decisions.

The chapter by Oxelheim, Randøy and Stonehill (Chapter 10) suggests how this can be done in terms of financial strategies. They argue that a firm's financial strength affects its ability to engage in FDI. Finance-specific strategies can be either proactive (efficiency-based) or reactive (arbitraging market imperfections). Proactive financial strategies generate OLI advantages that can be exploited through international production. What is missing from this chapter is managerial beliefs, which link the feasible set of strategies to the chosen strategy (see Devinney, Midgley and Venaik, Chapter 8). The juxtaposition of these two chapters, to me, suggests the way that OLI might better incorporate strategic decision-making. (The Oxelheim, Randøy and Stonehill chapter is important for a second reason: the OLI paradigm underplays the financial aspects of international production. The effects of exchange rate changes, over/under-valued currencies and international financial management are seldom examined through the OLI lens; this chapter is a rare and welcome exception.)

Another comment linking OLI and strategy should be made here. The 'bottom line' for strategic management scholars is firm performance; that is, how strategy affects financial and market returns to the firm. The links between OLI and firm performance are particularly slim in the international business literature. This is not surprising: OLI outlines advantages and disadvantages of international production; the link to firm performance is not straightforward unless one makes the simplifying assumption that the greater the OLI advantages, the better the performance. Benito and Tomassen (Chapter 9) begin the process of 'unpacking' the relationship between OLI and firm performance by applying the resource-based view to the OLI paradigm. This enables them to create an inventory list of performance implications for each component. The next step would be to operationalise this list and engage in empirical tests.

There have been few empirical attempts to explicitly link the eclectic paradigm to firm performance. Robins, Tallman and Fladmoe-Lindquist (2002) applied OLI to the performance of joint ventures in Mexico. Interestingly, they argued for and found evidence of some resources provided by the US parent to the Mexican joint venture actually *reducing* the venture's performance. This suggests that not all OLI factors must have positive impacts. Eden, Thomas and Olibe (forthcoming) also examined the influence of O and L advantages on the performance of US MNEs. They decomposed L into two components measuring firm depth (foreign market penetration, foreign production presence) and one measuring depth (country scope). The authors concluded that both financial and market performance of US MNEs over 1990–4 were positively related to O and L variables, but that breadth was more important than depth. Clearly, more research linking OLI and firm performance would be a useful addition to the literature.

Motivations for international production

Dunning (1988b) in his restatement of the OLI paradigm, recognised that the link between OLI and strategy could be made through firm-level motivations for international production. The O factors answer the 'why' question in terms of general Oa and Ot motivations. However, once one begins to think of MNE activities in terms of the value chain, individual plants, products and factors, the actual motivation – or *strategy* – behind establishing any particular value-adding activity abroad becomes important. Thus, focusing on motivations at the individual activity or investment level can be the link between OLI and strategy.

Dunning (1988b: 13) provided perhaps the first 'cut' at these motivations, outlining what he called the 'three main forms of international production': market seeking (import substituting), resource seeking (supply-oriented) and efficiency seeking (rationalised investment). A fourth category – strategic asset-seeking FDI – was soon added, reflecting the increased use of knowledge-based strategic alliances within OECD countries (Dunning 1991). By 1993, Dunning (1993a: Chapter 3) had identified four basic types of international production: natural resource seekers, market seekers, efficiency seekers and strategic asset or capability seekers.[17]

Throughout the 1990s, strategic management scholars wrestled with the research question: why do firms differ and how do these differences affect their performance? Perhaps the dominant explanation was the resource-based view, which argued that the long-run competitiveness of a firm depended on its resources and capabilities (Barney 1991). Initially, the resource-based view focused on domestic firms, but IB scholars soon began to make the link between resources and capabilities in the strategy literature and ownership advantages in OLI.

Dunning (1999) addressed the resource-based view in the context of globalisation of economic activity during the 1990s. He argued that the world was shifting from hierarchical to alliance capitalism, where, at the end of the twentieth century, knowledge, regional and global activities and intra- and inter-institutional alliances were of increasing importance. He recognised that the resource-based view and evolutionary theories of the firm were close relatives to the O in the OLI paradigm, but the strategic management theories took as their focus the creation and upgrading of these advantages, whereas the eclectic paradigm focused on their exploitation.

Recognising that the eclectic paradigm needed to encompass the growing importance of knowledge-based FDI, Dunning (1993a) again revised the paradigm. While the motivations for FDI (the 'why') were still seen as *exploitation* of the firm's O advantages, the purpose of the actual investment was now defined as *seeking* or *acquiring* either products (market seeking FDI) or factors (resource, efficiency and strategic asset-seeking FDI).[18]

Two chapters in this volume focus explicitly on OLI and the resource based view. Madhok and Phene (Chapter 5) argue that the OLI paradigm was useful when firms were beginning to internationalise, but is less useful now that many firms have multiple foreign affiliates. Now, the key source of competitive advantage is creating and managing a 'knowledge portfolio'. They see OLI as focused on explaining the home-country firm as an institution as compared to firms from other countries, rather than on a specific firm. The key issue, from their perspective, is knowledge management, both in terms of asset exploitation and asset seeking.

Maitland and Nicholas (Chapter 3) argue the resource-based view ignores location-specific differences in firm resources and capabilities because they are not unique to a particular firm. Their solution is to incorporate institutional theory into OLI. Reviewing the new institutional economics (NIE), the authors argue that NIE could replace OLI as an explanation of country and industry patterns of international production. That is, rather than incorporate institutional theory under the 'OLI big tent', the authors make the audacious proposal that NIE replace OLI!

Reconciling the tension between exploitation and acquisition within the motivations for international production, first identified in Dunning (1993a) has been a key preoccupation of his recent work. For example, Dunning (2002) broadened the 'OLI envelope' to encompass the resource-based view by including relational assets within the O variable. He split firm-specific advantages into tangible and intangible assets; intangible assets into intellectual and relational assets; and relational assets into private and social assets. Relational assets were defined as facilitating assets that had to be used jointly with the relational assets of another actor. After reviewing the implications for the OLI paradigm, Dunning concluded that O needed to be modified to include the creation, coordination and sustenance of relational assets; L, the presence/absence of networks of related activities; and I, a greater focus on cooperative non-equity economic linkages, particularly networks. Lastly, in this volume, Dunning (Chapter 2) now argues that incorporation of asset-augmenting FDI into the eclectic paradigm requires reconfiguring the traditional OLI variables, but leaves the overall paradigm intact.

Conclusions

My review of the OLI paradigm and its evolution over the past 30-plus years supports Dunning's contention that OLI can be and is an 'envelope' or 'big tent' for all theories that address the 'why, where and how' of MNE activity (Dunning 2000b). The paradigm has shifted its focus from explaining international production as the exploitation of OLI advantages through wholly owned foreign subsidiaries in manufacturing to a rich, complex analysis of globalised businesses at the beginning of the twenty-first century.

The chapters in this book point to ways in which the OLI paradigm is being interpreted and altered by international business scholars. Many of these contributions focus on strategic management and the need for OLI to better incorporate the resource-based view and managerial perspectives. In terms of my matrix of macro, meso and micro levels of analysis, the thrust of these extensions is at the micro level. One advantage of moving down the level of analysis is the increased ability to perform econometric analysis; that is, we can examine how OLI affect firm strategies and performance using firm-level data rather than FDI statistics. However, a fully fledged integration of strategy into OLI, even of the resource-based view into OLI, has yet to appear. The chapters in this volume point the way, but more work is needed here.

Second, I agree with the authors in this book that OLI should better integrate insights from institutional theory. The new institutional economics is a vibrant and thriving discipline, crossing management, economics, sociology and law. Institutional theory has brought new insights into international business literature, for example, liability of foreignness, relational capital, public corruption, and property rights. One can speculate how liability of foreignness and public corruption, for example, could quite easily be incorporated into the 'OLI big tent'. However, I do not agree that NIE should, or even could, replace OLI as the reigning paradigm of international business.

Third, the focus on international finance, bringing finance back in to OLI, is a welcome sign. As international business researchers increasingly adopt the tools of finance (e.g. real options theory, foreign exchange exposure, currency unions) the OLI model needs also to redirect its attention to risk and uncertainty in connection with the financing of international production.

To restate my initial thesis at the beginning of this chapter, the focus of OLI was and remains the 'why, where and how' of international production. The chapters in this book focus primarily on *deepening* the OLI in terms of these three questions by incorporating the micro level of analysis (the firm and its managers).

Let me suggest, as a counterpoint, that the eclectic paradigm might also usefully be *broadened* by an explicit focus on the 'when' question; that is, the issue of timing of international production.[19] International business and strategy scholars are increasingly preoccupied by timing issues; for example, *de novo* versus sequential FDI, FDI as platform investments, incorporating real options theory into IB research, the 'big step' hypothesis, the recent issue of *Academy of Management Review* on timing.

Perhaps it is also time to go back to the 'what' and 'who' questions. As the boundaries of the MNE grow ever more fuzzier, what is and is not an MNE? Cantwell and Narula (Chapter 1) also raise the 'fuzzy border' question. With the growth in non-equity cooperative alliances and networks, where does the hierarchy stop and the market begin? As ownership becomes less important, does control remain the critical dividing line? Earlier debates on this topic have been spirited. For example, Wilkins and Schröter (1998) examined free-standing companies, firms set up in one country for the purpose of doing business outside that country. The hotly contested debate among Wilkins, Hennart, Casson and Corley (in the Wilkins and Schröter volume) about free-standing companies is a wonderful way to introduce graduate students to the question of what is and what is not a multinational enterprise.[20] Perhaps it is time for a discussion about whether alliances and networks are MNEs and where inter-firm trade stops and intra-firm trade begins.

A second angle on the 'what' and 'who' question is the growth of multinationals from emerging markets and developing countries. Whether in the form of family conglomerates (*chaebol*, *grupo*), East–West or North–South strategic alliances or born global firms, MNEs from the East and South have been under-explored in the international business literature. Partly this is because 80 per cent of FDI comes from and goes to OECD countries; but the lack of high-quality statistics also hampers scholarly research. There are lessons from earlier work that could prove useful here. Kojima argued in the 1970s that MNEs from Japan were different from Western MNEs. This argument was refuted both by OLI and by later events that showed Japanese MNEs, as they matured, began to behave like Western firms (Kojima 1978; Dunning 1980). Giddy and Young (1992) also argued that developing country MNEs were non-traditional. My suspicion is that 'new-style MNEs from emerging market economies' will come to look increasingly like Western MNEs (e.g. shifting their corporate strategies from unrelated to related diversification), but this remains to be seen.

In conclusion, this chapter has attempted to do two things: provide an overview of the development of the OLI paradigm over the past 30-plus years and consider the chapters in this volume in light of that paradigm. In terms of my first goal, I have touched on only a small segment of Dunning's enormous volume of research in international business, and have perhaps missed several important contributions.[21] However, I hope that this chapter testifies to the resilience and robustness of OLI in the face of the theoretical and real-world storms that have occurred since Dunning first proposed the eclectic theory in 1976. In terms of my second goal, it is clear from the chapters in this book that the OLI paradigm has a life of its own outside of Dunning's work. William Arthur Ward once wrote: 'The pessimist complains about the wind; the optimist expects it to change; the realist adjusts the sails.' The OLI paradigm is a testimony to the ability of John Dunning and scholars such as the authors in this book to 'adjust the sails'. OLI is alive and well!

Notes

1 My latest purchase, the two-volume set of Dunning's selected essays, *Theories and Paradigms of International Business Activity* (2002b), is also a 'must have' since the first volume gathers together many of Dunning's OLI writings in one place from 1973 to the present. There is one problem with the *Selected Essays*: they have been updated from the originals in places, making it harder to trace Dunning's own theory development from one paper to the next. For example, Chapter 4, 'Explaining the international direct investment position of countries', published originally in 1981, contains references to 1993 publications. A second problem is that some of my favourite OLI readings (e.g. Dunning 1979; 1988a: Chapters 11 and 13) are missing.

2 For example, in this volume, OLI is referred to as a framework (Maitland and Nicholas) and as a theory evolving into a paradigm (Tolentino). Dunning in his earliest OLI writings (1977, reprinted in 1981, page 33) shares this uncertainty about where OLI fits; 'the eclectic model can be perceived as a general theory of international production in so far as it provides an analytical framework for explaining all forms of such production'.

3 *Laws* are generalisations about observed regular relationships between two phenomena, from which we can generalise about what we expect to happen (e.g. the law of one price or Gresham's Law).

4 According to Van Evera (1997), a good theory has large explanatory power, importance, parsimony and a wide explanatory range. A good theory is applicable to the real world, clearly framed and satisfying. A good theory should be, in principle, falsifiable. It should have prescriptive richness and explain important phenomena.

5 As a result, models allow complex systems to be understood and behaviours predicted, but may give incorrect predictions for situations outside the model's assumptions.

6 The first major statement of OLI as a paradigm is Dunning (1988b), which draws his previous writings on OLI together, shifting from eclectic theory to paradigm.

7 Dunning (1981b) pulls together his first eight years of publications on the MNE, organised into three topics: OLI, impacts of MNEs on home and host countries, and lessons for governments.

8 To recap the three conditions: (1) It possessed net ownership advantages over firms from other countries in serving a particular national market (O advantage). (2) It was more beneficial for the firm to use these net ownership advantages itself rather than sell or lease them (I advantage). (3) These net ownership advantages were more profitably exploited when used with factor inputs outside the home country (L advantage).

9 This distinction was important later for incorporating, first, internalisation theory, then strategic management and finally, sequential FDI into the OLI envelope.

10 Published as Narula (1996).

11 These included the spatial distribution of inputs and markets; prices, quality and productivity of inputs; transport and communication costs; government intervention and policies; infrastructure; psychic distance; and scale economies.

12 Personally, I also teach L as broken into E + S + P but use a different grouping that I find more intuitively appealing: E(economic), S (socio-cultural) and P (political-legal). This is closer in spirit to Guisinger's breakdown of the international environment (this volume).

13 This is a short chapter, but it has always been one of my favourites.

14 We now refer to these processes as investment-led and policy-led regional integration.

15 Piscitello (this volume) is also notable as an empirical piece using OLI to predict location.
16 And right side up too – upstream activities are at the top and downstream at the bottom, rather than moving from left to right horizontally! Thus, horizontally integrated activities can be shown horizontally (e.g. the component supply stage in Figure 13.1 on page 340) and vertically integrated activities in a vertical direction. See also Chapter 9 in Dunning (1993a) for a full discussion of value chains in MNE entry and expansion strategies.
17 Other motives (escape, support and passive investments) are also discussed although they appear to have been dropped from subsequent Dunning papers, leaving only the first four. In the OLI paradigm, the key verbs used to describe motivations are exploit and acquire; however, there are other verbs such as arbitrage (market imperfections) and avoid (risk) that now are seldom heard.
18 Originally cast as exploiting O advantages, the eclectic paradigm now includes seeking O assets. Note the shift from *advantages* to *assets*, which may come from the resource-based view. Dynamic capabilities should be next.
19 Although I do believe that paradigms do not have to explain everything, nor does the 'OLI big tent' have to shelter all theories under its umbrella.
20 See also Contractor and Lorange's new book where Dunning (2002a) appears.
21 For example, I deliberately omitted the *World Investment Reports*, where Dunning has served for many years as the lead intellectual scholar, and his work on MNE-state relations and the benefits and costs of FDI, which are tangential to the OLI paradigm.

References

Barney, J. (1991) 'Firm resources and sustained competitive advantage', *Journal of Management*, 17: 99–120.

Buckley, P.J. (1981) 'A critical review of theories of the multinational enterprise', *Aussenwirtschaft*, 36: 70–8.

Buckley, P.J. (1983) 'New theories of international business: some unresolved issues', in M.C. Casson (ed.), *The Growth of International Business*, London: Allen & Unwin.

Buckley, P.J. and Casson, M.C. (1985) *The Economic Theory of the Multinational Enterprise*, London: Macmillan.

Casson, M.C. (1987) *The Firm and the Market*, Oxford: Basil Blackwell.

Contractor, F.J. and Lorange, P. (eds) (1988) *Cooperative Strategies in International Business*, Lexington, MA: Lexington Books.

Dunning, J.H. (1973) 'The determinants of international production', *Oxford Economic Papers*, November, 25: 289–325.

Dunning, J.H. (1977) 'Trade, location of economic activity and the MNE: a search for an eclectic approach', in B. Ohlin, P.O. Hesselborn and P.M. Wijkman (eds), *The International Allocation of Economic Activity*, London: Macmillan.

Dunning, J.H. (1979) 'Explaining changing patterns of international production: in support of the eclectic theory', *Oxford Bulletin of Economics and Statistics*, 41(4): 269–95.

Dunning, J.H. (1980) 'Trade, location of economic activity and the multinational enterprise: some empirical tests', *Journal of International Business Studies* 11(1): 9–31.

Dunning, J.H. (1981a) 'Explaining the international direct investment position of countries: towards a dynamic or developmental approach', *Weltwirtschaftliches Archiv*, 117: 30–64.

Dunning, J.H. (1981b) *International Production and the Multinational Enterprise*, London: George Allen & Unwin.

Dunning, J.H. (1983) 'Market power of the firm and international transfer of technology', *International Journal of Industrial Organization*, 1: 333–51.

Dunning, J.H. (1984) 'Non-equity forms of foreign economic involvement and the theory of international production', in R.W. Moxon, T.W. Roehl and J.F. Truitt (eds), *International Business Strategies in the Asia Pacific Region*, Greenwich, CI: JAI Press.

Dunning, J.H. (1988a) *Explaining International Production*, London: Unwin Hyman.

Dunning, J.H. (1988b) 'The eclectic paradigm of international production: a restatement and some possible extensions', *Journal of International Business Studies*, 19(1): 1–31.

Dunning, J.H. (1988c) *Multinationals, Technology and Competitiveness*, London: Unwin Hyman.

Dunning, J.H. (1991) 'The eclectic paradigm of international production: a personal perspective', in C.N. Pitelis and R. Sugden (eds), *The Nature of the Transnational Firm*, London and New York: Routledge.

Dunning, J.H. (1992) 'The competitive advantages of countries and MNE activity', *Transnational Corporations*, 1(1): 135–68.

Dunning, J.H. (1993a) *Multinational Enterprises and the Global Economy*, Wokingham, UK: Addison-Wesley Publishing Company.

Dunning, J.H. (1993b) *The Globalization of Business*, London and New York: Routledge.

Dunning, J.H. (1994) 'MNE activity: comparing NAFTA and the EC', in L. Eden (ed.), *Multinationals in North America*, Calgary: University of Calgary Press.

Dunning, J.H. (1995) 'Reappraising the eclectic paradigm in an age of alliance capitalism', *Journal of International Business Studies*, 26: 461–91.

Dunning, J.H. (1997) 'The European internal market program and inbound foreign direct investment', *Journal of Common Market Studies* 35(1): 1–30 and 35(2): 189–223.

Dunning, J.H. (1998) 'Location and the multinational enterprise: a neglected factor?', *Journal of International Business Studies*, 29: 67–83.

Dunning, J.H. (1999) 'Globalization and the theory of MNE activity', in N. Hood and S. Young (eds), *The Globalization of Multinational Enterprise Activity*, London: Macmillan.

Dunning, J.H. (ed.) (2000a) *Regions, Globalization and the Knowledge-Based Economy*, Oxford: Oxford University Press.

Dunning, J.H. (2000b) 'The eclectic paradigm as an envelope for economic and business theories of MNE activity', *International Business Review*, 9: 163–90.

Dunning, J.H. (2001) 'The Key Literature on IB Activities: 1960–2000' in A.M. Rugman and T.L. Brewer (eds), *The Oxford Handbook of International Business*, Oxford: Oxford University Press.

Dunning, J.H. (2002a) 'Relational assets, networks and international business activity', in F. Contractor and P. Lorange (eds), *Cooperative Strategies and Alliances*, Amsterdam and Oxford: Elsevier Science.

Dunning, J.H. (2002b) *Theories and Paradigms of International Business Activity: The Selected Essays of John H. Dunning. Volumes I and II*, Cheltenham: Edward Elgar.

Dunning, J.H. and McQueen, M. (1981) *Transnational Corporations in International Tourism*, New York: UNCTAD.

Dunning, J.H. and Narula, R. (eds) (1996) *Foreign Direct Investment and Governments*, London and New York: Routledge.

Eden, L., Thomas, D.E. and Olibe, K.O. (forthcoming) 'Why multinationality matters: exploring the "L" in the OLI paradigm', in P. Gray (ed.), *Extending the Eclectic Paradigm in International Business: Essays in Honor of John Dunning*, London: Edward Edgar.

Giddy, I. and Young, S. (1982) 'Conventional theory and unconventional multinationals: do new forms of multinational enterprise require new theories?', in A.M. Rugman (ed.), *New Theories of the Multinational Enterprise*, London and Canberra: Croom Helm.

Itaki, M. (1991) 'Critical assessment of the eclectic theory of the MNE', *Journal of International Business Studies*, 22(3): 445–60.

Kojima, K. (1978) *Foreign Direct Investment: A Japanese Model of Multinational Business Operations*, London: Croom Helm.

Narula, R. (1993) *An Examination of the Evolution and Interdependence of Foreign Direct Investment and Economic Structure: The Case of Industrialized Countries*, Ph.D Dissertation, Newark, NJ: Rutgers University.

Narula, R. (1996) *Multinational Investment and Economic Structure*, London: Routledge.

Robins, J., Tallman, S. and Fladmoe-Lindquist, K. (2002) 'Autonomy and dependence of international cooperative ventures: an exploration of the strategic performance of US ventures in Mexico', *Strategic Management Journal*, 23: 881–901.

Rugman, A.M. (1980) 'Internalization as a general theory of foreign direct investment, a reappraisal of the literature', *Weltwirtschaftliches Archiv*, 116(2): 365–79.

Tolentino, P. (2001) 'From a theory to a paradigm: examining the eclectic paradigm as a framework', *International Economics*, 8(2): 191–210.

Van Evera, S. (1997) *Guide to Methods for Students of Political Science*, Ithaca and London: Cornell University Press.

Wilkins, M. and Schröter, H. (1998) *The Free-Standing Company in the World Economy, 1830–1996*, Oxford: Oxford University Press.

Index

An environmentally friendly book printed and bound in England by www.printondemand-worldwide.com

PEFC Certified

This product is from sustainably managed forests and controlled sources

PEFC

PEFC/16-33-415

www.pefc.org

FSC

www.fsc.org

MIX

Paper from responsible sources

FSC® C004959

This book is made entirely of sustainable materials; FSC paper for the cover and PEFC paper for the text pages.

#0065 - 250413 - C0 - 234/156/17 [19] - CB